# MEDICINES
# THAT FEED US

# MEDICINES THAT FEED US

## Plants, Healing, and Sovereignty in a Toxic World

STACEY A. LANGWICK

Duke University Press   *Durham and London*   2026

© 2026 DUKE UNIVERSITY PRESS. All rights reserved.
Project Editor: Lisa Lawley
Typeset in Garamond Premier Pro by Westchester Publishing Services

Library of Congress Cataloging-in-Publication Data
Names: Langwick, Stacey Ann, author.
Title: Medicines that feed us : plants, healing, and sovereignty in a
toxic world / Stacey A. Langwick.
Description: Durham : Duke University Press, 2026. |
Includes bibliographical references and index.
Identifiers: LCCN 2025026113 (print)
LCCN 2025026114 (ebook)
ISBN 9781478033226 (paperback)
ISBN 9781478029779 (hardcover)
ISBN 9781478061960 (ebook)
Subjects: LCSH: Traditional medicine—Tanzania. | Herbs—
Therapeutic use—Tanzania. | Alternative agriculture—Tanzania. |
Environmentalism—Tanzania.
Classification: LCC GN477 .L36 2026 (print) | LCC GN477 (ebook)
LC record available at https://lccn.loc.gov/2025026113
LC ebook record available at https://lccn.loc.gov/2025026114

Cover art: David Mzuguno, *In the Garden of Elden*. Courtesy of the
Mzuguno Family.

Publication of this book has been aided by a grant from the Hull
Memorial Publication Fund of Cornell University.

For my dearest

TSADIA

and Mama Helen's dearest

LINA

# Contents

## Introduction  *Healing (in) a Toxic World*

*Medicines That Feed Us* examines the relationship between toxicity and remedy in the face of the environmental and health crises shaping the twenty-first century. It locates its provocations alongside of, and in solidarity with, the innovative work of Tanzanians who are challenging the ways that "health" conceptualizes and governs the entanglement of bodies and ecologies. Together we ask: *What does it mean to heal in a toxic world*? How is that which counts as "therapeutic" shifting with the growing acknowledgment that the extractive relations fueling contemporary economies and animating modern life undermine possibilities for future survival? The double-bind defining our contemporary moment unsettles and disorients. It also has the potential to forge creative responses that reimagine the territorial and the corporeal, posing configurations of care that invite alternative forms of sovereignty in the service of both ecological and bodily healing.

This potential begins, I argue, with the recognition that modern modes of dwelling and the substantive changes that they have engendered in the matter of the earth and of the body have rendered nineteenth- and twentieth-century articulations of relations between toxicity and remedy inadequate. The narrow choice they seem to offer—either apocalypse or salvation—creates both intellectual and political claustrophobia. *Medicines That Feed Us* tells a story that reworks the pasts and the futures of the relationship between toxicity and remedy through healing in Tanzania. It is anchored in the hard work of both people and plants attending to the vitality of bodies and soils in the midst of the ongoing ecological and social violence wrought by the economization of life, labor, and land. It is dedicated to accounting for the radical potential of initiatives to redefine the times and spaces of healing both on and of the earth.

One friend and mentor, Helen Tibandebage Nguya, who you will meet throughout the pages that follow, proposed that we call the set of social and entrepreneurial projects on which this book centers *dawa lishe*—medicines that feed us. Mama Nguya is the founder of Training, Research, Monitoring and Evaluation on Gender and AIDS (TRMEGA), an innovative nongovernmental organization (NGO) that addresses health issues through land relations. We were driving together in my car, returning from a long day visiting a garden that TRMEGA had helped to seed at an orphanage in the dry volcanic plains west of Arusha, the fourth largest city in Tanzania, when she proposed the phrase "*dawa lishe*." Earlier in the day, as we drove out of the city increasingly farther from the forests of Mount Meru that stretch above it, the land flattened and vegetation became sparse. For many miles, the "road" was rather indistinguishable from the dry, sandy soil that stretched out on either side of it. Our path was less direct than it might have been, as we were forced to find ways around the huge erosion gullies that cut through the landscape. When we arrived at the orphanage, however, the garden was flourishing. During our visit, children bounded between the rows, showing us the plants and picking armfuls of greens for the kitchen. Mama Nguya and Jane Satiel Mwalyego, who worked with her, discussed with the gardener at the orphanage where they might best transplant the seedlings and cultivate the plant cuttings that we had brought from the TRMEGA gardens. Later, as we sat talking to staff, they drew our attention to the lemongrass Mama Nguya and Jane had previously brought, which was now flourishing under the window of the classroom and keeping the mosquitos at bay.

As we pushed to get ahead of the waning light on our way back to Maji ya Chai, we debriefed in the car. We three talked about the plants and the kids in the garden, as well as the politics behind the founding of the orphanage and the tensions around its leadership, before we turned to the conceptual questions about the work we were doing together that regularly shaped our discussions. In the midst of this, I confessed to them that I did not know what to call the sorts of projects TRMEGA and others were generating: projects that offered renewed relations between people and plants as an intervention into the prolonged depletion and ongoing injury of bodies and soils in postcolonial Tanzania. After a thoughtful moment, Mama Nguya suggested *dawa lishe*.

*Dawa lishe*, a phrase that merges the more official categories of medicine (*dawa*) and fortified or nutrient-dense foods (*chakula lishe*), was an offering, a proposition, and, as I have come to see, a theory. While it seemed to arise spontaneously in response to my grasping for language, it was animated by years of collective work with people and plants. As Dian Million, the Tanana

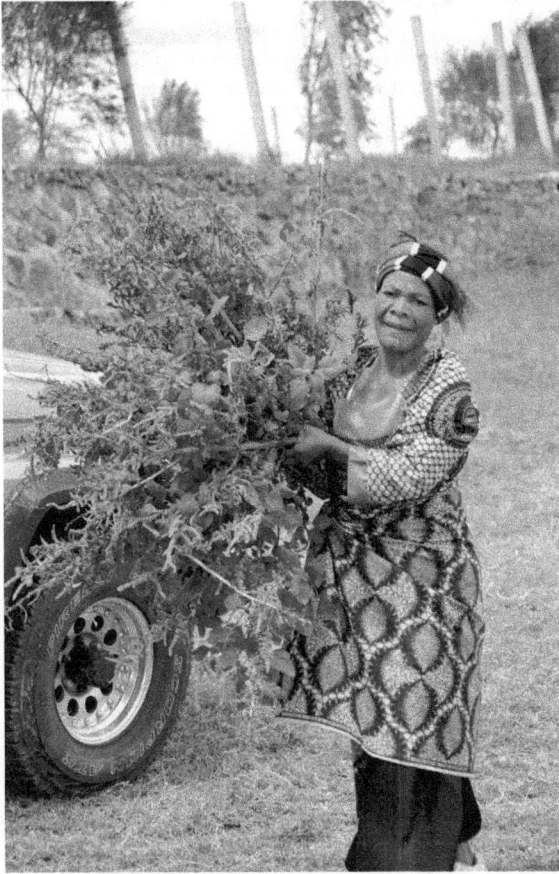

FIGURE I.I Jane Satiel Mwalyego carrying cuttings gathered from the gardens of Training, Research, Monitoring and Evaluation on Gender and AIDS (TRMEGA) and to be transplanted in the garden they support at an orphanage west of Arusha, Tanzania. Photo by author, 2015.

Athabascan theorist of indigenous feminist activism, teaches, "Theory is always practical first, rather than abstract."[1] Remembering this fact shifts the ground of our stories and the relations in which they emerge. It expands ideas of who is (already) theorizing, which conversations find traction in the academy, and how a vocabulary in service of decolonization is generated.

On that late afternoon, Mama Nguya and I were grappling with a sense that there is something worth distinguishing about an informal network of projects in northeastern Tanzania that live in the interstices of medicine and agriculture. In these projects, Tanzanians are simultaneously assessing the conditions under which contemporary life is attenuated, diminished, exhausted, or drained away, as well as experimenting with ways to intervene in these conditions. We saw the collective commitment drawing them together as something worth naming because it renders visible experiences of dispossession, assessments of healing and harming, and strategies for reckoning with the past in the service of a more caring, just, and equitable future. This book takes *dawa lishe* as a provocation. Every time the phrase *dawa lishe* appears in this book, it deserves a footnote to Mama Nguya; every reference embodies our relationship and her work. Yet, in generating new proximities, *dawa lishe* is also an invitation to think together with others, to invite *dawa lishe* to energize new lines along which to theorize, to build connections with projects well beyond the borders of Tanzania, and to find common cause.

The social-therapeutic-ecological projects that the phrase *dawa lishe* strives to draw together attend to bodies as effects of land relations. Plant(ing) remedies seek to intervene in the slow violence of colonial dispossession, land enclosure, extractive labor, and the insatiability of appetites for natural resources that drive them. Born of a moment in which Tanzanians are witnessing a dramatic rise in chronic disease, *dawa lishe* articulates these persistent illnesses not only as a consequence of irresponsible ecological practices but also as part and parcel of the derangement of the forces through which the physical capacities of bodies and lands come into being. Remedies work by feeding, fortifying, and strengthening bodies and soils. This book argues that in addition to describing a collection of remedies and a modality of care, *dawa lishe* proposes an imaginative and practical experiment in healing (in) a toxic world and fostering real possibilities for continuance.

The projects described in the chapters that follow share a kinship with the rise of various initiatives in Africa and throughout the Global South that combine local knowledge of how to support the flourishing of plants and people with a range of global ecological and health movements. The community organizers, NGO leaders, and entrepreneurs in Tanzania innovating these plant(ing) therapies share some concerns with those who work to "modernize" traditional medicine. At times, their initiatives draw on agroecological techniques and permaculture practices. At other times, they find resonance with environmental health efforts as they work to name the impacts of racialized capitalism. *Dawa lishe*, however, is an invitation to distinguish work that strives to address the limitations and erasures structuring even the most subtle and progressive work around traditional medicine, agroecology, and environmental health. It refuses the ontological division of bodies and lands that enables medicines to "work" without fundamentally attending to the histories of nature in which they are embedded and that enables agriculture to develop through technologies that describe negative consequences to human health as "secondary" effects. Remedies reorient the times and spaces of healing by attuning bodily senses. Addressing the lived relations of body and land provokes theoretical sensibilities that support alternative ways of living and dying,

FIGURE 1.2 *Mchaichai* (dried lemongrass) produced and packaged by Dorkia Enterprises, Moshi, Tanzania. Photo by author, 2016.

growing and decaying, composing and decomposing. *Dawa lishe* embodies what Michelle Murphy has called the "experimental otherwise."[2]

Take, for instance, the *mchaichai* (Kiswahili: lemongrass) produced by Dorkia Enterprises, a small-scale entrepreneurial initiative in northern Tanzania. The colorful label on this carefully packaged tisane announces: "It removes toxins from the whole body—those [toxins] that come from the food whose growth we have cut short, aluminum pots, steel wire, chemical medicines [synthetic pharmaceuticals, as well as synthetic fertilizers, herbicides, and pesticides], mental stress, and nicotine, which is from cigarettes and their smoke [and] is the cause of lung cancer. It cleans the kidneys; it removes all the residue stopping urine/plugging up the bladder (especially in elders). It returns a quicker memory and it puts the body in a good and lively state after using."

Here eating, drinking, cooking, healing, and breathing all pose threats as they regularly expose bodies to toxins. *Mchaichai* cleanses bodies that are stressed, clogged, and sluggish as a result of chronic exposure to substances that facilitate modern agriculture, medicine, and domesticity. *Mchaichai* offers a way to mediate the forms of contamination and complicity that are constitutive of twenty-first-century lives and bodies. It is not a targeted or singular cure. Indeed, how could there be for the relentless, low-level assaults on bodies described here? It kindles appetites that might reorient human-plant relations. It proposes an alternative to the sweet (and when possible milky) black tea that has grown to structure both private and communal moments in Tanzania since German settlers planted the first experimental tea at the Amani Research Station in 1904. As the lighter flavors of lemongrass gently divert postcolonial appetites, they invite bodies into livelier economies of taste.

*Mchaichai*, like other remedies described in the chapters that follow, attends to the prolonged depletions and recurring injuries intrinsic to a world in which toxicity has become a condition of life. The double-bind that compels this story forward is not uniquely Tanzanian. A walk through the supplements aisle in a local grocery store in the United States will reveal a wide array of remedies for the ills of modern life that purify polluted bodies, build individual defenses, and promise increased energy. The investment in herbal remedies and nutraceuticals has catalyzed rapid growth over the past two decades. Industry analysts estimated in 2022 that the global market for herbal medicine was USD135 billion and projected that it would reach USD178.4 billion by 2026.[3] The nutraceutical market is even larger, estimated to be USD317 billion in 2023 and expected to grow at a compound annual rate of 9.6 percent in the near term.[4] While these global markets both shaped and were shaped by middle-class concerns with "lifestyle" diseases, the hunger to harness ever-greater consumer spending power has driven companies such as GlaxoSmithKline, PepsiCo, and Coca-Cola to target the rural poor burdened by micronutrient deficiency.[5]

*Dawa lishe* is born of the frictions of this moment in which the therapeutic properties of plants are being rendered profitable as both middle-class health obsessions and humanitarian technologies. It seeks to name initiatives that disrupt the logics of these expanding international markets even while, at times, commercializing plant-based remedies and therapeutic foods. The social and entrepreneurial projects described in this book creatively navigate intellectual property regimes and trade agreements in part by moving between regulatory tracks for drugs and food. They find common cause through their efforts to undo the scalar forms of organization central to Big Pharma and Big Food. In aggregate, I argue, they open a space to redefine the efficacy of individual remedies in relation to their ability to hold the historical tensions over the properties of therapeutic plants rather than to resolve them through the magic of commodity relations.

Such provocations are not easy to sustain. Science, capital, and law are entangled in ways that incentivize mobilizing plants and plant knowledge as a resource for navigating life and countering harms. *Dawa lishe*, in contrast, is an invitation to work collectively toward a radical revisioning of health and a redefinition of care. It is in this sense aspirational. *Medicines That Feed Us* trains attention on efforts that are drawn together by this aspiration. This does not mean that the projects that follow are always able to live up to these aspirations by creatively reimagining the world in which they live. Nor does it mean that all individuals remain unwavering in their commitments to undo the scalar projects through which the accumulation of knowledge, capital, and credibility is generated. *Dawa lishe* as a theoretical proposition is refined through collective reflection on specific projects and reaction to individual products, as well as collaboration between producers emerging in response to the dramatic rise in chronic diseases, including diabetes, cardiovascular disease, rheumatism, and cancer, as well as the persistence of AIDS.

The edges of *dawa lishe* take form in eyebrows raised over a business owner who is taken in by an American dietary supplement and vitamin company, allowing his attention to be diverted from their more radical work as they redesign their clinical space to facilitate gatherings to promote the multinational's pyramid scheme. They are shaped in the subtle distancing from colleagues who, as their projects expand, slide into modes of production that treat land and labor as disposable or whose business practices come to more cleanly sever commodities from the assemblage of relations through which they are made. *Dawa lishe* is also distinguished by the more explicit critiques of a Kenyan man who rents space from a popular upscale bar owner, moving aside tables that gather festive groups in the evening to create a pop-up clinic in the morning. He buys plant products such as aloe juice and rosella in bulk in Nairobi and repackages it in small bottles with well-designed labels announcing his brand of Ideal Health. His consultations inevitably result in recommendations to buy these products and bills that frequently reached TZS 100,000 or more (approximately USD 50, a sum that exceeds half of the government's minimum monthly wage). Anxieties of ecological, economic, and social exploitation are folded into fears of the substances that circulate on streets, in buses, and in small shops in every town. No one suggests that all plant-based remedies are good or safe. Demanding that remedies be nourishing is an effort to recast how benefits *and* risks are evaluated and to hold medicine accountable to the slow violence of extractive ecological and economic relations organizing modern life.

Whether developed through a nongovernmental organization or a small business, *dawa lishe* marks projects committed to organic crop-management practices, composting, and seed sharing. Producers find alliances with food sovereignty movements to be more generative than biomedical collaborations driven by arguments about access to medicines. Products skip back and forth across regulatory tracks for drugs, food, and, separately in Tanzania, traditional medicine. In so doing, they trouble the institutional practices that fix objects of government control and work to expand the intellectually and politically cramped ontics policed by the state.

Anchoring an analysis in Tanzania broadens the space for political critiques and the options for decolonial work, as healing and sovereignty have been linked in this region long before the rise of the modern nation-state. Precolonial vocabularies articulate how harm accumulates in bodies and lands and reveal older practices to navigate the ways that harming and healing are entangled.[6] They provide analytical leverage in approaching twentieth-century struggles over healing. Colonial and postcolonial attempts to control healers reveal the mobilization of what David Arnold has called the "imperial pharmakon" as a technology of governance.[7] Modern notions of toxicity and the relations with remedy that inhere in it emerged through the effort to control people and plants and to harness their energies for empire. It has taken form materially and conceptually within the slow violence of global, racial capitalism—the same relations that gave rise to the modern nation-state. Therefore, problematizing toxicity and its relationship with remedy from Tanzania enables an account that takes the nation-state seriously without making it the foundational subject of the analysis, either implicitly or explicitly.

The social-ecological-therapeutic projects at the core of this book are not faithful to the epistemological and ontological commitments of Eurocentric philosophical and scientific practice, which offer biological bodies as the primary sites of healing, botanical plants as resources for innovation and therapy, and the environment as externalized context. Rather, by directing therapeutic attention to the ways bodies and land move through each other, *dawa lishe* articulates healing as a practice of dwelling. This focus on the relations that give rise to bodies, plants, and soils (re)spatializes concerns about health and sites of healing.

Refusals to forget the ways that colonialism, slavery, missionization, international development, and the extractive industries they privilege have drained—and continue to drain—the capacities of people, plants, pollinators, and other creatures expose toxicity's investment in a liberal form of bodily and territorial sovereignty. Remembering is facilitated by the vocabularies with which Tanzanians reflect on the violence that has "disabled ecologies," as well as deeply embodied practices that shape the ways sickness and struggle are lived and the ways that assistance, accommodation, and care are cultivated.[8] Insofar as precolonial vocabularies remain alive in everyday life, the leverage they provide is not reserved for scholars but also exploited by Tanzanians as they work toward healing in a toxic world. *Dawa lishe* is a product of these histories; it works across their differences to trouble the rhythms and forms of existence that frame modern health.

As *dawa lishe* names projects that apprehend sickness and debility by addressing disabling relations between people, plants, and soils, it also captures a felt sense of the limitations of the bodily and territorial forms of sovereignty that ground modern politics. Producers and users are beginning to build a lexicon (words, objects, practices, relations) through which to consider and (re)define the epistemic objects of healing and their connections to collective governance. They are beginning to articulate a space to debate, dismantle, and reinvent the forms of *therapeutic sovereignty* that have been central to national and international governance.[9] *Dawa lishe* offers an accounting of (post)colonial "struggles for control" through acts of care for people and plants—healing for communities and soils—that strive to nourish alternative forms of sovereignty through a redefinition of the therapeutic.[10]

*Medicines That Feed Us* examines social-ecological-therapeutic projects in Tanzania as they are reworking the scales, times, and spaces of "health."[11] These projects, at their most potent, reimagine the body through its incorporations and excorporations in order to raise critical questions about power and justice. They are, at their most radical, a reinvention of the forms of political sovereignty that have defined possibilities of independence and autonomy (through the nation-state and the body) in the past century.

## The Ground for Argument and Action

In 2018, I attended Slow Food International's biennial Terra Madre gathering in Turin, Italy, with several Tanzanian friends and colleagues. I sat with them during a moving conversation between Amitav Ghosh and the Indian environmental activist Sunita Narain on "Climate Change: How to Face the Biggest Challenges of the Coming Decades." I watched my friends' faces when a Kenyan man at the end of our row of seats stood up and asked the speakers, "Should we conclude that a country needs to become a powerful polluter in order to get a seat at the table discussing climate change?" The irony in the question and the frustration its dark humor released point to the frictions and fissures in the public sphere. There may be agreement that these are toxic times, but even in these progressive conversations, there is tension over whose histories and imaginations will shape global articulations of toxicity, remedy, and their relations.

Many of the solutions "at the table" are conceptualized through the language and logics of the imperial pharmakon—that is, of the particular notion of toxicity and its relationship with remedy that emerged through colonialism and became important to colonial governance. The identification and circulation of "poisons" through colonial networks fostered the development of the twin sciences of toxicology and pharmacology. Scientific practices of recognition in these fields came to reformulate relations between toxicity and remedy. As these fields displaced precolonial lexicons of harming and healing, they also displaced ways of articulating agency, framing problems, assigning responsibility, and designing solutions.

As toxicology and pharmacology have come to provide the epistemic grounding for medico-juridical regimes of governance, they have also come to shape the lexicon in which they can be resisted. For instance, these sciences implicitly authorize public health studies of environmental exposure. Such studies have proven important in illustrating the uneven distribution of harm by mapping disease prevalence data alongside the location of industrial contaminants and toxic waste. The data generated has supported the development of environmental regulations, a conceptualization of a truer cost of industrial production, the monitoring of industries, and (in the best of cases) the ability to hold them accountable to the people they affect. Effective resistance then becomes a question of scientific capacity and, by extension, political economy. The uneven production of knowledge about the harmful substances being released into the air, water, and soils creates spaces where resistance is possible and where it is rendered mute.

Air pollution provides a particularly telling, if singular, example of the toxicity of global racial capitalism that late liberalism obscures through a politics of substances resting on practices of scientific recognition.[12] The World Health Organization (WHO) acknowledges that pollution is the most significant environmental cause of disease and premature death in the world, and yet, as Gabrielle Hecht points out in her essay on "The African Anthropocene," it has no monitoring station in Africa.[13] South Africa is the only country in sub-Saharan Africa that has been able to consistently support an air-quality monitoring program.[14] Ghana has worked tirelessly to piece together a monitoring system from different shorter-term research grants.[15] The resulting gaps shape the invisibility of pollution in Africa, as do the gaps in other forms of monitoring lamented by the toxicologists striving to investigate and remediate the presence of poisons in Senegal described in Noémi Tousignant's *Edges of Exposure*.[16] The processes of chemical recognition and the institutions that lead to the global and state politics of security have been largely impossible to mobilize in much of Africa. This structural lack is obscured by the portrayal of Africa as a material instantiation of the world's preindustrial past and therefore essentially free of pollution and industrial waste. In fact, in some circles, this assertion has been used to argue that Africa's value is in its ability to absorb more toxicity.[17] This perverse addendum to the extractive logics of racial capitalism twists representations of Africa as a site of raw materials, in order to bolster new claims that its "nature" is perfectly positioned to absorb the toxic waste of the Global North.[18] These comments are not merely in bad taste; they also illustrate the cunning ways that rooting claims to environmental and health justice in a politics of recognition can be manipulated to reinforce structural inequities and keep Africans from having a "seat at the table."

Arguments about regulatory invisibilities can at times lead to a sense that more knowledge is all that is needed to manage the line between toxic and nontoxic substances, between unsafe and safe use, dangerous and acceptable exposure. The image of individuals free of specific chemical compounds provides both a ground for normative legal intervention and benchmarks for measuring success. In the process, it seems to suggest there is a *we* that can be held separate, and thereby safe, from the toxic matter of the world. A growing scholarship on toxicity and chemical exposure is challenging philosophical investments in the possibility of discrete, bounded objects and subjects that undergird the ideal of bodies that can be whole, independent, authentic, and pure, as well as their versions of politics, both progressive and conservative.[19] Expanding Bruno Latour's argument about the practices of purification that have been central to the project of modernity, Alexis Shotwell captures this argument in her assertion that *we* have never been pure.[20] The processes of purification built into scientific and juridical knowledge-making practices structure our forgetting of this fact. Recognizing toxicity as a condition of modern life means remembering. It means remembering that the work it takes to purify human bodies and synthetic chemicals is embedded in the work it takes to solidify a long chain of ontological distinctions between us and them, subjects and objects, and science and culture. It means remembering that this work has proven critical to the formation of contemporary hierarchy and privilege as well as the notions of contamination, corruption, and contagion that justify their maintenance.[21] Such remembering challenges any easy recognition of a *we* who has a right to live uncontaminated, of a toxin separate from and threatening this *we*, and of a neutral position from which to adjudicate this separation.[22]

Shotwell's research clarifies why purity politics offers only deeply fraught spaces from which to argue and act in Africa and beyond.[23] The forms of forgetting developed through practices of purification have justified violence and dispossession through colonialism, nationalism, postcolonial development, and humanitarianism. How we grapple with toxicity is thus a question of how we remember and whose remembering matters.[24] Postcolonial memories shape how Tanzanian publics hear both official news stories and circulating rumors about the continued spraying of DDT,[25] the large graveyards of e-waste from the United States and Europe,[26] the international trade in radioactive waste,[27] the use of cyanide and mercury in mining,[28] and chemical fertilizers, pesticides, and herbicides in large-scale agriculture.[29] Public debates highlight the ways in which the lack of regulations on known toxins is exploited as an asset throughout the continent. Capital's strategic use of regulation is accompanied by a strategic nonchalance to the leakage of toxins into informal markets, such as when the surplus pesticides used by the large-scale flower farms in Arusha made their way into small agricultural shops and were picked up as a cheaper way of protecting tomatoes in local kitchen gardens.[30] In this space, a politics of purity, an ideal of an uncontaminated body, is unthinkable both practically and politically. More evocative is the *mchaichai* that "returns a quicker memory and . . . puts the body in a good and lively state after using." More evocative is the invitation to take in a plant and attune to its transformative potential. This is not a nostalgia for tradition but a call for memory, for a remembering that relations between plants, people, and place have not always been as they are; that they were reorganized through colonialism and continue to be stabilized through large-scale (plantation) agriculture. As a result, liveliness might be found by inserting the body into alternative economies of people and plants. Another particularly powerful site for this memory work is *kitarasa*, an indigenous banana, whose orange sap animates efforts to rescale the therapeutic. In Kilimanjaro, bananas are more than a staple food; they are embodiments of long histories of human-plant collaborations in the making of home, lineage, and health.

For Tanzanians, such remembering also means that everyday toxicities are not only the result of capitalism's offloading of its harmful waste to Africa but also the social-material effect of efforts designed to address insecurity, poverty, and disease. The pesticides and herbicides in food, the growth hormones fed to "modern" chickens, tissue cultures injected into banana plants, the aluminum pots used in everyday cooking, the hybrid (at times, genetically modified)[31] corn whose reproductive strength decreases over generations, and the pharmaceuticals required to address chronic diseases (whether HIV, hypertension, or diabetes) and provide birth control are all held responsible for forcing modern bodies to bear complicated toxic loads. Approaching toxicity as a condition of life rather than an anomaly, however, does not mean accepting sickness, pollution, injustice, poverty, and death. It *does* mean that interventions focused narrowly on modes of recognition (and management) are not sufficient.

The absence of specific voices in conversations about climate change, national discrepancies in contributions to the production of harmful pollutants, uneven community exposure to the harms of toxins globally, and lopsided production of knowledge about the burden of these toxins on the bodies and lands are all co-produced. Together, these asymmetries trace the history of the dispossessions, both fueling racial capitalism and universalizing the European, imperial forms of reason that justified it. Liberal multiculturalism's call for more diversity at the table is at best an anemic response. Identity politics does not necessarily challenge the "epistemic line" that divides the rational from the irrational, the true from the false, the scientific from the mythical. In fact, it is often used to hold the line. This was the dark humor of our Kenyan colleague. His provocation lay not in why an African was not on the stage with Ghosh or Narain, or even how a similar conversation might happen with African intellectuals and activists, but his provocation lay in the bold assertion that polluted landscapes and bodies-at-the-table are of a piece.

## Healing in the Anthropocene

The growing recognition that the very processes through which modern life has developed are also the processes that threaten human and nonhuman survival has animated a search for ways to speak about and apprehend pollution and toxicity as processual relations, not just matters out of place.[32] One particularly charismatic space has been the flurry of articles, books, conferences, and art projects generated over the past two decades in the name of the "Anthropocene."[33] The ecologist Eugene Stoermer and the atmospheric chemist Paul Crutzen first proposed the "Anthropocene" in key publications in 2000 as one way of temporalizing how humans have altered the matter of the world.[34] They drew together diverse scientific work and consolidated a range of other terms that emerged over the late twentieth century, in an effort to argue that the biogeological dynamics of what has been referred to as the Holocene no longer accurately describe Earth's systems today. The changes in Earth's atmosphere (warmer), flora and fauna (less diverse), sea levels (higher), as well as carbon and nitrogen cycles (more rapid) and phosphorus cycles (downregulated) are, scientists tell us, irreversible and defining. Yet, locating change in geological sciences centers attention on markers visible in the rock strata. Debates over whether these markers signal an end to the Holocene or an event within it privilege geological time. Proposals to declare a new epoch demand a disciplinarily legible origin story.

Many have highlighted the political implications of any given geological marker indicating the beginning of an era in which humans fundamentally altered the composition of the world. What does it mean to identify the mid-twentieth century with the acceleration of industrial production, the rise of the chemical industry, and the testing of the first atomic bomb, versus the beginning of the Atlantic slave trade and the violence of turning both life and land into property, versus the European "discovery" of the New World and beginnings of settler colonialism, versus the first evidence of human agriculture that changed the course of life and nonlife on the planet? Origin points are never neutral. Furthermore, as others have demonstrated, all proposed origins of the Anthropocene, with their emphasis on the *anthropos*, manifest an ideological vision.[35] The Anthropocene is compelled by commitments to humanism, embedded in racialized and racializing economies of knowledge, and circumscribes notions of justice. As a proposed name for an era only beginning to be lived, the "Anthropocene" consolidates these ideological commitments in articulations of a future described through their impacts for millennia to come. In so doing, it also shapes the ground on which global justice might be articulated.[36] In response to these critiques, other terms, other starting points, other ways through this moment have rushed forth: the Plantationocene, the Capitalocene, the Chthulucene, the Ravenocene.[37] Each calls us to re-temporalize and re-spatialize accounts of the moment in ways that might enable us to imagine and work toward more just, habitable relations. Most continue to hold separate the biopolitical work of environment and health.

The inextricability of ecological and human health, however, is growing less possible to ignore as the harmful waste of human industry changes the matter of rock, soil, water, and air and as the products of human ingenuity transform plants, bacteria, fungi, and animals. Previously siloed experts in both the environmental and medical sciences have been driven to think together. They grapple with how to address their twinned challenges: as biodiversity and health,[38] as climate change and human survival,[39] as capitalism and newly innovated renewable economies, as One Health,[40] or Planetary Health,[41] or Sustainable Futures. All of these initiatives point to concerns about the ways that widespread environmental change is shaping both human health and the landscapes in which health is possible. All mobilize techniques for articulating "the problem" and strategies for intervening in "it," which are themselves relations of power. Most remain tied to Euro-American historical notions of what Donna Haraway has called nature-cultures and their commitments to the centrality of human needs.[42] Both the framing of, and increasingly the solutions to, the challenges humans face as a result of rising temperatures, increased carbon in the atmosphere, rising sea levels, salination of the oceans, widespread microplastics, emergence of new infectious diseases, and increasing rates of chronic disease are dominated by economizing logics that articulate the greatest threat as failing ecosystem services. The choice of grammar we use to describe this moment, the economies of knowledge we mobilize, and the histories we create promise to reinforce or destabilize the inequalities and erasures that have been critical to modern world-making practices.

In *The Great Derangement*, Amitav Ghosh argues that "the climate crisis is also a crisis of culture, and thus of imagination."[43] It remains unthinkable within the circumscribed forms of knowledge and politics born of the marriage of imperialism and capitalism. When imaginations are stuck maneuvering stiffly within consumption-driven logics, salvation from ecological and health crises leans heavily on dreams of spectacular technological solutions.[44] As Ghosh turns for inspiration to premodern forms of storytelling and poetry in South Asia for apprehending the exceptional and the catastrophic, the historian of public health and medicine Julie Livingston is one of few who offer narratives of our global moment that start with Africa. In *Self-Devouring Growth*, a parable set in contemporary Botswana, she entreats her readers to consider what rain-making might look like on a planetary scale. She asks what structures might create "the forms of collective self-agreement necessary to coax the climate." Her planetary parable suggests that contemplating ways of knowing deemed irrational, superstitious, and marginal by colonialism and enlightenment reason might "contribute to the unlocking of our collective imagination."[45]

I am interested in how the realities that shape ways of being, qualities of living, and possibilities of healing in Tanzania might re-situate our analytics of ecological and health crises and open up spaces for collective action. The languages and logics through which Tanzanians confront toxicity and remedy recognize the extractive relations that shape our contemporary geologies and sociologies yet exceed their imaginaries. *Medicines That Feed Us* argues that *dawa lishe* is a way of problematizing the present and shaping the space in which solutions might be formulated. It is a way of making a proposition about what is happening and why. That is, *dawa lishe* is a mode of theorizing, as well as a modality of care. It is a way of accounting for the relations of power that bear down on the present, burdening some bodies with toxins more than others, driving up rates of hypertension, diabetes, and other "chronic diseases" unevenly. In the pages that follow, I describe this accounting as it is emerging through the work of Tanzanians, several of whom were sitting with me listening to Ghosh and Narain at the Terra Madre gathering in Turin, Italy. They nodded affirmatively, supporting the Kenyan provocateur at the end of the row. After all, they do not offer *dawa lishe* as a celebration of Tanzanian (or African) specificity, or as a salve for postcolonial ills. Rather, they offer it, I argue, as a program for the dislocation of the imperial pharmakon and a reinvention of relations between toxicity and remedy through innovative work rooted in alternative relations between healing and sovereignty.

## The Politics of the Imperial Pharmakon

In *Toxic Histories*, David Arnold defines the "imperial pharmakon" as the specific configuration of science, law, and economy through which colonial administrators managed the constitutive ambivalence of the *pharmakon*—that remedies are also poisons and vice versa—in the service of empire. The term glosses the techniques through which contemporary notions of toxicity were forged as a solution to problems of knowledge, politics, and economy at the intersection of colonialism and capitalism. While toxicity is in this sense a "global" concept, universalized through the sciences, Arnold argues that it embeds itself differently in different places as it navigates the layered histories of local poison cultures. His interest lies in the place-based specificities of the rise of toxicity as a site of biopolitical governance. Ultimately, *Toxic Histories* excavates the place of science in the making of the modern state. Arnold situates his history in India and points to the diverse and layered poison cultures British colonial administrators sought to control. Poisons have a similarly complex history in East Africa. British colonial efforts to manage these histories were central to tactics of governing bodies and populations in colonial Tanganyika. In fact, techniques were shared across the empire as administrators (and plants) moved between territories in South Asia, Africa, and elsewhere.

Arnold makes a powerful argument that the imperial pharmakon produced "toxicity" as a solution to colonial problems of knowledge and politics and masked the violence of colonialism itself. In colonial Tanganyika, like in India, poison cultures attracted the attention of the colonial state insofar as poisons threatened to resist, complicate, or disrupt colonial rule. Scientific techniques for identifying the "toxic" in poisons facilitated strategies to contain or offload it, as well as efforts to (re)mediate and direct its effects. Medical and juridical infrastructures consolidated "toxicity" through forms of proof and kinds of evidence that located the problem and potential of the toxic in substance, thereby obscuring the dispossessing relations through which racial capitalism systematically depleted and disabled bodies and lands. Other ways of organizing and of being in the world—ways that enable people to articulate power and sovereignty otherwise—were rendered illogical, mythical, and ignorant. The silences built into this marginalization continue as the imperial pharmakon shapes postcolonial infrastructures, grounding both state control and resistance to it.

Arnold's recognition of diverse poison cultures frames his argument that the centering of the imperial pharmakon required the decentering of other ways of healing and harming. His historicization of the imperial pharmakon as a technology through which self-identical substances ground biopolitical governance, however, stops short of questioning the metaphysics of poison. Asserting "India"—the emerging nation-state—as the historical subject that grounds his account leads him to narrow his focus around colonial and post-colonial knowledges and institutions that articulate toxicity as frozen in substance. Poison is apprehended through the articulations that developed with the rise of toxicology, and these then organize a reading back into precolonial South Asian history. Toxicity as a modern concept and the nation-state as the legitimate form of modern political sovereignty not only arise together but also reinforce each other, making both seem inevitable: Toxicity is an intrinsic component of evolutionary change, and the state is the universal conclusion in the evolutionary development of complex societies.

While I have learned from Arnold's careful attention to the forms of inclusion and exclusion through which participation in the state is organized in the name of toxicity, *Medicines That Feed Us* stays attentive to the erasures of other ways of knowing and insists on "provincializing" European configurations of bodily and territorial sovereignty. I explore the imperial pharmakon as an "epistemology of unknowing," in the sense that Vimalassery, Pegues, and Goldstein articulate in their essay "On Colonial Unknowing."[46] Practices of "unknowing" render slow violence and dispossession invisible. In the service of these occlusions, they obscure forms of sovereignty that are not rooted in the nation-state and its concept of citizenry. This obfuscation can feel ironic, such as when the politics of the imperial pharmakon is pushed into the future through the category of "traditional medicine" and projects of "integration."

Through my work in Tanzanian clinics and hospitals since the mid-1990s, I have been confronted regularly by biomedical and public health specialists (Tanzanian and not) who assert that traditional medicines poison people. Public health initiatives strive to convince traditional midwives and healers to stop administering medicines, and trainings continue to develop healers and midwives as a referral network for the clinic. The rhetoric advancing "integration" suggests that traditional healers and midwives are uniquely valuable because people trust them, not because they know things. They offer a solution to the labor shortage insofar as they limit their work to convincing the sick and the pregnant to attend the local clinic or health center sooner than they otherwise might. As trainers repeatedly emphasized, however, healers and midwives were *not* to administer medicines; they were not to do the work of healing. Chapter 3 traces the infrastructure through which healers' integration systematically dismantles their power to manage tensions between healing and harming.

In *Epistemic Freedom in Africa*, the historian and theorist Sabelo J. Ndlovu-Gatsheni identifies the recognition of infrastructures that support practices of unknowing as the first step to what he calls "epistemological decolonization." A second is the recognition that while other ways of knowing, modes of attention, and techniques of world-making were marginalized, they did not disappear. The forms of knowledge and politics rendered unspeakable through the imperial pharmakon receded from the working of formal state institutions, but they are alive in other spaces. Today, the frictions generated by layered histories of knowing and forms of embodiment shape the ways that Tanzanians approach healing in (and of) a toxic world and how they are reimagining sovereignty when toxicity is a condition of everyday life.

Given that toxicity as a modern concept has been co-constituted with particular notions of sovereignty, a rigorous engagement with questions of justice demands rethinking toxicity. *Medicines That Feed Us* joins a rapidly growing body of work emerging at the intersection of science, humanities, and art that is committed to rendering visible toxicity's relations and its entanglements with dispossession.[47] In such work, toxicity is recognized as a quality of relations rather than limited to a quality of substances. This work defines the toxic as that which is beyond remediation, that which has been forcibly torn from its place, unearthed and alienated, and that which must be mediated through exposure and dosage.[48] In the process of formulating *dawa lishe*, my colleagues in Tanzania extend this conversation as they recast the toxic as depleting, injurious, dissociative, dismembering, attenuating, barren, infertile, and exhausting. Focusing on toxicity as a quality of relations sedimented in the making of substances reveals it as the accumulation of deeply unequal and unjust racialized economies.[49] The complex and shifting notions of consumption and growth driving these economies have transformed landscapes, or in the words of those advocating for the "Anthropocene," they have transformed the substance of the planet. Yet, while toxic spaces are debilitating and sometimes deadly, others have argued that they can also open a (compromised) space for forms of existence that had been rendered impossible. When toxicity names not only the poisons in the land and water as the result of extractive industries but also capitalism's hunger for new frontiers once the desired resources have been exhausted, these toxic spaces hold out the possibility of creative alterlives, if not freedom.[50] As Elizabeth Povinelli has argued, toxicity may in these instances forge a space for limited survival and the development of new sovereignties.[51]

The anxieties and potentials that animate the "toxic" fuel (alternative) medicine, agriculture, and cultural life. Some of the most provocative interventions in art and activism, as well as in scholarship, suggest that the toxic demands new forms of sensing and sensibilities and perhaps alter(ed) bodies.[52]

This conversation across disciplines and publics points to a theoretical proposition structuring this book: Toxicity has become the "ethical substance" of our epoch-in-transition (whether we label that with an -ocene or not). This proposition draws on Foucault and his articulation of "ethical substance" in the second volume of the *History of Sexuality (Uses of Pleasure)*, as the matter that raises the most impactful moral questions of a historical moment.[53] He develops this ontological argument through his genealogies of Greek and Roman sexual ethics, and he concludes that bodily pleasure is the ethical substance that came to define the ethics of modernity. If reflection and labor over the aphrodisia generated the dynamics through which bodies and selves came into being in the nineteenth century, then, I argue, reflection on and labor over toxicity are dismantling and reinventing the ontics of the body and self in the twenty-first century.

### Toxicity in Translation

Confronting the overrepresentation of the imperial pharmakon clears a space to see other locations for thinking and acting in response to the moral problematic posed by modern toxicity. Ndlovu-Gatsheni argues that while such efforts toward "provincializing" Europe are necessary, they are wholly insufficient. Epistemological decolonization also requires what he refers to as "deprovincializing Africa."[54] I take Ndlovu-Gatsheni's call as an invitation not only to attend to the ways that the pharmakon has been thought (and its ambivalence managed) otherwise in Africa but also to take them as legitimate locations from which to interpret modern notions of toxicity and remedy. Such efforts, he notes, push beyond work to develop "theory from the South," as they refuse to be restricted to the forms of speech and ways of thinking recognized as "theory" by the Global North.[55]

One way to theorize while expanding the limits of theory is through the frictions and fissures of translation. Souleymane Bachir Diagne argues that all thought is generated in translation, that thinking is the process of working across languages and the ideas, practices, styles of comportment, landscapes of relating, and possibilities for being in which they are entwined.[56] Tanzanians work across English, Kiswahili, and a range of local languages every day. Among the people who animate the following chapters, these include multiple dialects of KiChagga as well as KiMeru, KiPare, KiArusha, KiKagera, and KiMaasai.

Communication in the region is layered with explicit and implicit translations. Reducing these translations to better or worse—innocent or noninnocent—efforts at drawing equivalence misses their power. Translating dissolves some, and surfaces other, modes of existence. It stages interruption and invites invention in the "in-betweens" created through the forms of difference-making that structure postcolonial modernity.[57]

*Dawa*, a word that will appear often in this text, illustrates the ways that translation composes and decomposes worlds. As the dictionary from the Institute of Kiswahili Research (TUKI) at the University of Dar es Salaam identifies, the dominant English terms used to translate *dawa* are medicine, medication, medicant, or drug. Yet, *dawa* as it lives in everyday life is never only medicine and never only material. *Dawa* refers to healing plants and pharmaceuticals, as well as to rat poison, fertilizers, pesticides, herbicides, holy water, Qur'anic verses, Christian prayers, and evil looks. *Dawa* is swallowed, rubbed on skin, showered over bodies, worn bound in amulets, whispered to people, spoken over plants, sprinkled on dirt or in food, buried at the crossroads, or tucked above door frames. *Dawa* is a liveliness that intervenes in the forces that body forth humans and nonhumans, the vital and the elemental. *Dawa* has brought rain and stopped it; injured and turned the forces of injury back on the one who initiated them; brought wealth, love, and fertility; and depleted bodies, lands, patience, and fortitude. *Dawa* may be experienced as *kali* (fierce, impactful) or *baridi* (cool, gentle), but its capacity to be therapeutic or injurious is not (primarily, at least) a quality intrinsic to it. The effects, a *dawa*'s efficacy, are a product of the relations that it mobilizes and that mobilize it.

Connoting relations far more expansive than those of the English "medication" or "drug," *dawa* might be more faithfully translated as pharmakonic substance, yet only if the metaphysics of substance itself is taken as a field of inquiry. *Dawa* in Kiswahili captures the constitutive ambivalence of the pharmakon while holding together the qualities rendered incommensurable through the materialism of the imperial pharmakon (e.g., the mythic, religious, or fraudulent). For this reason, it has generated and continues to generate friction in medical and juridical infrastructures. That which is referred to as *dawa* is regulated through laws concerning medicine, agriculture, religion, and witchcraft. As *dawa* ranges across the categories of modern governance, it threatens to destabilize them. It troubled colonial—and continues to trouble postcolonial—efforts to manage substances through the epistemic and ontological settlements of the imperial pharmakon.

The relations of *dawa* are more thoroughly grasped within the complex philosophical relations of healing and harming that inhere in *uganga* and *uchawi*, terms often translated as medicine and witchcraft, respectively. These common translations are themselves, however, acts of decomposing precolonial worlds and composing colonial ones. Colonial officials advocated for "witchcraft ordinances" to control healers (*waganga*) whose work destabilized colonial authority. Officers had neither the resources nor the inclination to use the witchcraft ordinances to join together with communities to address the harms of those accused of *uchawi*. Rather, these ordinances provided a legal mechanism through which to reshape healing in allegiance with the colonial state.[58] As colonial scientists formulated the imperial pharmakon (in no small part) through botanical studies of local plants, they interrupted the ways that *dawa*'s constitutive ambivalence was managed through the lexicon of *uganga* and *uchawi*. Investigations in field stations and laboratories, such as the Amani Research Station in northeastern Tanzania, elucidated plants' potency through the materialism of the sciences. Colonial ethnobotanical field studies, botanical gardens, and later postcolonial phytochemical efforts internalized the efficacy of plants, whether for medicine, food, pesticides, building, or other uses. Such ways of knowing and mobilizing plants were effective in conceptualizing them as botanical resources critical to the wealth of the state and the livelihoods of its citizens. In part, because scientific articulations of plants-as-botanical were linked to state projects in medicine and agriculture, they threatened to overwhelm other ways of working with and living through plants in this region.

Yet scientific articulations of plant potency have never been exclusive. East African healing traditions do not take a plant itself to be intrinsically medicinal; rather, the relations into which plants are called can invite healing (or harming) effects. As others and I have argued, herbal remedies in Tanzania and Kenya do not locate efficacy inside the internal matter of the plant— for instance, in secondary metabolites produced through the plant's stress responses, which phytochemists often identify as active. Rather, a plant comes to engender effects that might be deemed therapeutic when taken up through a set of relations that include a healer, their ancestors or spirits, the patient (and their ancestors or spirits), the forces causing illness or debility, and the dynamics that influence the plant growth (and their relations). A plant heals, in other words, through its work of relating and in being related to, its ability to gather and disperse, and its movements that bind together or bypass and leave alone. Whether or not an individual says that they "believe" in traditional healing, these alternative ways of approaching plants contribute to the rich vocabularies through which Kiswahili speakers continue to access diverse historicities of plants.[59] Descriptions of healing and harming in Tanzania are not limited by a vocabulary that fixes self-abiding substances inside fields of social and institutional power. Accounts of healing are accounts of bringing into being that which has capacity, agency, and endurance. Accounts of harming are accounts of bringing into being that which is depleted, disabled, and infertile.

References in Kiswahili to something as *uganga* or *uchawi*, then, are *not* assertions of fixed categories of practice or expertise available for government regulation (medicine and witchcraft, respectively). Rather, these speech acts are mobilized as a judgment on the ethics of the dynamic relations that give rise to the subjects (bodies) and substances (medicines, poisons, written words, etc.) at hand. Through *uganga* and *uchawi*, Kiswahili (and other Bantu language) speakers maintain a lexicon that does not forget, that cannot forget; that matter itself is a relation of power. By dissolving substances back into the relations through which they emerge, this lexicon holds out a space for articulating an alternative metaphysics of substance as actionable. In so doing, it works against the scientific and juridical practices that render dispossession invisible by freezing the toxic in discrete substance (whether at the scale of the botanical or the chemical).

This book tells the story of Tanzanians who are rethinking modern problems of toxicity through the rich, multiple histories of *dawa* alive in Tanzania today and reflecting on the practices through which *dawa*'s constitutive ambivalence is negotiated. *Dawa lishe* strives to name their efforts to navigate the incommensurabilities built into the lexicons of healing and harming in Tanzania. As producers work with the research institutes, regulatory bodies, and clinics, they shape their interventions by troubling tensions between the "traditional healer" and the "scientist" (alternately: the ethnobotanist, phytochemist, and pharmacologist). When engaged as contemporaries, the healer and the scientist each consolidate a different set of discourses, affects, and tactics used in shaping the relationship between remedy and toxicity, healing and harming. In so doing, these figures render visible continuities in precolonial and colonial relations of power. The healer is not marginalized as an anachronism or engaged as a living archive of primitive traditions that might be exploited by the botanist or their scientific colleagues. Rather, the healer's co-presence with the scientist denaturalizes any one depiction of relations between bodies and environments, as well as any one way of linking substance and sovereignty. By "staying with the trouble," as Donna Haraway would say, *dawa lishe* providers cultivate their access to the different modes of existence that each of these figures proposes and the human-plant-soil arrangements through which they emerge.[60] They illustrate that refusing to explain away epistemological and ontological difference, through a social evolutionary logic captured in bioprospecting's assertion that healers point to plants most likely to evidence phytochemical activity, does not necessarily mean a descent into relativism. Rather, by maintaining the incommensurabilities that the "healer" and the "scientist" index, they strive to conceive and intervene in the inequity that sediments in bodies and lands over time.

This book also illustrates the ways that some Tanzanians are exploiting the unruliness of *dawa*, as well as the frictions in everyday translations of healing and harming, in order to (re)formulate and (de)compose the version of toxicity produced through the imperial pharmakon. The argument that follows, then, is not limited to the ways that the modern concept of "toxicity" is localized. I am suggesting that *dawa lishe* responds to the inadequacy of nineteenth- and twentieth-century notions of toxicity by turning toxicity itself inside out. It is a way of intervening in thinking that has been overwhelmed by the imperial pharmakon, of dislocating colonial ontics, and of experimenting with ways of being otherwise.

Toxicity, as it is remade through *dawa lishe*, undoes the scales through which we understand agency and reorients the boundaries of life. What this means practically is that when toxicity is not (only) a material quality essential to the identity of a substance, it cannot be effectively managed by mediating the thresholds of acceptable exposure to substances (whether considered active ingredients or poisons). Theoretically, this means that remedies are not another strategy for reasserting bodily sovereignty against capitalism's stealthy trespasses through the air, water, food, and medicine (i.e., they are not mobilizing indigenous plant knowledge for postcolonial ills). Rather, *dawa lishe* names efforts through which sovereignty itself becomes a site of creative reinvention. What this means for the writing of the stories that follow is that bodies and plants, institutions and land, fade in and out of view. A sense of unevenness can emerge from this effort to unsettle the forms of vision and corresponding aesthetics of storytelling that require ontological solidity to naturalize the scalar logics of modern scientific, legal, and bureaucratic regimes. Toxicity and its relations with remedy and memory become a site to experiment with collective action that disrupts objects of analysis and politics. In order to surface the multidimensional, heterotemporal harms attenuating the strength of bodies and depleting the capacities of land in Tanzania today, these stories seek to surface intimate land relations that have been rendered invisible through analyses that hold tightly to botanical plants and forms of justice possible in their wake.

By attending to the interpenetrations of people and plants, and their (de) composition into soil, Tanzanians explore "the enmeshment of flesh with place" and its implications for modes of healing that recognize our ontological inseparability from the world.[61] This starting point was brought home to me by a woman working at one of the herbal clinics where I have been conducting research since 2013. Through our many hours together in the face of the pain of those seeking help at the clinic, in the camaraderie and joys of collective efforts to support them, in the rush of multiple demands, and in the boredom of long hot afternoons, Romana and I had grown very fond of one another. One afternoon, she shared her concern about the toll my trips back and forth between Tanzania and the United States would inevitably take on me. The dis-ease of my body, as it had to adjust and readjust to these different environments, would benefit from some attention. To cultivate this dual orientation, she recommended a strategy that she had learned from a German missionary in the area. When I traveled back to the United States, my friend told me, I should take a little dirt from Tanzania. On arrival, I should mix it with a little water from the United States and drink it. On my return to Tanzania, I should do the opposite: bring US dirt, mix it with Tanzanian water, and consume it. This way, my body would be constituted in the interstices of these two lands; it would be of both places and could make the corporeal translations and shifts necessary to my constant returns.

The vulnerability of a body is implicated not only in exposure to toxic substances but also in the very movements that give rise to the conditions and labors through which strength is constituted. In his effort to account for the toxic in North America, Nicholas Shapiro has argued that "bodies are sites of both actively absorbing the world and being put into motion by its constituent medley of human and nonhumans."[62] *Dawa lishe* resonates with such efforts to rethink toxicity and extends them. Healing is not necessarily limited to managing what the body absorbs. The phrase offers a way to call out the times and ways that producers and providers heal by intervening in human–nonhuman relations to affect how specific bodies are put into motion, how they dwell. *Dawa lishe* retheorizes the entanglements of bodies and ecologies by reorienting what it means to heal in a toxic world. Remedies are less dedicated to harnessing the internal properties of self-referential objects (body or plant) and more focused on cultivating ways of dwelling, which might make it possible to respond to ongoing disruptions of the possibility of being well together. Justice is less restricted by efforts to manage the boundaries of ontologically stable entities (institutions or land) and more concerned with mediating the knot of relations they hold in place.

## Note on Method of Un/knowing

Although anthropologists talk more than many others of the importance of relationships and building rapport, too often this "rapport" is in the service of extractable data and authorial economies. Within the genre of the monograph, a claim that this book is a product of my ethnographic work between 2008 and 2018 in Tanzania could slide by relatively uncontroversially. Yet, a commitment to co-create accounts in the service of decolonization means working against the grammar of such a claim. The reference to my agency and to disciplinary labor directs attention to the techniques of data collection, to the data itself, and to the distribution of property rights. Even the dates belie anthropology's constitutive ties to an economy of knowledge that locates author as producer and owner of knowledge based on extractable data as the decade highlighted refers to the time frame of the grants that funded me for the travel and for the research that is officially connected to this book (rather than the previous one or the next one, or someone else's). During the research that animates this book, "data" was not the goal, and in the analysis, "property" is being explicitly troubled. Recognition of the fact that the "ethnographic work" in any book is neither solely the author's, nor exactly "work," is too often relegated to the acknowledgments section rather than encountered as a methodological (and writerly) challenge.[63] For this reason, while I have many to acknowledge and much for which to be grateful, these final sections of the introduction draw the people and plants who made this work possible beyond the acknowledgments into the body of the text in order to think our co-laboring as method and the resulting text as relation.

*Medicines That Feed Us* is the product of a series of invitations offered in the midst of friendships, collaborations, a few explicit disagreements, and more subtle refusals. They were often incremental, emerging during everyday tasks, small collaborative projects, strolls along the road between events, and lunch or tea together as we fortified ourselves for more work as well as through the work itself. Each invitation came as part of a process of confronting the ways that my own expertise is embedded in histories of colonial unknowing and defined through practices of seeing, sensing, speaking, and writing that have supported positioning Africa as a resource for first colonial expansion and then postcolonial development. Each invited me (sometimes together with the one who offered the invitation) to push against the categories and practices—dispositions, vocabularies, and styles of engagement—that render the violence of dispossession invisible. As the research has been an unfolding of invitations (rather than discovery), writing has been a process of recollecting the thinking and acting possible in the wake of the invitations (rather than my findings or data). *Medicines That Feed Us* is offered in the service of our ongoing work together and what my colleague Mama Mtweve calls our "coevolution."

The invitations that mattered started in the late 1990s, when Binti Dadi took me under her wing, a full decade earlier than any grant through which I conducted formal research for what has become this book. As a graduate student, whose Kiswahili was then nascent and whose articulation of a research project was still hazy, Fatma Dadi, or Binti (daughter of) Dadi, as she was known throughout southeastern Tanzania, invited me into her home and family and became my teacher and my mentor. My indebtedness to her still grounds all my relations in Tanzania and many well beyond. What I understand of Tanzania as a place, how I engage with people and plants, how I imagine the possibilities and impossibilities of these relations, and how I sort through my own complicity in structures of global inequality have been and I imagine always will be shaped by Binti Dadi and her family. Binti Dadi once told me that her *majani* (those familiars that guide her life and healing) may find me one day. I do not yet know if they will, but her spirit presses upon me, my work, and all I do. As I write, my WhatsApp is ringing with incoming self-portraits of Binti Dadi's youngest grandchild, who was born only a few weeks after I began my doctoral work and is now grown and accomplished in her own right, having recently finished teachers' training college. My first book, *Bodies, Politics and African Healing*, which was dedicated to this grandchild, shares what I understood at the time to be my learnings in my most intense years of research with Binti Dadi. To the government, Binti Dadi was known as a traditional birth attendant. During the years that I regularly spent long days in her home, she attended only a few births, and it is unclear if in this part of southern Tanzania on the Makonde Plateau, there was ever a tradition of all births being attended by such "experts." For Binti Dadi, healing involved listening through Islamic forms, attending to plants, remaining sensitive to relations with ancestors, being climbed on by spirits, and learning to engage bodies already shaped by biomedicine. Her days focused on farming as well as what health professionals might call "reproductive health." She worked to help others conceive, maintain pregnancies, give birth, and welcome energetic newborns, as well as navigate threats to connection, reproduction, and vitality. Her therapies also focused on feeding and nourishing these newborns, securing them to body and land so they would not slip away too early. While registered with the state, she felt that district-level efforts to integrate traditional healers and birth attendants into the health care system were rather anemic. She did not waste energy shunning them or stretch to participate in them. She did, however, wonder out loud about the possibilities of collaborating with scientists. So, this book might be said to have been kindled by Binti Dadi's curiosity.

Wanting to explore the spaces where Binti Dadi saw possible collaborations, I spent six months in Dar es Salaam in 2008 investigating the "modernization of traditional medicine." I located myself within networks of scientific work at the University of Dar es Salaam, Muhimbili Medical Center, and the National

Institute of Medical Research. Dr. Ken Hosea's microbiology laboratory, investigating traditional knowledge, offered the most dynamic site. At the time, he and his graduate students were particularly interested in the antiviral and antibacterial properties of two traditionally fermented "foods": *idundu*, a moldy banana cultivated by the WaPare in northeastern Tanzania for postpartum women, and the stirring sticks Chagga on Mount Kilimanjaro traditionally used for brewing banana beer. I hung out and made myself as useful as possible in the lab, learning to handle Petri dishes and cultivate bacterial and viral growth, as well as developing an eye discerning enough to count different kinds of growth by shape and color. I read drafts of grant proposals and thesis chapters. I spent long hours talking about the wide range of pressures shaping the kinds of work Ken could do and strategizing ways to overcome the obstacles with him. He shared some of his efforts to offer his expertise on questions around genetically modified organisms as an active member of the President's Biosafety Commission. Through his work and the kinds of close ties built through elite secondary schools in Tanzania, he knew many in government and generously introduced me to colleagues who were grappling with questions of public health, scientific ethics, and legal technologies from a range of professional positions. Only in retrospect did I appreciate that it was Ken who first illustrated the power in dissolving the hard boundary between food and medicine, even as his vocabulary differs from my own. He used *idundi* and the beer stirring stick to open up ways of asking what sorts of food might be understood as therapeutic and how. Interestingly, Ken identifies as Chagga himself. When he wanted to investigate if genetically modified organisms (GMOs) had entered Tanzania illegally, he went to the seed shops in Kilimanjaro over Christmas break to ask for the best seeds to buy as a gift for his mother. At that time, I did not know that I would come to spend many years in Kilimanjaro and come to feel the impact of the affective force that pulls so many Chagga back in December. Ken's subtle sense of the relations through which knowledge is generated and through which matter could be therapeutic or harmful made him brilliant to some and dangerous to others. His struggle to generate rigorous scientific research and thoughtful public debate needed to address pressing questions in science, policy, and law was inspiring. It also highlighted the narrowness of the space in which scientists might articulate their work.

Binti Dadi and Ken Hosea both pointed the way. They shape this book's scope, even if much of my work with them falls outside of the main throughline that drives *Medicines That Feed Us*. They taught me where to look and how to frame questions when I got there. During this time in Dar es Salaam in 2008, I also met John Ogonidek and Victor Wiketye as well as their Head of Department (the late) Gloria Mbogo, in the Traditional Medicine Research Unit of the National Institute of Medical Research (NIMR). John and Victor became good friends and trusted colleagues of mine over the years. You will come to know them in the pages that follow as their scientific knowledge, curiosity, and compassion gave shape to this project, and their friendship has long grounded me. In the time that this research unfolded, Victor married. We all had children. John named his youngest Stacey, and he affectionately referred to her as "the professor." In the early days in Dar es Salaam, they helped me to understand the layout of the state's scientific interest and increasing investment in traditional medicine. The negotiation around NIMR's acquisition of a research station for traditional medicine developed by an Italian nongovernmental organization was already underway. John and Victor not only worked to facilitate this transfer but also soon found themselves reassigned to Arusha in order to manage and direct the new research station. Their move proved a major factor pulling me north. In addition to the infrastructure of the research station, NIMR inherited collaborative relations with five healers. I saw John and Victor's vision of good science most clearly through their drive to generate good relations with these healers and the collective work they sustained with three of them. The ways John and Victor opened up new horizons of what it might mean for healers and scientists to produce knowledge together, as well as their interest and care in the professional and life trajectories of these healers and their incorporation of the healers' treatments into their own lives, suggest ways that traditional medicine might be developed as a decolonial science. Yet, most of this was not, perhaps could not be, the official work of the NIMR research station—even though these were often the most substantial, generative engagements happening there—given the political battles and jealous fighting over resources that left the research station's phytochemistry lab understaffed and underfunded.

As I moved to the northeast, I hoped to work with John and Victor as well as with an interested group at the major teaching-research hospital in the region, Kilimanjaro Christian Medical Center (KCMC). I was compelled by questions of how property itself might be innovated within research methodologies concerning therapeutic plants through the dynamics of carefully built collaborations. While I was welcomed into both lab and clinic, the funding for research in both settings was driven by development projects with goals around specific public health or clinical interventions. The limited efforts generated from evening or weekend work and leftover resources did not constitute the conditions under which my colleagues at the research station or in the hospital could sustain projects with therapeutic plants that expanded property relations. Yet, they

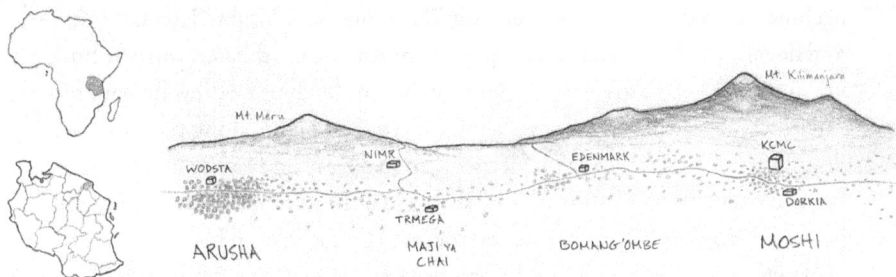

FIGURE I.3 *Upper left*: Map of the African continent showing Tanzania. *Lower left*: Map of Tanzania showing research area. *Horizontal map*: Valley between Mount Kilimanjaro and Mount Meru, where the stories in this book unfold. Drawn by Margot Lystra and Stacey Langwick, 2025.

folded me into the work they did, and I examined how traditional medicine emerged and moved through both settings. I also began (sometimes with John and Victor, sometimes following up on a reference made by a patient in a clinic interview) investigating what at the time I called the emerging herbals industry.

In 2013, I followed a vague recommendation from an outpatient at KCMC to see an herbal clinic at the main intersection in Bomang'ombe, a town halfway between Moshi (where KCMC is located) and Arusha, where John and Victor's research station is located. I grew curious, realizing I must have passed it many times over the previous months. EdenMark was in a strip of shops tucked behind a row of fruit and vegetable sellers at this busy crossroads. Yet, EdenMark's brightly colored van and sandwich board inviting people in for acupuncture, among other treatments, jumped out at me once I was looking. The storefront sat adjacent to a more traditional pharmacy. In many ways, they looked interchangeable, with their glass doors opening onto glass counters filled with boxes and canisters of "medicines" and staffed by women in white coats. Indeed, as I came to see, it was not uncommon for someone to descend from a *daladala* (minibus) across the street and mistakenly walk into Eden-Mark asking to fill a prescription they had just received from the hospital.

Alex Uroki, the driving force behind EdenMark, wants to work at scale. His knowledge of plants grew from conversing with elders on the mountain, reading international research into therapeutic plants and functional foods, attending permaculture workshops, participating in pranic healing gatherings, studying acupressure books, and making connections among peers in Kenya, South Africa, and elsewhere. He travels widely, maintains diverse relations, cultivates eclectic interests, and exudes boundless energy. Plants, people, and machines moved in and out of EdenMark. An intensely curious, lateral thinker (and doer!), Uroki's interest was piqued by an American affiliated with the hospital interested in plants. Our relationship began slowly as he drew me into his activities and relations. He tested my knowledge of plants along the roadside. He gifted me remedies to try, inviting me to make my body a site in collaboration and experimentation, as well as sharing how he approached his own body this way. We would talk as he assessed the large burlap bags in his storage room filled with dried bark, leaves, or tubers; as he examined the plant matter in his large industrial dryer; or as we drove into Moshi together after I had spent a day at the clinic. He wanted to know what I knew and if it would be helpful for him to know. While he came to introduce me to his network of relations, he was not interested in being a traditional research subject. He pressed to be introduced to my network in exchange, expressing interest in people he could work with, or through, or alongside. I brought John and Victor to EdenMark, and they developed their independent relations with Uroki. We traveled to India together with others for a Cornell "partnership" meeting. Uroki's generosity and energy not only opened up a world of activity to me but also reshaped how I might be invited into that world as he sometimes patiently, sometimes impatiently, worked with me to orient my engagement toward building common projects.

The relations central to *Medicines That Feed Us* and the trajectories along which this story is told all unfolded from and/or were folded into EdenMark. I spent many hundreds of hours in EdenMark facilities over the years. The two women who worked in Uroki's main clinic, Romana and Jenipha, drew me into the intimacies and rhythms of healing with "modern traditional medicines." It was from EdenMark that I originally followed plants (e.g., green bananas to Dorcas Kibona, chapter 5), addresses on canisters of medicine (e.g., a remedy called *imarisha yako* to TRMEGA and Mama Nguya and her network of people, chapter 1), and phone numbers on posters (e.g., the glossy poster *Dawa za Asili katika Nchi za Joto*, Traditional Medicines in Tropical Countries, with its sixty colorful thumbnail photos of therapeutic plants, to anamed, the international NGO that produced it). It was Uroki's engagement with the government offices—his registration as a healer and efforts to register others—that first rendered visible frictions with the state. In addition, it was in his office that I met the Slow Food Vice President Edie Mukiibi, who played a pivotal role in the development of the 10,000 Gardens in Africa project and from whom I learned of Mama Nguya's leadership in food sovereignty projects in the region.

Uroki embodies one way of working within the postcolonial, postsocialist, and neoliberal pressures shaping the possibilities of action in Tanzania today: the entrepreneur. His work, however, was always in relationship with the other subject position from which Tanzanians can find political leverage in the first quarter of this century: the NGO. The entrepreneur and the NGO director are deeply entangled subject positions. NGOs in Tanzania support entrepreneurship, and entrepreneurs tap into the organizing and work of NGOs. Neither are outside the state, or the market, or each other. In fact, it was a canister of one of TRMEGA's remedies on EdenMark's shelves—an herbal formula that Romana and Jenipha sold to those with symptoms that indicated immune disorders—that drew Mama Nguya and Uroki together. He had first picked it up at a national agricultural fair. This is the same remedy that drew me to Mama Nguya, who not only conceived of, founded, and directed TRMEGA, as mentioned above, but also served on a range of NGO boards (some for decades) and was appointed the Slow Food coordinator for northeastern Tanzania.

Mama Nguya has a gift for elevating others. She fostered the leadership of Rose Machange and Jane Satiel Mwalyego, and she invited me into relation with them. Both Rose and Jane are skilled community organizers, although they originally found their footing in different spheres. Rose came up through Women Development for Science and Technology (WODSTA), an NGO started in 1990 as the issue of gender drove international development agendas. This women's membership organization recognizes farming as the basis of many women's livelihoods and has come to focus on sustainable agricultural practices that will support women and their environment. Early on, the more elite women who founded the organization enlisted Rose as a "grassroots woman." Over the past twenty-five years, she has come to be a primary animating force for WODSTA's social projects. Mama Nguya has long served on WODSTA's advisory board. Admiring Rose's intelligence, honesty, and hard work, she has advocated for her to be recognized not only as an exemplary "grassroots woman" but also as a leader in the organization. Rose now holds the professional title of WODSTA "community mobilizer." When Mama Nguya founded TRMEGA, she pulled Rose and her women's groups into their work. Rose's agricultural skill and knowledge of therapeutic plants, as well as her strong connections to women's cooperative organizations, have shaped the agenda and impact of TRMEGA's projects. Jane began developing her community organization skills in the early 2000s, several years after Rose, first through participation in and then through the mobilization of support groups for people with AIDS. Her combination of charisma and compassion continues to incite generative connections and sustain collective work, even as the health development funding for social infrastructure around AIDS has attenuated with greater access to pharmaceuticals.

Many others helped me understand efforts to (re)kindle relations with plants in order to redefine health and the forms of governance that might be promoted in the name of health: healers, herbal producers, shop owners, governmental officials, women's cooperatives, intellectual property lawyers, and food sovereignty advocates. John, Victor, Uroki, Dorcas, Mama Nguya, Rose, and Jane, however, form the loose collective at the heart of *Medicines That Feed Us*. They were working in relation to each other before my research but came to engage each other more deeply through it. Each recognized in the other's work an effort not only to respond to individuals seeking help with discrete illnesses but also to the slow violence and everyday disruptions that worked against the possibility of being well together in northeastern Tanzania (and beyond). *Dawa lishe*, the phrase offered by Mama Nguya, seeks to name this recognition. Such efforts were not something I could see or speak of during my time in Newala in the late 1990s and early 2000s, nor was *dawa lishe* as such visible to me in Dar es Salaam in the mid-2000s in the laboratories of the University or the Medical College or the offices of the National Institute of Medical Research. The experiments to which Mama Nguya gave language likely did not exist as a collective effort in the late 1990s. The abandoned social and material infrastructure of the AIDS support groups on which *dawa lishe* later came to lean were then only coming into being, and their articulation of these support groups with older peasant rights organizations was only just being organized. Slowly, after I shifted to the Kilimanjaro-Arusha area in 2010, the practices and projects that define *dawa lishe* started to be pulled into my view. The process of moving toward *dawa lishe* as a therapeutic modality, as a theoretical proposition, and as the central organizing principle of the book was a kind of ethnographic practice—being invited into the story and into the collective work through progressive invitations to see outside of the categories of knowledge that marked the original questions.

The relations that animated these invitations also gradually remade my own body and environment. Plants moved into the garden around our home in Moshi, joining the growing grove of avocado trees that our daughter tirelessly sprouted from seeds and planted in the backyard. I was regularly gifted cuttings and seeds, which drew my deepening relations into a dense green circle around me and my family. Friends and collaborators thinning their own gardens piled us with various plants to transplant at home. When we first rented our home in Moshi in a residential neighborhood near our daughter's school and walking distance from the hospital, the flat uniform lawn was broken up only by an occasional ornamental plant. Over the years, it grew lush and active. We composted along the back fence, behind the banana trees, in an open pit, sharing with the animals (but at a distance). We snacked on the moringa leaves from a tree gifted by Mama Nguya, as well as other fruits and vegetables that came to flourish in the garden plot near it. Lemongrass was planted under the window to fend off mosquitoes, and a neem tree sent out branches over the gate that were fragrant with purple flowers in season. A dear friend who kept bees lent us a hive and valiantly tried to teach us to attract a wild colony. Coming to see *dawa lishe* was also a process of coming to be in place and of coming to understand healing as a quality of lushness in everyday life.

## Plant(ing) Reproductive Justice

When I started to think about writing this book, I imagined that I would write narratives around individual plants. Several specific plants have come to signal that producers or users may be invested in the sort of commitments that *dawa lishe* glosses, such as banana (especially *kitarasa*), avocado, lemongrass, moringa, ginger, and pumpkin. Some or all of these plants scaffold the gardens I trace. Between them, people cultivate their favorite greens, tomatoes, taro, rosella, and orange sweet potatoes, as well as beans less commonly found in the market and often some fruit trees. Medicinal herbs sprout between these foods—sometimes intentionally cultivated, having been gifted from kin and neighbors, and sometimes as "spontaneous" offerings that are afforded space to grow during weeding. Each plant signaling a possible engagement with the commitments of *dawa lishe* has a complex history within colonial and postcolonial agriculture. They troubled plantation logics through indigeneity (*kitarasa*), difficulty in harvesting (moringa), wildness (ginger), or waste (seeds of the pumpkin and of "traditional" avocados), and their therapeutic potential seemed rooted in this capacity to hold space for more-than-economic relations. Yet, as I began to organize my notes and recordings, storying through individual plants started to feel inadequate. Too much was falling between or outside of the imagined chapters. As described above, being invited into the proposition that *dawa lishe* poses was also an invitation into ways of evaluating agency that cannot be articulated through the internal workings of plants, and ways of evaluating efficacy exceed engagement with individual plants. That is, the material substance of the plant was not the only or even primary unit through which producers and users were experimenting with ways of supporting nourishment and flourishing.

*Dawa lishe* attends to the forces generated between and among plants, people, animals, soil, and other elements as they co-labor. The efficacy of interventions rests in planting as well as in plants. This involves, but also exceeds, what agricultural experts refer to as "intercropping." As chapter 1 illustrates in detail, these gardens are about not only strategically staging discrete interactions but also nurturing a density of relations that offer rich possibilities for surprise and a broad platform for future response. Even producers who source their plant material from elsewhere talk extensively about the style of planting and care by those growing the plants they are purchasing. In producing remedies, they take the sourcing of plant material as an opportunity to self-consciously foster protection and strength through proximity with human and nonhuman others. Many gather their material through community-based collectives. All eschew synthetic pesticides, herbicides, and other products used for maintenance and turn attention to the co-laboring of people and plants.

Those who are extending their plant(ing) remedies through the sharing of seeds and cuttings recognize that land relations are explicitly multispecies affairs. Many of the gardens cultivated through these networks of people and plants include beekeeping, especially the keeping of African stingless bees (*nyuki wa dogo*, literally small bees). Local beekeepers identify eight different types of stingless bees. All prefer to build their hives in the hollowed-out logs passed down through generations. These ancestral logs used to be hung in the forest, and those with the knowledge worked to attract bees to their hives. As the forest has shrunk from deforestation, and the national park has excluded people from such activities within its boundaries, the hives have moved into home gardens. The honey they produce (Kiswahili: *nyori*), with its distinctive notes of lemon, can be found on the shelves of herbal shops. A women's cooperative that Rose leads has trademarked their brand of stingless bee honey, and it has been elevated to a food worth saving in the Slow Food Arc of Taste.[64]

*Nyori* does not lend itself to large-scale industrial production. The ancestral hives of stingless bees require reclaiming an intimate knowledge of dwelling well with others through everyday practices. The therapeutic value of their honey is not only attributed to its chemical content but also generated as beekeepers afford space for the hives, care for the bees, and tend to the relations with the past they foster. This honey intervenes in forces that have rendered particular bodily vulnerabilities durable and injuries chronic by (re)kindling alternative land relations. Similarly, *dawa lishe* producers, like many other Tanzanians, have a deep commitment to, and taste for, the sinewy meat of traditional chickens (*kuku wa kienyeji*): those raised without medicines or hormones, among the household, eating corn, scraps, and bugs in the courtyard, fertilizing the soil with their waste and producing eggs for human consumption until they pass their reproductive prime. Home gardens envelop pens for goats, pigs, and cows, and their manure is used to enrich the soil. For community gardens, manure is brought in, usually from members' own animals.

Many in the pages that follow have found common cause with global others through this attentiveness to multispecies relations. They have, for instance, participated in permaculture workshops and drawn on skills they learned while there. Yet, their gardens are not designed to mimic discrete ecosystems, nor are these gardeners interested in fidelity to the closed loop of a particular plot. This is at least in part because the temporalities of *dawa lishe* reach beyond the temporalities of such an ecosystem economy. Ancestors do not figure in permaculture theory, even if goats and bees have a role. *Dawa lishe*, however, takes the ways that past relations are alive in the present as a primary site of therapeutic work.

Chapter 4 explores *dawa lishe* as a search for the times and temporalities needed to address the durability of bodily vulnerabilities and weaknesses glossed as chronic disease and seen as a symptom (if not also an index) of modernity. Together with my interlocutors, I push back against dominant biomedical notions of "the chronic," suggesting that such notions limit the spaces through which persistent diseases might be addressed. Sustaining in the midst of chronic injury, they assert, is not healing. Rather, healing requires the hard work of addressing the slow violence of dispossession and toxicity by dismantling the pasts alive in, and continuing to undermine the liveliness of, bodies in present-day relations.

*Medicine That Feed Us* describes social-ecological-therapeutic projects that mobilize plant(ing) remedies. Drawing attention to planting emphasizes the practices and relations that put plants in motion with each other and with multispecies others. It surfaces these movements as the forces defining their therapeutic properties (rather than internalizing their agency through phytochemical elucidation). In Kilimanjaro, the *kihamba*, the "Chagga home garden," offers a lexicon for healing (*uganga*) as an assessment of efforts to interrupt forces attenuating life and to nourish those extending it. The *kihamba* feeds those who grow from its soil; on average, half of the bananas produced are consumed by the household, and half go to market.[65] Yet, the description of it as a "globally important heritage agricultural system" by the Food and Agriculture Organization of the United Nations does not fully capture its dynamism.[66] Tending to the *kihamba* as a space of everyday healing (not only as an ecological form or an agricultural strategy) illustrates how people apprehend the therapeutic capacity of plants through assessments of the relations that put them in motion. As a mode of dwelling, of attuning senses and cultivating sensibilities, of orienting to the co-laboring of plants and people and others, the *kihamba* gives rise to genealogical or reproductive temporalities that define notions of healing being brought forward by *dawa lishe*.

I saw this when I returned to Tanzania in the summer of 2023, for the first time after the COVID-19 pandemic. The story of COVID-19—of how people coped, survived, and grieved during the pandemic—was carried in these home gardens. Although the impact of the virus was greatest in 2021, two years later, the stunted growth of many lemon trees still bore stories of the demand for a tea made from their leaves and the networks among kin and neighbors that their circulation sustained. Plants used for respiratory ailments and immune boosting continued to be afforded more space. Young people who had returned home from the cities talked enthusiastically of the uses of weedy herbs as they toured me through parents' and grandparents' *vihamba* (Kiswahili: plural of *kihamba*, more than one Chagga home garden). The graves of those who passed away during the early 2020s were covered with plants used to ease the distress of their last days and months. While hospital statistics—and debates over them—tell a story of global politics, national tensions, insufficient resources, and inadequate health care service infrastructure, the *vihamba* tell stories of rekindling relations with land and with others, of the people who were tied together in loose networks of support, and of the flow of plants and care through households, communities, and broader kin. They mark passings in the soil and hold space for the pasts that still animate the present.

Below, I offer a brief sketch of the *kihamba* as a lexicon for meaningfully grounded understandings of reproductive justice located in the long arc of plant relations that define living well in the region.

### Home Gardens on Kilimanjaro

Literature in the environmental sciences defines the *kihamba* as a banana/coffee home garden. This agroforestry system, however, preceded the introduction of coffee by German colonists.[67] Bananas defined the precolonial *kihamba* not only as a dominant species of flora but also for their intimate co-laboring with people to generate more life.

Knut Christian Myhre's evocative ethnography among the Chagga on the eastern side of the mountain that borders Kenya (Rombo district, where Romana is from) excavates the precolonial relations animated by the *kihamba* in this area of Kilimanjaro.[68] He argues that in the nineteenth century, when land was more abundant and coffee not yet central to livelihoods, *vihamba* moved between women along patriarchal lines. A mother gifted her *kihamba* to a son's (first) wife upon marriage. The older woman moved herself and her children to a new *kihamba* above the one her son's wife then occupied. The young wife would therefore inherit a fully functioning *kihamba*. In this area where polygamy was common, men would cycle between their wives' *vihamba*. Although a *kihamba* extended a husband's lineage, it was the space of a wife's power and authority. Her presence invited its flourishing. Her labor channeled the forces of reproduction and continuance. Youngest sons' wives inherited the parents' final *kihamba*, and when these elders passed away, they were buried in the *kihamba*. The son's children grew up with grandparents' and sometimes great-grandparents' graves holding down the courtyard. The *kihamba* was, and was more than, a plot and a style of planting; it was the cultivation of a dense node of reproductive energy that potentiated lineage and land. It engendered a particular kind of lushness marked by a density of relations—human and non-human, animate and inanimate—that supports the possibility of children and harvests, that is, of ongoingness.

Coffee slowly entered these home gardens during the first half of the twentieth century. Not until the 1950s, with the rise of global coffee prices and the successful political organization of the Chagga through coffee cooperatives, did the relationship between bananas and coffee come to thoroughly redefine the composition of the *kihamba*. The mid-twentieth-century *kihamba*, as a banana/coffee home garden, both indexed and animated changes in social organization, modes of dwelling, and forms of trade. Coffee as a cash crop offered a way to respond to economic pressures and shifting social priorities, from colonial taxes to mission school fees. It also fueled Chagga political organization, as growers petitioned colonial administrators to resist pressure from white settlers who wanted to limit African coffee production to boost their own claims to land.[69] The history of the emergence of the banana/coffee "home garden" tells a story of the incorporation of coffee into the reproductive energies of the household and the broader community. This contemporary formation is also a manifestation of the broader changes in land relations in response to colonial enclosure and the forms of sovereignty through which it was levied.

The coffee bush facilitated, and its extension was facilitated by, the forces that destabilized precolonial social, ecological, and economic organization. Settler farms and mission compounds alienated land from those whose continuance had long been tied to the mountain. As churches advocated for the end of polygamy and the schooling of children, they shifted the allocation of labor within the household. Fathers came to live with one wife permanently rather than moving between wives' homes. Their authority grew as they consolidated their lives and livelihoods in a singular *kihamba*. This focus settled in the land through the planting of coffee bushes to which men claimed ownership. Gendered imaginings of the "family" propagated by the church reinforced models of patriarchal ownership.[70] Children were simultaneously less available for labor on and off the *kihamba* as missions established schools.[71] The styles of homes changed as modern aesthetics discouraged drawing cows into the house at night to sleep.[72] The desire for more permanent houses increased the investment made in the home space, and parents grew less willing to gift their *kihamba* to the son's new wife. Upon marriage, fathers began giving sons a piece of land on which to begin cultivating a *kihamba* (rather than mothers gifting sons' wives a fully functioning *kihamba*). These plots might be a section of the parents' *kihamba*, former grazing land, or area in the lowlands. Colonialism disrupted the intimate relations through which plants and people tended forces generating life in myriad ways.

Demographic pressures continued to drive more land under permanent cultivation. As uncultivated land grew scarcer, young men's access to land grew scarcer. Grazing land next to *vihamba* for pregnant cows and those with calves shrank. Cattle had to be taken farther for grazing or provided fodder. Plots grew increasingly fragmented, encouraging the further intensification and diversification of the *kihamba*.[73] Long histories of irrigation, manuring, and terracing supported farmers' ability to cultivate the steeper and more inaccessible areas and to plant continually through three growing seasons while still attending to the fertility and capacity of the soil.[74] The contemporary banana/coffee "home garden" not only captures the incorporation of coffee into the reproductive energies of the household but also holds the tension of these shifting land relations and the reconfiguration of gendered power, modes of production, and practices of dwelling.

The twenty-first-century *kihamba* is shaped by, even if not fully defined by, both colonial and postcolonial dispossession. Land continues to be alienated from smallholder farmers through commercial farms (often foreign-owned), forest plantations, conservation areas, and rapid urbanization. Today, smallholder parcels average .5 ha on Kilimanjaro, and access to additional land for cereal crops and grazing is much more difficult. Fewer people have access to even small parcels, and this land scarcity reinforces existing inequities and vulnerabilities. As the advocacy of women's rights groups such as the Kilimanjaro Women's Information Exchange and Consultancy Organization (KWIECO) teaches, customary and colonial law systematically supported men's rights to land over women's claims. More recent land reforms have been inadequate to ensure equity, even as the urgency of land rights has been exacerbated in the wake of AIDS.[75] The *kihamba*, then, is a site of both inspiration and struggle in efforts to keep alive a mode of staying close to soil, plants, and ancestors.

While environmental historians describe the *kihamba* as a form of early agricultural intensification, ecologists celebrate the biodiversity that inheres in the *kihamba*.[76] In a 2006 study, Andreas Hemp found that each *kihamba* consisted of over 500 vascular plant species, including 400 noncultivated plants; that is, approximately 80 percent of the plants identified had taken and/or were afforded space, but they had not been intentionally cultivated by humans for a specific or immediate use.[77] The *kihamba* entwines histories of agricultural intensification with stories of human cultivation that foster the distribution of indigenous species by increasing habitat diversity, but it cannot be fully captured in these narratives. Both perspectives engage land and plants in the *kihamba* as resource, one to fuel economic systems and the other to animate ecological systems. In so doing, they obscure the logics and labor that reveal the therapeutic potential of land as relation and the densification of these relations as a rich space of response.

The density and diversity of the *kihamba* resonate more with the generative space-making that Isabelle Stengers describes in *Capitalist Sorcery* as "casting the circle" than with the logics of resource extraction and industrial production.[78] Stengers draws on the ancient European traditions of sorcerers and healers who gather the forces of change by creating "the protective space necessary to the practice of that which exposes, of what puts at risk in order to transform." The boundaries of the *kihamba* are delineated less by keeping out than they are by pulling in; its strength grows through an intensification of relations and its power through the transformative potential of new exposures.

### Healing (as) Land Relations

*Vihamba* are dense gatherings of plants and people working together to amplify reproductive potential. Through and amid the tensions that compose the contemporary *kihamba* and its exclusivities, *dawa lishe* providers explore the kinds of land relations that might be experienced as healing. They do not offer the *kihamba* up wholesale as a model garden, but rather, by working within its layered histories, incommensurable economies, and unexpected proximities, they offer a space to reimagine what it means to engage healing as an act of working for reproductive justice in the broadest sense. But it is not the only one.

The *kihamba* has never been the sole site for cultivating reproductive energies. It is one of a diverse array of interlocking configurations of more-than-human relations. The specificity of the land relations engendered in the *kihamba* was forged in coordination with other areas: grazing land for cows and goats, lowland fields for cereals and beans, highland fields for growing fodder, and forests for wood, wildlife, and foraging.[79] Precolonial mountain communities took advantage of the different ecological zones and microclimates in the region to enhance reproductive possibilities and generate sufficient food. They dedicated labor to accumulating excess across these spaces in order to foster exchange and negotiate continuance with others through village markets.[80] The *kihamba* emerged through, and continues to exist in, complex relations to the field and to the market. The multiple ways of being and being with plants in and beyond the *kihamba* have long afforded farmers dexterity as they assess and engage regional and global economic forces and care for the health and well-being of those that grow from its soil.

Furthermore, today in contemporary northeastern Tanzania, the *kihamba* has been joined by other configurations of multispecies liveliness. The desirability of the volcanic soils, the growth of the urban centers, and the quality of secondary schools in this region have drawn many to the area since independence. Those who hail from further afield bring different histories of planting and land tenure. They are pressed by differently gendered responsibilities to human and nonhuman kin and rights to access land (as well as obligations to that which they can access). Mama Nguya, for instance, was born and raised in Kagera, a region in the northwestern corner of Tanzania running between Lake Victoria (to the east) and Rwanda and Burundi (to the west). Among the Bahaya in Kagera, life, land, and lineage are similarly interwoven with bananas. The ecologies and the social movements that shaped land relations in this region, however, have been quite different. Trade has moved most easily around Lake Victoria, building connections across what is now Uganda and Kenya. In Kagera, men traditionally owned the banana trees surrounding homes, and this ownership determined property relations, whereas women planted seasonal gardens in the grasslands. The diversity of ways that garden plots have enacted and continue to enact relations across Tanzania has become a source of creativity for those striving to heal contemporary bodies and lands. The differences serve to highlight the garden as a complex form of enclosure and of ecological choreography.

*Dawa lishe* plays with all these differences as it moves in and out of home gardens, through collective gardens, opening to the sorts of fields that would allow for increased cultivation. This multiplicity offers frictions through which to dismantle and reinvent these gardens and the social relations of which they are part. Plant(ing) remedies remember that colonial land tenure was set up to manage global capital for empire. Postcolonial land tenure was set up to manage national resources for national (and after the 1990s and the fall of socialism, increasingly corporate) development.[81] Modern medicine and agriculture emerged in relation to the formation of land tenure invested in the vitality of productive citizens and productive ecologies. Neither colonial nor postcolonial land tenure was designed to maintain ancestral graves and sacred groves. Neither were compelled to generate forms of growing, eating, and healing that foster the viability of lineage and land.

Those innovating *dawa lishe* recognize that precolonial, colonial, and postcolonial land tenure systems have *all* been sites of inequality and violence, as well as constant struggle over responsibilities and obligations to others (human and nonhuman, animate and elemental). Powerful forces external and internal to households and to communities have striven to accumulate wealth and have often resisted the distribution of resources. For the past four decades, Tanzanian activists have fought for women's ability to own, control, and manage land, and they have highlighted the urgency of land rights as AIDS has loosened the bonds through which land and lineage form.[82] Female-headed households have increased; sisters and grandmothers find themselves taking in children from kin. Widows are dispossessed of land, or their rights to use land are increasingly challenged. Many find themselves displaced to periurban areas, in small rented houses or rooms.

The plant(ing) remedies in *Medicines That Feed Us* do not offer a magic bullet but rather something more modest: invitations into, and strategies for building up, a density of relations as a powerful social-ecological place from which to respond to persistent injury and chronic depletion. I write to extend the work of Tanzanians who are exploring new, hybrid vocabularies for reflecting on how harms accumulate in bodies, lineages, and lands, and who are developing practices to intervene on those harms. Their efforts to formulate remedies through current legal and medical regimes are fraught. The avenues for intervention are narrow. The pressures for their work to be absorbed into the economy are intense. Yet, by drawing attention to the co-laboring of plants and people, they teach us how we might attune (and continually reattune) to relations that offer times and spaces for decolonial reinvention of health and new possibilities for healing.

Therapeutic Sovereignty

The next chapter explores how plant(ing) remedies reach out to those increasingly marginalized and abandoned through histories of alienation from land and labor. Neither I nor my colleagues elevate one style of planting or one historical moment as a solution in and of itself. Nor do those experimenting with *dawa lishe* as medicine and theory romanticize the fact that African healing has long recognized land relations as central to both sickness and health, depletion and restoration, for healing has been a site of power as well as resistance and therefore entangled in these inequities. There is no pure place to stand. No innocent ground for argument. But there are more and less subtle, elaborate, and impactful ways of thinking relations between bodies and environments. There are modes of healing that render visible and those that render invisible the ways that violence, dispossession, and economism exhaust some bodies more than others. And there are efforts to intervene—through both argument and action—in the relentless depletion and chronic exhaustion that have left people and land worn down, more vulnerable, and less fertile. *Dawa lishe* seeks to recognize and, by doing so, draw together those people and plants in Tanzania that are innovating plant(ing) remedies to address such body-land injuries.

*Dawa lishe* emerged as a proposition linking the work of farmers' collectives, women's cooperatives, imaginative entrepreneurs, and resourceful AIDS support groups in Tanzania that are experimenting with ways to care for depleted bodies and disabled ecologies through innovative projects that (re)kindle relations with plants and (re)formulate ways of dwelling together. Their efforts insist on remembering the ways that toxicity has become a condition of modern life. In an effort to expand the scope of how to address the urgency of the environmental and health crises defining the twenty-first century, I offer a story of *dawa lishe* as a provocation for collaboration among projects experimenting with ways to dislocate the imperial pharmakon and reinvent relations between toxicity, remedy, and memory. Drawing on African healing, these projects open a space to articulate therapeutic sovereignty as the power to determine the terms through which reproductive justice is articulated in a toxic world.

The projects described in the pages that follow invite people and plants into the collective labor that might create times and spaces for healing and repair. They are creative gestures in the face of environmental and health crises that have a momentum of their own. Recognizing the extent to which dispossession has exhausted some bodies and lands, draining them of the relations that enable response to injury and loss, these plant(ing) remedies start small: a canister of herbal medicine given to a woman confined to her bed with AIDS and hunger, abandoned by family and tormented by the pain of not being able to care for her children; a bag of dandelion greens to stimulate a taste for bitterness and awaken the knowledge that plants which fortify can be found along the road and in abandoned lots; seeds shared for a container garden near the door of a small room rented by one pushed to a periurban area by poverty; cuttings and saplings brought to schools and orphanages in a refusal to abandon those rendered hungry, weak, or unproductive. These projects locate remedy in reparative, transformative acts that challenge claims to environment and health justice rooted in ownership. They insist on kindling other ways of being close to land and to each other.

Some take the garden itself as an intervention, highlighting that healing is about land relations. *Dawa lishe* gardens extend from one to the other through the exchange of seeds, cuttings, and knowledge. In so doing, they work directly against the continued alienation of plants from small-scale farmers, as Tanzania has bent to increasing international pressure to comply with strict inter-pretations of plant breeders' rights.[83] They also trouble deeper commitments to plant–people relations rooted in long histories of the economization of life, labor, and land and their enclosures. Remedies rekindle relations with, and ap-petites for, plants in an effort to remind people of body-land relations that are otherwise. The innovative projects that describe *dawa lishe* strive to render vis-ible the slow violence and accumulated burdens of environmental degradation and economic exploitation, by cultivating appetites for alternative times and spaces for going on through (and with) plants.

Therapeutic possibilities are generated in the dynamics of growing and eating, gathering and composting, drying and burying. Attending to the dynamics of composition and decomposition, *dawa lishe* focuses on not only the powers that determine life and death but also those that animate the ways that life and nonlife move through each other. It shares a kinship with concerns over how to read, name, and act on changes in the substances that make up our planet, both the lively and the inert, and how to evaluate the ways that these substances move through each other, pooling in bodies and earth, finding new lives and life cycles, and changing the compositions of humans, animals, plants, soils, waters, and air. Producers and users locate the therapeutic in reproduc-tive capacities that animate forces of continuance that exceed human lifetimes, as well as forms of agency and animacy that refuse the ontological separation of body and land (a separation critical to apprehending soils, plants, and ani-mals as economic resources). In the process of reimagining the times and spaces of the therapeutic, they hold our notions of corporeal and territorial sover-eignty accountable to the work of healing. *Medicines That Feed Us* strives to capture the provocation that interrupting the rise of chronic illness in Tanzania requires revisioning the times, spaces, and scales through which harm is articu-lated, responsibility is delineated, and obligations to the bodies and lands that are harmed are held.

This book comprises five substantive chapters through which I propose *dawa lishe* as a mode of collectively moving toward an answer to the ques-tion: What does it mean to heal in a toxic world? Each chapter is both an ethnographic examination of practices and a theoretical provocation inviting collaborative possibilities in developing modes of attention, techniques of ob-servation, ways of storytelling, and forms of active engagement that hold thera-pies accountable to nourishing the life force through which lineage and land are bodied forth. I offer a notion of "lushness" to conceptually capture the way that the innovative projects described in the pages that follow are reorienting notions of health and healing. In the process, I argue that they draw out and innovate on versions of sovereignty that support healing and continuance through rich ecological relations. Chapter 1 roots this argument in the work of TRMEGA. I account for this NGO's co-laboring with plants and the extension of gardens as interventions into the persistent injury and chronic vulnerability that define periurban spaces. Forms of care and composting draw people closer to the soil and to each other in ways that intervene in the ongoing depletions and slow violence of extractive economies of land and labor.

Chapter 2, "Efficacy of Appetites," takes up appetites as desires that drive body-land relations and that energize lively response. In Tanzania, as elsewhere today, appetites are under scrutiny and have become the focus of national and international public health efforts. Plant(ing) remedies—as well as the large social projects of which they are part—challenge the forms of knowledge that authorize therapeutic efficacy and elevate interventions into the palate as a ground for politics.

In chapter 3, "Registers of Knowledge," I take a short detour to elucidate why *dawa lishe* cannot be faithfully engaged or managed as traditional medicine. This argument requires tracing the institutionalization of traditional medicine and identifying its origin in the epistemological and ontological settlements forged by colonial policies that separated African therapeutics into herbalism and witchcraft. *Dawa lishe*, as an invitation to collective labor and an incitement to contemporary theory, troubles the settlements made through the forging of traditional medicine as a modern category of knowledge and practice. Plant(ing) remedies denaturalize now well-institutionalized answers to the questions: "What are plants?" "Who can know them?" and "What counts as knowing and knowledge?"

Chapter 4, "Work of Time," is inspired by how the Senegalese philosopher Souleymane Bachir Diagne mobilizes his work on time as that which is conceived in action. I ask what times or temporalities are needed in the face of the toxicities shaping the African Anthropocene. For many Tanzanians with and without an HIV diagnosis, antiretroviral therapies have become a generative object around which to contemplate what it means to live in toxic times—that is, to reflect on this historical moment in which life (individual and communal) is only possible through the "toxicity" of such pharmaceutical regimes. *Dawa lishe* strives to enroll bodies in rhythms that respond to, but also expand beyond, the pharmaceutical logics and the tempos of biomedical institutions.

The banana offers a generative example of the rhythms created in the plant(ing) remedies. Chapter 5, "Properties of Healing," looks most closely at *kitarasa*, a banana said to be indigenous to the Kilimanjaro region and a particularly charismatic actor in the rise of the social-ecological-therapeutic projects indexed by *dawa lishe*. The temporal and spatial horizons toward which *kitarasa* strives to orient care challenge global health's current ways of formulating relations between bodies and their environments. The interscalar capacity of *kitarasa* incites the theorizing of relations among toxicity, healing, and memory in ways that challenge the properties of therapeutic and economic value that drive the pharmaceuticalization of health.

I conclude the book with a meditation on how *dawa lishe*, as a practice of fostering real possibilities for continuance in a toxic world, begins to compose forms of political and therapeutic sovereignty that support a dispersed ecological body, broad notions of reproductive justice, and innovative notions of property rooted in an ethics of hospitality. In so doing, these practices decenter the imperial pharmakon and reinvent toxicity and its relation with remedy through vocabularies forged in African histories of healing.

These efforts to re-story chronic illness within the persistent depletion, injury, and loss of postcolonial Tanzania find connections beyond the country's borders. I invite you to trace them in the service of ongoing work. What modes of attention, forms of care, spaces for repletion, and times for regeneration will support reflection on and labor over toxicity in the service of healing? What relations will hold us responsible for the kinds of lushness that support the ability to accommodate injury, attend to loss, and move through illness?

I

# Futures of
# Lushness

What struck me first was the lushness. I had just stepped through the gate and into the compound of TRMEGA (Training, Research, Monitoring and Evaluation on Gender and AIDS). The organization sits just a few hundred meters off the east–west road that carries local commuter traffic in and out of Arusha as well as large commercial vehicles running between the coast and Tanzania's northern and western borders. As I stepped away from the dusty muted browns of the road and into the intensely, varied greens of the garden, the air felt cooler. Passion-fruit vines covered the fence. Neatly labeled medicinal plants packed narrow beds along the edges of the compound and squeezed between the corners of buildings. Chard and a range of other dark green, leafy vegetables grew in a kitchen keyhole garden and burst from sacks placed around the courtyard. The moist, rich brown soil smelled earthily sweet. A brightly painted sign reading "Maji ya Chai Slow Food Community Garden" sat at the head of a winding path that invited walkers through plots of taro, sweet potato, watercress, and tomatoes.

FIGURE I.I TRMEGA garden as viewed upon entering the compound at Maji ya Chai, Tanzania. Photo by author, 2015.

I had followed the address on the label of an herbal formula, *imarisha afya yako* ("strengthening your health" in Kiswahili). I initially encountered a canister of *imarisha* in Alex Uroki's EdenMark pharmacy about 30 kilometers west of Maji ya Chai in Bomang'ombe. Sitting on one of the glass shelves, stacked next to Uroki's own plant-based formulas and those he had gathered from friends and colleagues throughout Eastern and Southern Africa, it had contributed to my early impression that these commodified remedies were examples of an emerging herbal industry in Tanzania. My fieldnotes and photo archive from that first visit to TRMEGA document that I found other carefully dried and packaged plant remedies stacked on the table against the back wall of the office. Yet, what has stayed with me most powerfully is the sense that, upon stepping through the gate, the stories that I had anticipated finding began to recede. The canister, the plants inside it, their chemical properties, presented too narrow a scope to locate the efficacy of this place; the notions of property that inhere in the idea of an "herbal industry" were too cramped as concepts to contain the relations through which remedy unfolded in this garden.

TRMEGA apprehends the efficacy of *imarisha* and other similar herbal formulas through its ability to stimulate appetites that catalyzed other gardens and other economies of knowledge, configurations of kin, versions of the body, and notions of healing. Each canister of *imarisha* reaches out, carrying its

FIGURE 1.2 *Imarisha afya yako* (strengthening your health), a TRMEGA herbal formula made from four different plants, Maji ya Chai, Tanzania. Photo by author, 2014.

plants to bodies far beyond the garden's borders and even beyond the homes of TRMEGA members. The material properties of individual plants are pulled into a broader conception of the work of herbal formulas. By attracting people's attention, developing familiarities, and rekindling appetites, the dried herbal formula cultivates the senses and sensibilities necessary to recognize healing as land relations. It works by catalyzing a thickening of relations, nourishing density, and cultivating a lushness that makes flexible and fulsome responses to the depletions, injuries, and hungers of everyday life possible.

On that first day, when *imarisha* drew me to TRMEGA, and every day I have entered the gates since, I have been prompt to walk around the garden, to feel my feet on the paths, to notice plants that are flowering, fruit that is ready, newcomers that have emerged from seeds carried from afar, and weedy herbs that have been afforded space, as well as the level of water in the irrigation channels, the texture of the soil, and the transformations of the compost. Sometimes I am invited to taste the sweetness of a ripe cherry tomato or to tend to a scratchy throat with a bitter leaf. During periods when we spent a great deal of time together, working to extend this garden to others, this walk was a brief greeting, a grounding, an orientation to the day's work. On occasions of reuniting after I had been away teaching in the United States for several months or during COVID for even longer, we spent more time lingering with the plants, for it was with them that the stories of the time apart could be told: the networks that were created or sustained, the relations that took root in and through the garden, the ways that the densities of the space composed and decomposed themselves through stresses that had defined that period. The plants held stories of remedies tried, strengths tested, capacities returned, and people lost. As we moved between them, noticing growth, touching bark, picking leaves, these stories were coaxed out of the textures, tastes, and smells and given time to circulate between us. In remembering, we reconnected, surfacing the intimacies of our friendship and naming our entanglements. We brought our bodies back to the garden and it into us. In these moments, our bodies sensed histories folding in one another. We tasted, smelled, heard, saw, and felt the dynamic forces forming and reforming through people–plant relations. Our connections to each other and to the land strengthened and multiplied.

During the time I wrote this book, COVID-19 moved across the region. From TRMEGA, I saw how this pandemic became entangled with the HIV/AIDS epidemic. The chronic conditions that render bodies more vulnerable to COVID-19—diabetes, hypertension, lung disease, and cardiovascular disease—are also comorbidities associated with HIV/AIDS. The gender inequalities and economic inequities that exacerbate the intensity and impact of the first epidemic shape those that follow. Entangled in these layered histories, individual medical diagnoses and the vocabulary that accompany them are taken up in service of describing bodies disconnected from their sources of growth, traditions of nourishment, and forces of ongoingness. To render illness actionable as an effect of a manifestation of long histories of dispossession, remedies intervene in the affective and material relations that concretize people's alienation from land in everyday life. Healing requires not only attending to the dynamics inside individual biological bodies but also building intimate relations with plants and a close connection with land.

Vocabularies forged in African histories of healing find in such a lushness of human and nonhuman, social and ecological, relations a manifestation of what might also be called "health" and describe the context that makes being "healthy" possible. Bantu metaphysical lexicons juxtapose healing (Kiswahili: *uganga*, that which nourishes, strengthens, and fortifies, causing a thickening of relations and resulting in lushness and fertility) with harming (Kiswahili: *uchawi*, that which depletes, exhausts, and weakens, causing an attenuation of relations and resulting in wasting and barrenness). This analytical frame is not opposed to public health analytics of inequities that surface political economic tensions as driving the pattern of disease and debility, but it does shift the focus to the ethical and relational tensions driving the patterns of disease and debility. This alternative vision is consequential. The imperial pharmakon—that configuration of science, economics, and law that shapes the modern relation of toxicity, remedy, and memory—is central to the articulation of the political economic and its link to notions of bodily and territorial sovereignty. TRMEGA's extension of plant and human communities, their offering of gardens as a therapeutic response, decenter the imperial pharmakon and reinvent toxicity and its relationship with remedy. *Dawa lishe* names their and others' efforts and identifies projects whose work nourishes alternative forms of sovereignty.

The next four chapters describe *dawa lishe* through experiments that nourish collectives (human and nonhuman) through the process of remembering flavors and building appetites, of remaking time and tempos needed to address the wounds of dispossession and embodiments of structural violence, of generating knowledge arising in action rather than locked objects, and of invigorating styles of dwelling that push against the limit points of private property to reinvent relations between economy and ecology. In this chapter, I consider how this work comes to be felt in a fight for the futures of lushness.

## Healing (as) Land Relations

Over time, with growing trust, deepening relation, and hundreds of hours of co-laboring, Mama Nguya offered me a more private story as an origin point for the TRMEGA garden and as a touchstone for navigating forms of knowledge and economy that threatened to limit articulations of efficacy to the material properties of plants. It began with her physical collapse.

The stress of more than a decade of demanding work and long hours as a director of social service programs for the Roman Catholic Church in the northern part of Tanzania, combined with two intense years during which she organized and worried over the care of her children back in Tanzania while she was in Europe earning a master's degree, exhausted Mama Nguya. She was depleted. From northern Europe, her social network was too attenuated to bear the shock of the deaths of her mother and sister when they came. She collapsed. She could not stand or feed herself. She returned to Tanzania. Her kin nursed her in her marital home of Maji ya Chai. Nguya's healing slowly drew her back to the land. First, as she sat in the banana grove that surrounded her home, and then, eventually, as she put her hands in the dirt, she felt that she began to heal. Or, as she tells it, she started to remember those things that had fed her, had given her vitality when she was younger. While Mama Nguya is not a traditional healer by her own or anyone else's account, her (re)tellings of this story echo a narrative genre of African healers who locate their knowledge and expertise in a challenging illness that forced them to change their path in order to heal. Remembering this origin point allows TRMEGA and its remedies to resonate in multiple registers: as home garden, as nongovernmental organization (NGO), as entrepreneurship, and as healing.

The vision that eventually grew into TRMEGA emerged from Mama Nguya's efforts to reimagine a life in which she could do meaningful work and sustain her strength. Among her extensive portfolio of social service programs was a church-supported HIV clinic. While visiting the clinic one day, she met a man whose story moved her. He confessed to her that through the treatment he received in the clinic, he had come to feel capable of continuing his life with AIDS, that his diagnosis was not a death sentence. She remembered and reached out to him. Together, they began connecting with others who had been diagnosed with HIV/AIDS. Through their initial efforts in 2009, Nguya began to see a particular set of needs and possibilities. She explained,

> I realized that some had problems, and I didn't have the funds to help. They had clothing needs and food needs. So, I told my family and friends. Some contributed clothes. Some gave me shoes. Some donated domestic utensils. Some donated food. They came and they saw [my work]. They were amazed and gave me maize flour and whatever they had. I taught those who gathered about nutrition. We had already planted some plants. I taught them how to cook them. They came here [to my garden in Maji ya Chai]. We ate together and participated. This made them more active. After that we saw that it would be better to teach them about herbs, and that they should not only practice it here, but also at their homes. And those who had land around their homes, we gave them things to plant.

Mama Nguya's personal experience of healing in her home garden inspired a broader, pedagogical intervention. The garden became an object of therapeutic practice, the site of everyday acts of healing and justice.

Gardens are not exactly a solution to the problems that Mama Nguya perceived but something more as well as something less. For TRMEGA, gardens— organic, cultivated without synthetic pesticides and fertilizers, full of therapeutic plants, spreading from one to another and growing up around homes and schools—are remedy and relation, resource and momentum. They constitute the very material of individual, institutional, and community extension. Gardens offer a response to the attenuation of life; their answer is a thickening, a density, a multiplicity that one might experience at times—like when walking through the gate of TRMEGA—as lushness.

Today, the garden at TRMEGA's headquarters serves as a community garden for the members of TRMEGA's support groups and as a Slow Food demonstration garden to introduce new members, school groups, and other visitors to the plants and principles of sustainable agriculture. Experiments with recycling kitchen water, sack gardens, and composting are always underway. Bees have been introduced, both larger honeybees and smaller, stingless African bees. Medicinal plants are neatly labeled in both Swahili and the local language of the area where Nguya learned of their therapeutic potential. In the shade of the pavilion, medicinal herbs as well as therapeutic foods are dried and processed into TRMEGA remedies. This garden also generates the cuttings, seeds, and saplings used to stimulate other gardens.

The contents of the Maji ya Chai garden embody a lifetime of living and working with plants. Both in terms of its shape and the plants that populate it, Mama Nguya's garden innovates older ways of organizing landscape through the residential farms of her childhood. Mama Nguya grew up in Kagera. She helped her grandfather as he tended banana trees, which not only supplied the primary staple crop but also structured the spatiality of village life and the masculine forms of land ownership that drove it.[1] Her grandmother, like other Buhaya women, would have cultivated a garden in the grasslands outside the village.[2] The location of these plots shifted and did not embody ownership claims as the bananas did, but the harvests from these seasonal grassland gardens did supply a diversity of vegetables for household consumption and for sale. Mama Nguya also learned to identify and collect medicinal herbs with her grandmother, who was valued in the community as a traditional midwife. Mama Nguya has transplanted some of these plants into the Maji ya Chai garden. Others she continues to get from Kagera. Sometimes she goes to collect them herself, and at other times, she asks a nurse in the health center to collect a desirable herb for her and send the harvest by bus.

The novel collectives fostered through the rapid expansion of NGOs in Tanzania since the 1990s have enabled Mama Nguya to combine the ways that the gendered landscapes of Buhaya villages structured reproduction, nourishment, and expansion with the work of global projects. While her home in an adjacent lot is surrounded by banana trees, her TRMEGA garden is not. The contents of TRMEGA's Maji ya Chai garden might be better seen as an innovation on the grassland gardens Buhaya women cultivate. The plants Mama Nguya learned from her grandmother grow together with fruits and vegetables popular in her childhood, with those commonly eaten by people with deep roots in the area where she has settled, and with those introduced by international nutrition initiatives and health development projects. The Maji ya Chai garden stages a conversation between older social-political practices of dwelling and newer global practices. As plants and seeds extend through TRMEGA's projects, giving rise to other gardens, members are invited to be active participants in this conversation and to restage it in their own gardens, further multiplying engagements with other local forms. The TRMEGA garden is not a simple reenactment of older plant–people relations or a model garden to be replicated in the plots it seeds. Rather, these gardens are each an effort to cultivate a dense space of healing—that is, lushness—for bodies and lands left depleted by the relentless extraction of labor and resources central to the histories shaping periurban landscapes.

Mama Nguya's knowledge of agroecological models initially grew out of her participation in the NGO, Women Development for Science and Technology Association (WODSTA). In the 1980s, Mama Nguya worked with a group of prominent women leaders to found WODSTA. In those early days, President Julius Nyerere's wife, Maria Nyerere, joined their efforts. The association has always focused on "grassroots women," meaning rural and often disenfranchised women. Sustainable agricultural practices were central to the mission of the organization from its inception.

Because of this work, Nguya was first invited to the biennial Slow Food gathering in Turin, Italy, in 2004. She returned in 2008 and has gone to each subsequent gathering. These days, she generally goes with Rose Machange, who runs the gardens and does outreach for WODSTA. When Mama Nguya and Rose travel, be it locally to gardens or internationally to Kenya, Uganda, Italy, Ethiopia, or India, seeds often return in their pockets. Even after decades of friendship, they rarely leave one another's gardens without something in their hands. Knowing them both for over a decade has allowed me to witness not only their attentiveness to plants but also how they move through the world with plants.

Each time seeds are shared, cuttings are gifted, and transplants are taken, they are re-membered in and through the land and engage a complex past. Rose, for example, identifies as Chagga and lives in the Meru District of Arusha. The economic and social life of the Meru in Maji ya Chai and other areas of the Meru District, as for the Chagga on Kilimanjaro, originates from their *vihamba* (Kiswahili: plural of *kihamba*).[3] In fact, archeological evidence suggests that the Meru were likely descendants of those who settled early on Mount Kilimanjaro. Approximately three hundred years ago, they crossed the valley moving west away from Kilimanjaro and settled on the slopes of Mount Meru.[4] The Meru continued for at least two centuries outside direct pressures from the chiefdoms in Kilimanjaro. During this time, they grew in relation to the Arusha agro-pastoralists.[5] The Arusha share a past with the migratory Maasai, who have long moved across the steppe to the north of Mount Meru, over the plains, and up the western slopes of Kilimanjaro. During their migrations, the lush forests drew the Maasai for medicines and hunting. As TRMEGA extends to, and through, Rose's home garden and on to the groups with which she works, the Kagera modes of dwelling and global practices are brought into conversation with the *kihamba*, its multiple pasts, and the complex connections it is able to make in the present and offer to visions of the future.

Many of the gardens that grow out of TRMEGA also come to participate in the 10,000 Gardens in Africa project, one of the flagship initiatives of the Slow Food Foundation for Biodiversity. In 2011, while in Tanzania to celebrate the first gardens (one in Dar es Salaam and one in southern Tanzania), Italian staff members visited Nguya's compound and saw the TRMEGA garden. They ate with her and talked about their vision for 10,000 Gardens in Africa, asking if she would volunteer to be the *Convivium* Coordinator for the northern region. Although coordinators are not paid, Mama Nguya accepted this position because it would enable her to solicit limited funds for material enhancements to expand the network of gardens, such as hoes, wheelbarrows, watering cans, and fencing. In addition, the foundation strives to train *Convivium* Coordinators and local leaders and to support their travel to meet other communities. Mama Nguya drew Rose into Slow Food as well. They traveled together and participated in projects, such as the Arc of Taste, a Slow Food initiative to catalog indigenous foods and recognize groups whose work ensures their endurance (see more details in chapter 5). A women's group that Rose leads, called UMANGU, succeeded at having their African stingless bee honey accepted into the Arc of Taste. This group also oversees projects like the new traditional foods shop that sells various products from the women's support groups Rose runs.

Throughout its existence, TRMEGA has grown by extending its work through a variety of plants and people, and opening diverse, fertile spaces to experiment with ways of living, forms of organization, and styles of dwelling that make life possible in a toxic world. Originally, Mama Nguya's wide-ranging connections with the Catholic diocese and their projects animated TRMEGA's formation. They continue to provide fertile ground for growth. Mama Nguya has also folded some of the projects and energy of the Slow Food Foundation for Biodiversity into TRMEGA's work. These institutional and personal entanglements have been further extended and reinforced by Mama Nguya's connection to WODSTA and her friendship with Rose. The interpenetrations of these institutional trajectories also carry with them certain orientations toward history—here, Catholic social justice agendas, a global movement for culinary diversity and food sovereignty, and a postcolonial, nonaligned socialist drive for self-sufficiency. Each movement carries the inequalities and exclusions of its formation as well as the drives and biases of those individuals involved. Connections are partial, situated, and embodied, built through personal histories of resilience, friendship, and striving, as well as of inequity, abuse, and sickness. It is into this density of relations (institutional, human, and botanical) and these storied spaces that TRMEGA welcomes people. The gardens offer an architecture for rethinking the therapeutic; their lushness offers a felt sense of the transformation that drives healing.

## Storying Plant Remedies Otherwise

When I first entered TRMEGA's compound, I felt the lushness of the space, saw the dark chocolate brown of the soil, smelled the moist coolness, heard the hum of bees, and tasted the taro offered with lemongrass tea. It took me a long time, however, to understand this lushness. It was another two years before Mama Nguya drew together the official categories of medicine (*dawa*) and fortified, or nutrient-dense, foods (*chakula lishe*), inventing the phrase *dawa lishe*, medicines that feed us. With this word-work, she opened a space to recognize and cultivate alternative strategies for storying healing through remedies such as *imarisha*.

The four individual plants comprising *imarisha afya yako* came together over time in response to the needs of patients at a nearby HIV/AIDS clinic. TRMEGA proposes using this plant formula alongside antiretrovirals to strengthen bodies, enabling them to endure the pharmacological treatments. With *imarisha*, members of TRMEGA describe building their stamina for antiretrovirals.[6] For some, this is a story to be told through the nutritional sciences. Plants in this formula have dense nutritional profiles that authorize claims to a range of biochemical effects with therapeutic implications.[7] Mama Nguya and other trainers at TRMEGA frequently translate one of the ingredients in the formula, *mlonge* (*Moringa oleifera*, family Moringaceae), through its nutritional assessments: seventeen times more calcium than cow's milk, fifteen times more potassium than bananas, twenty-five times more iron than spinach, seven times more vitamin C than oranges, and ten times more vitamin A than carrots.[8] Mama Nguya borrowed these figures and this line of argument from a food justice organization she visited in Nigeria. Nutrients story remedies in ways that build "partial connections" with the objects of other ways of knowing and engaging this tree.[9] Mama Nguya mobilizes the nutritional data pedagogically in conversations with members. In fact, in 2022, she had a colorful mural painted on the wall of TRMEGA's new training center promoting this translation. She also uses the figures with state officials to legitimize the value and suggest the safety of this formula.

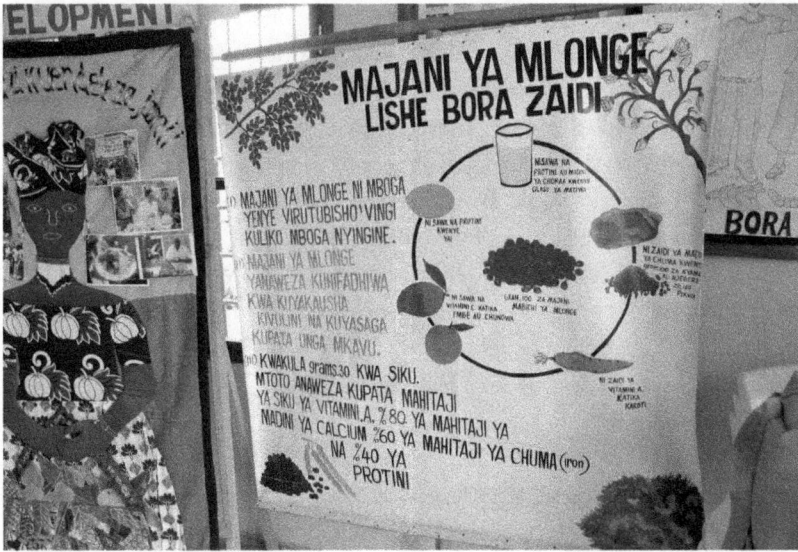

FIGURE 1.3 Banners TRMEGA used to teach members about the benefits of *mlonge*, known in English as moringa (*Moringa oleifera*), Maji ya Chai, Tanzania. Helen Nguya developed the banner on the left after being inspired by those used to promote *mlonge* in Nigeria. Photo by author, 2014.

Although nutrients story TRMEGA remedies in ways that are legible to biomedical health practitioners, relevant to chronic disease treatment, and active outside of the disciplining of the pharmaceutical science, the nutritional sciences obscure other aspects of *imarisha*'s potency. Arguments centered on nutritional value neatly bracket off toxicity as a side effect: a quality to be controlled for consumer safety perhaps but only relevant as matter out of place (i.e., effects of contamination or residue). This storyline narrows the space of effect and the temporalities of efficacy, situating them beneath the skin. From this location, it is not possible to see *imarisha*'s work within the double bind forging *dawa lishe* more broadly: that is, that toxicity is a condition of life in the twenty-first century. Our historical moment is defined in part by a recognition that the very substances used to support life also deplete the vital forces from which it comes.[10]

Producers' commitments to regenerative agriculture—in particular, to the use of organic crop-management practices, composting, and seed sharing—locate the concrete practices through which *dawa lishe* articulates toxicity as an attenuation of the forces of liveliness and the focus of ethical engagement.[11] By attending to the dynamism of soils and seeds—as well as to their communities, economies, aesthetics, and politics—Tanzanians reflect on and labor over the relations that constitute toxins and those that constitute remedy. The sense that toxins saturate everyday life compels *dawa lishe* producers to expand the spaces and times of therapeutic work. They address the effects of toxins in individual bodies and also by attending to how the plants in their remedies are grown and processed. As toxicity becomes re-spatialized and re-temporalized in this way, reflecting on and laboring over it is a process of reflecting on and laboring over that which is left behind as the very ground through which new or alternative life might then be nourished, coaxed, and cultivated—or cut off, deprived, and attenuated. The work of people and plants, as well as the myriad organisms, both human and nonhuman, that make up healthy soil and fertile communities of pollinators, constitutes a central focus of TRMEGA's interventions.

The garden, then, reframes the story of *imarisha*. Like TRMEGA's other colorfully packaged dried-plant remedies, *imarisha* draws people, as it drew me, to TRMEGA. The ease with which such remedies can be sold and gifted facilitates introductions to TRMEGA's work. Focusing on the path carved by their circulation draws attention to the dynamics of commodification and can render subtle articulations of the ways that Tanzanians shape and reshape possible futures through it. Yet, if this focus is held too tightly, it threatens to reinforce "rampant economism" even in critique by overwhelming stories that articulate ecology as anything other than that which is in excess of economy. From the garden, commodified herbal formulas are invitations to co-labor with other people and plants, to be remade within complex interventions into the forces that foster vital human and nonhuman communities. In fact, TRMEGA's garden explicitly destabilizes the commodities produced from regular harvests through the sharing of seeds, cuttings, and saplings across networks of support groups. Among *dawa lishe* producers, individual commodified formulas fade in and out of view. The efficaciousness of herbal products exceeds any properties of individual plants and is entangled in the ways that gardens cultivate healing through the compositions and decompositions that make places, times, and bodies livable again (and again).

*Dawa lishe* as an analytic, and as a theoretical provocation, names that which motivates solidarity among a set of social and therapeutic projects committed to co-laboring with plants in ways that disrupt the racializing, gendered, classed dynamics of exclusion and disposability that overdetermine which bodies and lands are repeatedly unsupported and unaccommodated. The familiar motivations that draw them to each other cannot be reduced to an interest in developing culturally appropriate ways of articulating the value of nutrition or in tapping local botanical knowledge to aid in addressing the burdens of global disease. Rather, these projects find common cause in a recognition of the disfiguring relations through which postcolonial, extractive economies compose and decompose lifeworlds. Their remedies draw, explicitly or implicitly, on histories of African therapeutics that oppose healing to the injustices driving attenuation, depletion, foreclosure, and barrenness (*uchawi*) and locate it in fullness, maturity, transformation, extension, and endurance (*uganga*). Relevant notions of efficacy emerge through the collective work of attending to the giving and taking, forming and transforming, that animates ongoingness. Healing means responding to dispossession and enclosure and reckoning with the continuing injuries caused by the world it has enabled.

Through *dawa lishe*, Tanzanians seek to not only address the attenuations, barrenness, and depletions constitutive of the extractive relations that undergird "development" but also unsettle the chemically saturated forms of lushness forced to thrive in the name of economic growth. Here I think alongside Emma Shaw Crane and her articulation of the "lush aftermath" of colonialism, dispossession, agricultural enclosure, and militarization.[12] Writing in the wake of the genocide in Guatemala, Crane traces how the US invasion in support of banana companies' insatiable desire for cheap labor targeted indigenous Maya, forcing them to flee. In Miami's suburban periphery, she finds migrant communities whose economic and legal precarity have rendered them available for dangerous work in ornamental plant nurseries. Their stories not only underscore the dispersed geographies of war and militarization but also illustrate how particular forms of lushness are forged through these displacements and the exploitative labor relations they enable.[13] In Kilimanjaro and Arusha, bananas may be the next frontier of radical enclosure. Whether or not Tanzania is able to resist the dangers of transnational banana companies, however, the landscape is already shaped by alienation of both land and labor in the name of commercial development. Sugar plantations, coffee estates, industrial-scale green bean farms, commercial timber forests, and export-driven cut-flower nurseries have restricted local farmers' access to fertile plots and undermined their rights to water while simultaneously increasing their exposure to chemical fertilizers, herbicides, and pesticides. Forged in this aftermath, *dawa lishe* emerges as a proposition that invites solidarity in efforts to reinvent the labor, aesthetics, politics, economies, and communities through which lush possibilities are experienced. The lushness that struck me as I walked through TRMEGA's gate describes the vibrancy and vitality that is *dawa lishe*'s remedy to the cramped space and vulnerable life produced in the wake of dispossession and enclosure. This lushness names the sensory experience of being pulled into the dynamic liveliness— caught in the composition and decomposition—of Mama Nguya's garden.

Many have cited the witticism, famously attributed to Fredric Jameson, that "it is easier to imagine the end of the world than to imagine the end of capitalism." In part, this failure of imagination is an effect of what (and who) we choose to think with. These gardens of therapeutic foods and medicinal herbs are not outside modes of capitalist extraction and world-making, but they do seek healing through ways that alter them and are alter to them. *Dawa lishe* offers lushness as a quality arising amid shifting relationships and ontological reorderings from plants to soil to microorganisms, from flowers to pollen to bees, from the one who feeds to the one who is fed. TRMEGA's gardens extend this density of relations through the sharing of seeds and cuttings. They infiltrate areas that have been depleted, drained, and dried out as they have been pushed into the margins by long histories of resource extraction in support of uneven capital accumulation. As they spread, one out of the other, TRMEGA gardens and the gardens that they catalyze are, I argue, complex experiments to reinvent what constitutes the therapeutic.

## Stigma as the Attenuation of Lushness

Jane Satiel Mwalyego described feeling lonely, isolated, and angry when she first met Mama Nguya. In her own words, she was "living with much stigma" (*unyanyapaa*, literally to exclude someone).[14] Jane had stopped taking her antiretroviral drugs in the midst of these pressures and the demands of providing for her children as well as her nieces and nephews, whom she had taken in after her brothers had died from AIDS. She tells a story of decline and searching, and of visiting a range of healers. Eventually, she lost the strength to farm. At times, she could not even rise from bed to cook. Both of her parents had passed away. She ran out of money and found herself without enough to be able to feed the children in her household or herself. The vitality that had sustained her life and that of others grew more and more attenuated; the courtyard of her home grew sparse. Jane marks 2010 as the turning point, when she met Mama Nguya and gradually began to repair and extend her relations with people and plants.

With Jane, Mama Nguya first learned to respond to the attenuation of relations around those with HIV/AIDs. She identified the primary problem as the attenuation of ties to others. Exclusion, marginalization, abandonment, and barrenness were not just obstacles to health but the opposite of healing. TRMEGA formed through an engagement with stigma as more than a social impediment to medical treatment. Programs apprehended stigma as a wounding of the forces that drive connections. As one member recounted, "In the earliest phase, we focused on people who were not infected, in order to be able to reduce the stigma of the disease. By reducing stigma, we were able to reach a point in which we could advocate for diverse forms of support and accommodation, such as taking loans from banks. The people from the Meru Commercial Bank came to see us and they advised us on how we can apply for loans and other things that might be possible. So, by that time, we were able to mix with other, different people."

The site of healing started with people who were *not* living with HIV. The threat posed by this virus sat not only in lowered CD4 counts but also in the marginalization of people with HIV/AIDS. Bodies grew thinner as relations strained. TRMEGA started by cultivating a community receptive to initiatives focused on the connection with and integration of people with HIV/AIDS.

TRMEGA quickly expanded its scope of work to include teaching people who were living with HIV to "leave aside stigma" (Kiswahili: *nawafundisha waondokane na unyanyapaa*). The organization sought to create spaces where people with AIDS could, in its own terms, "get used to each other," gain the confidence to talk with one another, and "share what is in their hearts." This language resonated with that of HIV/AIDS support groups. Indeed, Nguya did receive some external support for such work. Most prized at the time were bicycles that would ease outreach efforts and enable more connections.

Yet an account of TRMEGA's approach at this time must acknowledge that *stigma* here was in translation, a sort of boundary object that often emerges as NGOs appeal to global initiatives in efforts to design local projects and take up medicalized concepts to support interventions into everyday life. The way that TRMEGA actualizes stigma differs from the way that stigma is actualized through psychosocial research and public health programming. The histories of discrimination and social sanction that structure stigma (*unyanyapaa*) are most accurately accounted for through material instantiations. However, for TRMEGA, stigma glosses a thinning of social relations and all matter of life entangled with them—human, plant, and otherwise. Rather than seeing stigma as a psychosocial phenomenon, a judgment of the mind that can be fixed by addressing ignorance, TRMEGA takes up stigma as a relational phenomenon. It entails a turning away or attenuation of involvement that must be fixed by addressing the density of interconnection through which life is generated. While TRMEGA gladly folds education into its agenda, the efficacy of its efforts lies as much or more in the gatherings convened than in the correction of false beliefs or the transmission of information. Addressing stigma requires a series of encounters that interpolate one into vital, mutualist relations with things, people, and plants. In these early years, Mama Nguya introduced Jane to *mlonge*—the Kiswahili word used for many species of moringa, but here for *Moringa oleifera* in particular. She remembers that Jane's dramatic response to the plant was the first time that she "saw the power of *mlonge*."[15] Jane gained strength, resumed her antiretrovirals, and started gardening again. She was able to support her children and her brother's children. Her physical and social life thickened; the landscape around her house grew denser. This experience strengthened the role of *mlonge* and the centrality of the gardens as remedy in TRMEGA's work.

When I first came to know Jane, her life had grown lusher. She had come to embody the transformations possible through TRMEGA and its gardens, and to serve as a catalyst for others. In 2016, when an anonymous donor gave TRMEGA a plot of land so that the organization might increase its production, Mama Nguya asked Jane to live on the property and tend the gardens. In addition to a house, Jane received space to grow some of her own food. Using seeds, saplings, cuttings, and tubers from other gardens, she built a second demonstration garden and developed plans for a much larger production garden. Jane's daughter and grandchild moved in with her. As a new Slow Food community garden, this second TRMEGA garden attracted some additional investments. Jane started a chicken coop and envisioned acquiring goats. One of the first luxuries at the house was a long, thin wooden bench where TRMEGA members and others could gather, where Jane could teach, and where collective work could be organized.

TRMEGA turns an orientation to healing as the extension and expansion of relations into a pedagogical project. While Jane grew plants to be included in TRMEGA's remedies, and leaves could be dried in the shade of her overhang for inclusion in *imarisha* and other formulas, their efficacy required gathering. The bench so central to Jane's vision was to facilitate collective work and collective learning with plants.

## Infiltrating the Periurban

Much of my fieldwork with TRMEGA involved Mama Nguya, Jane, and me filling my vehicle with plants and distributing them together. Many of our best conversations arose as we drove through the region delivering cuttings, seeds, and saplings—as we extended TRMEGA's garden outward. Our activities catalyzed other patches of lushness in orphanages, street children's homes, elementary schools, and the homes of people planning community gardens of their own. Plants were delivered, along with a great deal of advice from Mama Nguya and Jane as we tromped around the spaces to be planted. Mama Nguya instructed on composting, organic pest and fungi management, irrigation and drainage, intercropping, and more. The form and content of her teaching fostered communities, human and nonhuman. She did not privilege the didactic dissemination of information through which knowledge is mobilized to facilitate the accumulation of capital. Rather, she brought both nutrient-dense vegetables and medicinal herbs to build rich, lush relations centered in and through place. After these long and detailed walks through the gardens, conversations regularly turned to health, to the therapeutic benefits of plants. Over tea, we would speak of the ways they cured and strengthened. Such meetings never stood alone; they always marked moments in long and involved relationships. Some had started a decade or two earlier through Catholic Church projects, while others emerged from newer, secular connections. Yet all gardens required tending and ongoing relationships.

We brought plants to a TRMEGA member living on the outer reaches of Arusha, to contribute to the community garden that he was beginning on his land. Other TRMEGA members were welcomed to join. His neighbors would drop by to explore ways that they might use plants to support the health of their families. He cultivates plants for medicinal teas to ease his parents' ailments. After years in which they attended to him as he struggled with HIV/AIDS, his pride in his ability to care for them was palpable.

Other gardens are larger. For instance, the garden at the Watoto Foundation, a home for male street children run by the Catholic Church, is quite established and substantial. The residents eat from the garden and make products to sell. Many of these boys go home during breaks. They have been encouraged to bring *mlonge* saplings home and to plant them at their parents', relatives', and neighbors' houses.

The Maji ya Chai garden and the gardens it has seeded are scattered around the city's edges. Through seeds and saplings, membership and mentorship, they infiltrate the patches of overly compacted, dusty earth disconnected from the births and deaths of lineages. These are not *vihamba*, the home gardens on Mount Kilimanjaro, or their relatives on Mount Meru in any traditional sense—that is, lands woven in and through the inheritance of a lineage, passed from parents to children, where the placentas of newborns are placed in the roots of a banana tree and elders are buried. In Tanzania, in the areas where land has not (yet) been rendered scarce by large-scale commercial interests, local leaders distribute plots to adults to farm. The fertile soils of Kilimanjaro and Arusha, however, have been seen as lucrative since early colonial and mission settlements, and local leaders have no land to distribute. Prices are high. Even old established and relatively secure lineages in the area find that the plots that parents are able to pass to children have grown smaller and smaller. Others are pushed toward the cities as they search for jobs. Those with resources send their children to secondary schools, most of which are located in urban centers. Youth with and without secondary education feel drawn to the city in search of livelihoods and liveliness. Ever-growing numbers rent or squat in the peripheral spaces on the outskirts and woven through the interstices of the city. Urbanization has transformed social, ecological, and architectural landscapes. Arusha town, for example, has grown exponentially since 2000 and is now the third largest city in Tanzania.[16] Maji ya Chai, where the TRMEGA garden is located, was once a distinct village but now feels contiguous with the city.

TRMEGA works in and through the periurban, a distinctive social and ecological form forged through intersectional forces of dispossession, urbanization, capital accumulation, environmental change, and state structure. Here ecological relations have thinned with dwindling access to land, overuse of resources, and declining soils. Yet, driven from rural areas by poverty and escaping both droughts and floods, people keep coming. Not only geographers and city planners but also public health specialists, environmental advocates, and a range of scholars recognized that settlements in and on the margins of cities are hybrid spaces, harboring both urban and rural characteristics and catalyzing new dynamics of their own. Settlement has often been too fast for the state to plan or control. Overwhelmed sanitation systems and inadequate water contribute to findings that periurban areas can exacerbate health inequalities. Some public health scholars have found that these dynamics are catalyzing a "major transition in the human–microbe relationship" and situate the periurban as the forefront of the coming era of new pathogens.[17] These zones of mixing and transformation, more-than-human migration, and experimentation are critical sites through which to understand intersectional inequalities as they are emerging in this moment and the radical environmental shifts reshaping human possibilities.

Periurban areas have also long been dynamic spaces, of diversity and mixing, new ideas and creativity. TRMEGA's work takes the periurban not only as site but also as method insofar as these dynamics compel particular modes of attention and engagement. Movement and migration are steady states. Members stress acts of gathering in and of reaching out. They track translations, transplantations, and transformation. Remedy and repair counter land alienation. The ethics cultivated through TRMEGA projects rest on practices of bringing together, of drawing closer, of cultivating entanglements and interdependencies.

Many TRMEGA members—having been separated from their *nyumbani kabisa* (Kiswahili, literally true home) where their ancestors are buried—rent their plots and homes in such spaces around the city's commercial center. When they move in, the soil in their courtyards is often densely packed. It has not been turned, nourished, or planted in years. Plots are too small to be imagined as agricultural. These homes are often considered temporary. Healing intervenes in this alienation. It works against the disabling imaginaries that prevent attachments to and care for the soils. Remedies draw people closer to land. They offer concrete practices of growing and composting that rekindle relations and open up horizons that cannot be fully defined by either private property or public commons. TRMEGA gardens jumped—through seeds and cuttings, lessons on double digging and intercropping, assistance with sack gardens and making liquid compost in buckets—across this uneven landscape of the periurban. In this way, they respond to what Anna Tsing calls the "patchy Anthropocene," the spatial configurations emerging from global capital's intentional and unintentional (re)shaping of more-than-human relations.[18]

TRMEGA gardens also spread to organizations that spring up in these urban and periurban areas to support those cast out in the fray of capital's drive to reshape urban and rural economies—children whose parents have died or who do not have relatives capable of caring for them, young pregnant women and new mothers who have no kin to support them and few skills to support themselves. To make TRMEGA legible to external funding agencies, Mama Nguya describes their work as focusing on "vulnerable people." At times, this has meant those with HIV/AIDS or, alternately, women and girls, orphans, or people with disabilities. TRMEGA is less interested in fidelity to particular categorizations of persons, however, than it is in disrupting the forces that systematically thin attachments, drain the liveliness of the relations, and render some bodies and land consistently more vulnerable than others. Their own attention moves with plants and soils.

When we were visiting Italy together for a Slow Food gathering, Mama Nguya and Rose were interested in permaculture demonstrations. Indeed, permaculture workshops in Tanzania have grown increasingly popular since the 2010s. TRMEGA and WODSTA experiment with many of these techniques. They find resonance with the dynamism of relations and the focus on movement. They are less likely to commit, however, to a closed economic loop animated by recycling. This is perhaps why the excitement about fishponds, iconic in the permaculture landscape, does not seem to generate sustained attention. I found fishponds scattered through the area, usually in demonstration gardens such as the Maji ya Chai garden at the TRMEGA offices, but fishponds do not move in the ways that seeds and saplings do. The complication is not that fishponds are not "traditional" in this area, but rather that they are not particularly skilled at infiltrating the alienated plots of the periurban. Fishponds, with their equally iconic rabbit house perched over an edge, are permaculture projects that can theoretically be "scaled up" and offer nutritional diversity to those living in cramped spaces. To do so, however, they require relations with land and control of water that cannot be assumed for many. They work within the logic of a closed-loop economy but also highlight the relations that enable such an imaginary. By offering a dynamic conversation with such global models rather than championing a particular form, TRMEGA's gardens orient members toward the movements in and out of the dense spaces that designate their gardens. Like the *vihamba*, these plots are conceived, in part, through their relations with the market and the field. They are nodes in complex and emergent transformations compelled by the people, plants, animals, and ideas that move through. They articulate a lushness whose rich biodiversity is distinguishable from the more uniform green of the plantation (or the industrial greenhouse) and whose relations exceed the circular economies of permaculture science.

### Beyond Cultivated and Uncultivated, or Plant Agency

*Mlonge*, the tree that was so powerful in Mwalyego's story, animates TRMEGA's current work and their Slow Food gardens. The saplings sprouting in small plastic bags that line the wall of the garden in Maji ya Chai lend themselves to travel, manifesting extensions, and initiating new spaces. Given that the nutrient-dense seeds, when dried, are light and easy to pocket for snacking, the species could have migrated from the Himalayas in northwestern India to the East African coast along numerous routes, involving both land and sea. In a 1976 essay that helped to catalyze interest in East Africa's role in maritime linkages across the Indian Ocean, Ed Alpers offers a Portuguese reference to three 200-ton vessels from Cambray at Malindi in 1500 as unambiguous evidence of Gujarati traders on the Swahili Coast by the turn of the sixteenth century.[19] Muzaffar Alam and Sanjay Subrahmanyam, in expanding Indian Ocean research to Arabic archives, however, allow us to wonder if *mlonge* may have found its way through the complex maritime travel linking South Asia to what is now known as Tanzania much earlier.[20] Or while we are speculating, perhaps the first seeds did not initially come by water but rather overland along trade routes coming down from Egypt or up from South Africa. Or maybe *mlonge* waited and came with the more substantial settlement of Gujaratis in East Africa in the nineteenth century. The British brought Indians as indentured laborers to work on colonial projects, such as the Kenya–Uganda railway, which began construction in Mombasa in 1886 and finished in Kisumu in 1901. The railway compelled some in the diasporic community to develop businesses farther inland.

In more recent times, *Moringa oleifera* has extended its reach through international nutrition and water purification projects. Today, many producers of

FIGURE 1.4 *Mlonge* seedlings, known in English as moringa (*Moringa oleifera*), sprouted and prepared to share with members, schoolchildren, and others to whom TRMEGA extends its gardens through collaborative projects, Maji ya Chai, Tanzania. Photo by author, 2014.

therapeutic plants dry and grind the leaves to be added to tea or porridge. At TRMEGA, it is mixed with three other plants (including the herb from Kagera discussed above) to make *imarisha*. TRMEGA shares seedlings with its members and sells them to support the organization's work.

As *mlonge*'s possible travels suggest, the therapeutic plants in these gardens are caught up in long histories of migration, human and nonhuman. Producers are no more invested in the purity of indigenous nature than they are in the authenticity of the healer or the discreteness of institutions. Local Slow Food festivals do celebrate indigenous foods. Some gardens share space with more recent hybrids, such as banana stock from the USAID Tanzania Agricultural Productivity Program or the many sorts of orange sweet potatoes that have made their way into the country through nutrition projects that have introduced a variety of nutrient-dense cultivars.

Relations with plants are cultivated based on the ways that collaborations with them enable particular modes of exchanging, growing, eating, tasting, healing, communing, and decomposing. The qualities of the plants themselves matter. *Mlonge*, for instance, does not lend itself easily to plantation agriculture despite the ongoing efforts of many. The leaves are small and difficult to harvest en masse; they wilt quickly. This recalcitrance led to the failure of large-scale efforts outside Tanga, 400 km to the southeast, where row upon row of trees stand unattended and unharvested.[21] *Mlonge's* value emerges when it is enrolled in human projects by different means. In collaboration with food sovereignty initiatives, Tanzanians work with *mlonge* in home and community gardens to cultivate alternative ethical-political possibilities. Through a careful parsing of local plants as "therapeutic," they experiment with ways to build common cause with efforts to protect indigenous knowledge.[22]

Relations among plants, as well as between plants and people, prove critical to decisions about which plants are brought in, cultivated, and catered to and which plants are not. A garden's capacity to heal works not only through the fostering of biologically mutualist relations—nitrogen fixing, pesticide management, pollination, and so on—but also through its ability to hold surprises and remember even when people do not. Mama Nguya and Rose illustrated this one day when walking through a garden belonging to a member of one of the women's groups. They saw a vine growing among the more purposefully tended plants and asked the woman what she did with it. She told them that it was "just a weed." Mama Nguya revealed that she could eat the fruits of this wild cucumber plant, and Rose added that the leaves made a good vegetable. The woman tasted the plant; later, she ate the fruits and cooked the leaves. Soon, she made a small income selling the leaves and seeds back to Mama Nguya and Rose for Slow Food gardens and to others in local markets. These gardens are weedy. Often blackjack or other medicinal herbs will be growing in the middle of a bed of another vegetable or on the side of a tended row. Such "weeds" are left. These gardens are not plantations. The plants brought in must make affordances for, if not explicitly support, other life. The practices through which particular plants are folded into the space of these gardens cultivate plant relations, and therefore a *plant politics*, that allow for surprise, not only recalcitrance.

I have often found myself returning to moments such as those with the wild cucumber—mulling over stories of plants that have been afforded space, their agency recognized as neither subordinate to nor symmetrical to human agency. Attending to the wild cucumber surfaces the forms of attention and affordance central to the *kihamba's* lushness. It was not uncommon that as I walked through home gardens with people, they would stop and wonder, often with humor, about a plant that self-seeded. I was most intrigued when the person neither knew what the plant was nor was particularly inclined to speculate (out loud at least) on its potential use-value. The plant might not be particularly interesting or attractive. No one seemed to know if it was edible or nitrogen fixing. Bees or other pollinators might or might not seem to enjoy it. At times, the grower might observe or wonder at it, and at other times, they would ignore it. They simply allowed it space. They did not weed it out when they removed other plants. This willingness to be moved by weedy charisma points to the dynamics at the heart of the *kihamba* as a mode of dwelling.

The logics of the *kihamba* in which the wild cucumber flourishes suggest modes of attention that exceed the relations of owner and resource generated in modern land relations. Yet, the alternative they offer cannot be captured by critical plant studies' fascination with experiments in plant communication or other modes of ascribing human-shaped agency to botanical beings. Rather, these modes of attention add depth to the findings of botanical studies by Andreas Hemp that these coffee–banana plantations contain on average 520 vascular plant species, including over 400 noncultivated plants.[23] He and those who draw on his research offer this as evidence of the *kihamba*'s ability to "maintain a high biodiversity" and therefore argue that this style of planting is important to environmental conservation efforts.[24] There is particular interest in the way that the architecture of the *kihamba* resonates with the multiple layers that define a tropical montane forest with trees, shrubs, lianas, epiphytes, and herbs, and thereby continues to afford space for species that have lost most of their former habitats. While the lushness of the *kihamba* certainly finds important points of connection with environmental scientists (and many of the rest of us) concerned about biodiversity loss, I am struck by the ways this vocabulary (again) articulates the *kihamba* through fundamentally economic logics. Plants are the objects. People are their cultivators. Biodiversity is its use-value.

The wild cucumber helps us dwell for a moment on other modes of attention and forms of agency that both support the unfolding of, and are themselves unfolded through, "400 non-cultivated plants." Wildness, here, is not a property of forest pitted against farm. It does not mark that which is outside of the human. The wild cucumber plant helps us to envision something other than ecology's resistance to the economic. Rather, wildness strives to recognize relations that slip through the binary of the cultivated and the uncultivated and, in so doing, points to the improvisational choreographies that are possible in the density of these gardens. The wildness of this cucumber is in invitation to—borrowing Agamben's words—"keep ourselves in harmonious relationship with that which escapes us."[25] That which moves and migrates and co-creates spaces for survival. It speaks to a subtleness and dexterity that is other-than-mastery to guide modes of care and styles of dwelling. And in so doing, it suggests a strategy of living with and through that locates lushness in radically different relations of economy and ecology. These are modes of attention and forms of agency that make it possible to locate alternative forms of sovereignty in the processes of composition and decomposition.

# Lushness in (De)composition

The surprises that occur in *dawa lishe* gardens defuse a politics of purity and, in so doing, the fantasies of salvation (or apocalypse) such a politics supports. Projects such as TRMEGA's make much more modest proposals about what it might mean to develop spaces with the capacity to nourish alternative futures of lushness. It requires making arguments for care and healing, for taking positions and initiating action from some ground other than crisis.[26] Composting incites a theorization of this ground as densifying or thickening relations.[27]

For centuries, smallholder farmers in Kilimanjaro and Meru added organic matter to their fields. The *kihamba* oriented relations with fields and markets in ways that focused attention on the nourishment of the forces through which people and plants, lineages and soils, were bodied forth. Farmers worked together with plants to conjure rich, moist soils. Colonial dispossession and the epistemological and ontological architecture structuring postcolonial development obscured this orientation to being of, and through, the land. Since the 1960s, as the population has grown, pressure for food and fuel has taken a toll, driving deforestation, the omission of fallows, and soil erosion. The social-economic strategies narrowed around land's role as resource—a context for livelihoods—and left soil vulnerable to overuse. Practices through which organic material was returned to the soil waned. When less organic material was added to the soils than was being removed during harvesting, the vitality of the soil (and that which it supports supports) decreased.[28]

In the TRMEGA gardens, composting is both a practical consideration and a form of re-membering that counters this attenuation of vitality. Parts of the past that have been left behind (plant residues and food scraps) transform through specific entanglements into something other than themselves and—if carefully tended—into the components of rich, healthy soil that grounds growth and other life. Compost embodies a process of living-through. It takes movement—flows in and out of the compost pile—as a steady state and, in so doing, focuses analytical attention on ontological dynamism rather than the properties of stable objects. Compost is not only a practice worthy of description but also an incitement to theory.[29] Compost's re-membering pushes against the forms of forgetting underlying the metaphysics of purity.[30] In so doing, it opens a space to reimagine the modern concept of toxicity and expand its relations with remedy.

Yet, there are stakes in this remembering. Compost constitutes a material, social, and ethical relation with the past. Not all things compost and not all compost well. In their guide to "compost politics," Abrahamsson and Bertoni challenge the "cozy versions of conviviality" that they see in much of the scholarship celebrating more-than-human approaches to interspecies relations.[31] The stakes rest in an ongoingness, not in niceties, symmetries, or authenticity. They rest in both the relations and the nonrelations—the attachments and the detachments—that support "making livable again."[32]

Some of the most involved conversations between Mama Nguya, Jane, and the others with whom they collaborate concern composting: what to compost, where to put it, how to tend to it, and how to use it. In my first interview with Mama Nguya, she emphasized compost as central to the identity of her garden, stating emphatically that they never use *dawa za dukani*, literally "medicines of the shops," referring to synthetic fertilizers and pesticides available at the market. Composting, which in practice includes recycling crop residues, food scraps, and animal waste, orients toward a "collective continuance" of the vital forces that support ongoingness.[33] In the process, it problematizes that which attenuates and diminishes these forces over time. On entering each of the gardens that we visited together, Mama Nguya attends to the signs of composting: the pits or piles of organic matter in slightly shaded spaces, in different stages of decomposition, bearing evidence of regular turning. She asks questions and advises on how to foster the sorts of relations that lead to soil that can go on—systems of replenishment, practices of nourishment. As she builds new relations or starts to become more deeply involved with a garden, establishing composting routines and ensuring practices that regenerate the soil's fertility often demand significant time. Successful composting requires reflection and attention, even in her own garden and in those of longtime companions.

The practice demands intimacy, engagement, and nonrepresentational modes of growing closer to and being with land and its many forms of liveliness.[34] It decenters knowledge as mastery and offers a capacity to keep trying, responding, and growing more attuned.[35] "Composting," as Afonso and Imbassahy note in their article on composting and compost in an indigenous garden in Rio de Janeiro, "is a constant maintenance process, in which the balance is never completely stable. There is no better guidance for making compost than understanding the experience of other practitioners and creating intimacy with your own specific compost, learning to manage its elements, and understanding how they combine."[36] Mama Helen teaches composting as a process of observation, interaction, and response. Being knowledgeable about composting involves having developed a sensitivity to changes in matter and a commitment to a centering of the practice in one's garden.

Composting involves cultivating thick relations. The greater the content of organic matter, the greater the capacities of the soil to retain water and nutrients, to make proximate those entanglements that support plants and a community

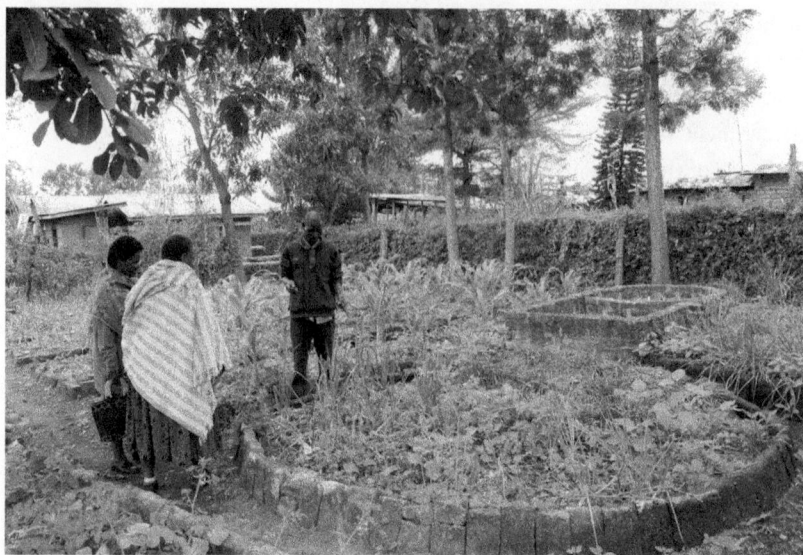

FIGURE 1.5 Talking about compost at TRMEGA, Maji ya Chai, Tanzania. Photo by author, 2017.

of decomposer microorganisms. When these relations are attenuated—when plants are harvested and taken away, when plant matter is not turned back into the soil, when there has been a thinning of relations—then the structure and porosity of the soil become compromised. Composting proposes an ethic of return or, at least, of a regular turning. The past is turned in and under to cultivate an environment dense with matter and with the potential for transformative relations. From the compost pile, relations of remedy with memory are given a concrete form, and a rethinking of toxicity as an effect of thinning, abandonment, and infertility is possible.

The pleasures of composting, as well as its particular form of lushness, rest in these interconnections, interpenetrations, and transformations. The stuff of the world must alter from itself, change from what it was: corn stalks, bean vines, fallen rotting tomatoes, spinach stems, and other debris from the garden and scraps from the kitchen left after human consumption. Exposure to living decomposers like bacteria, fungi, protozoa, algae, invertebrates, and insects triggers long and complex sequences of physical transformations: decomposition, mineralization, and humification. Composting calls for an ethics that rests in ontological dynamism. Our stakes in each other cannot be articulated through a loyalty or faithfulness to what one is; rather, the stakes consist of a commitment to becoming other—to becoming that through which other living might happen.[37]

## Conclusion

Over the past decade, the vibrancy of the TRMEGA compound in Maji ya Chai has expanded, changed, and taken new forms. When I visited in 2023, the dreamed-of training center had been completed. Colorful murals decorated the walls. Behind it stood a new shade structure with two long tables for gathering to work, eat, and share. Air potato vines grew up the poles and wound around the crossbeams, their heart-shaped leaves waving gently in the breeze and loose clusters of small fragrant flowers dangling playfully. Upon entering, we walked slowly around the garden, Jane and Rose detailing each change and describing how, why, and to whom each plant mattered. They stroked the leaves of a recently acquired medicinal herb, perched nonchalantly on the end of a bed of greens. Through each story, we began winding our way back to each other across the too long absence forced by a global pandemic and extended by the challenges with which it became entangled. When we eventually settled around the tables, we shared the fears and losses of COVID and the other injuries that those years had borne. We offered our condolences, our reassurances, our humor. They wrapped me in a new *kanga* and plied Sia and me with fresh hibiscus juice, lemongrass tea, hand-roasted peanuts, and sweet banana chips. I urged them to open bags with maple syrup, blown glass bulbs for container gardens, and seeds of golden beets and rainbow carrots.

In this chapter, I have taken up TRMEGA's work as an argument for the need to reconceptualize healing as land relations. TRMEGA is both caught up in and responding to a widely held concern over the ways that bodies today are forced to bear complicated toxic loads not only because of the dumping of capitalism's harmful by-products but also because of the very products that facilitate domestic life (e.g., plastics, kerosene), agriculture (synthetic pesticides, industrial fertilizers), and health (antiretrovirals, contraceptives). The toxic is social (as with stigma) and environmental (as with agricultural runoff). TRMEGA members are thinking about the conditions under which life becomes attenuated, diminished, depleted, exhausted, drained away.

Remedies include herbal formulas, such as *imarisha*, but "not only."[38] In raising gardens as remedy, TRMEGA shifts attention from the individual embodiments of plants and people to the ways that plants collaborate with people to make vital spaces that enable healing. Canisters of herbals fade in and out of view. Plants are not only agents with active ingredients, although both herbal producers and users are happy to fold in such articulations if available. Plants are not only resources to be capitalized on, although they do at times constitute a latent commons that offers the possibility of commodification. They are therefore not (or not only) anticommodities, even as they resist the forms of politics possible through stable trajectories of commodification.[39] In these gardens, therapeutic plants are never only resources to be harnessed for human need (corporeal, aesthetic, or economic); they are also collaborators in the making of nourishing and nourished spaces.[40] In fact, plants become therapeutic as they move and work with people to create habitable spaces, pockets where people and plants mutually reinforce the cultivation of vitalities over time.[41] Health is about lushness, density, diversity, and the mutualisms they enable.

Although agricultural policies in Tanzania over the past fifty years have focused on cash-crop productivity as a way of navigating the country's position in the global economy, the gardens at TRMEGA disrupt the forms of scalability so critical to these dominant notions of development. They are designed for ongoingness, rather than scalability and maximum extraction. Gardeners eschew notions of transcendence through continual growth; they work within the impermanence, dependencies, and decompositions of plant and human life. Organizing their gardens around compost, they question how we problematize toxicity by pointing to the ways that people and plants navigate value within the dynamics of composition and decomposition. They do not ignore other framings—environmental destruction, chronic disease, or food insecurity—but they unsettle them by recognizing and folding in the temporalities of a wider range of urgencies. Such gardening projects think expansion with their plants; reproducibility here is about seed sharing, cuttings, and transplants.

TRMEGA's engagement with food sovereignty movements, then, raises an interesting question about the sorts of sovereignty the organization imagines. At one level, these gardens appear to democratize the means of therapeutic production. Indeed, knowledge of the ingredients in one's medicines and the ability to make them stand in stark contrast to most people's relationships to pharmaceuticals. These gardens also expand access to particular therapies, reclaiming the right to alternative options for treatment. By sharing cuttings, seeds, and transplants with each other, TRMEGA members and affiliates might be said to create and extend a form of the commons. In this reading, the commercialization of plant-based medicines hovers as the ultimate, if not inevitable, goal. Sovereignty would seem to emerge through people's ability to determine what sorts of therapies are available, what they are composed of, how they are produced, and how they are distributed.

At another level, however, the extension of these gardens results in more than a democratization of the means of production for herbal therapies, because the gardens themselves move the therapeutic value of the plants beyond the process of commodification. Gardens cannot be explained away as apothecaries for the poor. In this reading, TRMEGA's work is not about control over the means by which therapeutic plants are turned into resources for the alleviation of human need, but rather about the creation of spaces that facilitate dynamic relations among plants, people, soil, and decomposers of various sorts, as well as the interpenetrations that might transform them. Plant relations may be organized through commodified formulas such as *imarisha* for a moment, but this mode of organization does not fully confine them. The plants in the formulas also live in TRMEGA's garden, and from there they reach out, fertilize other plants, move beyond the boundaries of the garden, seed other plots, and develop new relations. None of the gardens stands as a model of some original or imagined garden. None is merely a concept. Rather, each constitutes an actual extension of other gardens, an extension of their social relations, physical matter, and storied histories.[42] Not only does this challenge a particular notion of ownership, but, as diverse temporal relations fold in on one another, it also troubles the very notion of an owner. Opened up to the indeterminacies of both subject and object, this version of sovereignty suggests the ability to shape the conditions of possibility for thriving and becoming—that is, the conditions of possibility for lushness.

Indeed, thinking with the edges of capitalism means, in part, thinking through toxicities as a condition of life to be navigated, shaped, and engaged, rather than as apocalypse or salvation. The ethics of these gardens grow in the relations and logics of compost. They do not lend themselves easily to dystopic or utopic visions. They push against these "cramped spaces" for argument and action.[43] They are experiments in creating spaces for a new politics of habitability, one that continually asks: Which kinds of lushness can be cultivated in twenty-first-century Tanzania? Which relations enable bodies and landscapes to grow ampler, denser, more productive, and more potent? Whose ongoingness and which forms of continuance do our gardens support?

# 2

# Efficacy of
# Appetites

In Tanzania, Nane Nane, or "eight eight" in Kiswahili (for August 8), also known as "Farmer's Day," is a national holiday celebrated across the country during weeklong agricultural exhibitions. With over 65 percent of households involved in agriculture, this celebration of the farmer's role in the national economy feels personal to many.[1] Hopes for greater financial security are fueled by such public recognitions of the centrality of their land and labor to the generation and accumulation of national wealth. The signature event brings together hundreds of exhibitors from the private sector, public institutions, academia, research organizations, policymakers, and advocacy groups who are invested in agriculture as an engine for economic growth. The gathering is opened each year by the Minister of Agriculture and attended by the president. The organizers showcase agribusiness efforts in Tanzania, animating visions of a profitable future through increasingly industrialized agriculture and highlighting the technologies and producers that embody these visions.

Occupying a table at the fair, the nongovernmental organization TRMEGA (Training, Research, Monitoring and Evaluation on Gender and AIDS) introduced *imarisha* and other herbal formulas into this excitement for agricultural development and growth. Their presence not only pointed to less well-known botanical resources that Tanzanians might exploit to gain traction in local and global markets but also quietly insisted on a connection between agriculture and medicine, land and bodies, growing and healing. Their table drew others whose products moved across established modernist divisions held in place by institutionalized distinctions between food and drugs. One of these people was Alex Uroki, an entrepreneur who founded and ran EdenMark Nutritive Supplies in northeastern Tanzania, not far from TRMEGA.

Recognizing in TRMEGA's products a shared commitment to valuing food as healing and medicine as nourishing, Uroki purchased several canisters of *imarisha afya yako* at the 2014 National Nane Nane Agricultural Exhibition in Dar es Salaam. When he returned to Bomang'ombe, he placed these canisters on the shelves of his herbal clinic next to those holding EdenMark's plant-based medicines. Each of the EdenMark labels echoed the refrain splashed across the front of his clinic: *Chakula chako ni dawa yako.* "Your food is your medicine." Indeed, this proposition may feel familiar to readers in North America, Europe, and elsewhere.[2] It may also resonate with students of Ayurvedic and Chinese medicine.[3] Products, books, and scientific studies about the therapeutic effects of food move between North and South America, Asia and Africa. Yet, the occasions that this proposition describes—and catalyzes—unfold differently across these diverse international arenas. People live into and through the idea that food is medicine in concrete, local ways. In Tanzania, the phrase *chakula chako ni dawa yako* draws on African therapeutic traditions, local plant life, and ways of dwelling to expand notions of eating well and healing effectively.[4]

Uroki returned with more than a new product to diversify his sales, however. He had also begun to cultivate a connection with Mama Nguya. By placing *imarisha* on his shelves at EdenMark, he drew that new relationship into his clinic. Their connection provided invitations into the larger web of relations in which Mama Nguya worked. For instance, Mama Nguya suggested that the regional leaders of the local food sovereignty organizations in which she was active connect with Uroki. He became a member of Slow Food International and their 10,000 Gardens in Africa project. This chapter takes up the provocation of *dawa lishe* as an effort to name such gestures of solidarity and thereby invite attention to the forms of healing through which Tanzanians are unsettling the dreams of a new Green Revolution. I am interested in the ways that plant and planting remedies concretely address disease and debility as embodiments of ongoing social and ecological abandonment, justified by the alienation of labor and land, central to the reduction of the natural world to natural resources. TRMEGA's and EdenMark's therapies stimulate desires for eating in ways that elaborate alternative land relations, forms of embodiment, and conceptions of environment. Their social-ecological experiments have led me to see the work of plant-based medicine as being, at least in part, the work of reimagining appetite as a site of intervention for global health.

Over the past thirty years, appetites have come under increasing scrutiny by clinicians in a wide range of specialties concerned with the impact of "lifestyle" on health in response to the rise in noncommunicable disease. "Modern" appetites are taken up in public health campaigns responding to rising rates of diabetes, hypertension, and obesity on the continent. Appetites are implicated in the hunger that challenges compliance with, as well as the efficacy of, antiretroviral drugs. Appetites lie slightly to the side of medicine or medical sciences and sit more solidly in the purview of public or community health. Appetites are to be managed so that they do not interfere with medical treatment regimes. For clinicians as well as nutritional scientists, appetites are subjective experiences that either propel or prevent contexts for the biochemical processes necessary for their cures to work. However, when appetites emerge as a problem to be managed biomedically, the subtler politics that work through desire are often truncated, glossed as "culture," and labeled in flowcharts as an obstacle. This move narrows the relations for which medicine is held responsible. In this context, *dawa lishe* names initiatives that resist being pigeonholed as (only) community-based projects focused on overcoming cultural obstacles to eating well, and it draws attention to initiatives that apprehend appetite as the desiring force that motivates lush social and ecological relations.

In redefining appetites and their roles in remedying individual and collective health challenges, TRMEGA and EdenMark are also redefining the terms according to which the efficacies of remedies are evaluated. Insisting that medicine feeds, nourishes, and fortifies explicitly disrupts a public health lexicon that relies on binary ontological claims—that food nourishes and medicine cures—to evaluate the value of plant(ing) remedies.

I see TRMEGA's and EdenMark's strategy as an effort to disentangle what Eli Clare describes as the "knot of contradictions" that solidifies the ideology of cure in global health. Clare describes how biomedical notions of cure anchor a lexicon that draws attention deep into the body and away from the ways of living—practices of dwelling—through which bodies come into being. Cure emerges from both the construction of a biological body that has been severed from social and ecological relations and the commitment to the dominance of this body as the epistemic object of medical practice.[5] Efforts to re-embed the biological body in its social and ecological relations reveal the double-bind in which cure simultaneously heals and harms. As Clare writes, "Cure saves lives; cure manipulates lives; cure prioritizes some lives over others; cure makes profits; cure justifies violence; cure promises resolution to body-mind loss."[6] I am interested in the ways that producers and users work through the contradictions of "cure" as they develop and assess—grow and distribute—their remedies. I am particularly drawn to the ways in which their remedies refuse to track the division of body and land (healing and growing), on which the distinctions between medicine and agriculture depend. I explore how they advance the language of nourishment in order to expand the range of injuries, depletions, and illnesses that medicine is held responsible for addressing.

The provocation that inheres in *dawa lishe*—that medicine be nourishing— breaks up the horizon against which the political phenomenology of bodies is generated in Nane Nane's commercial festivities or public health–focused charitable campaigns. It identifies shared orientations and invites solidarity among social-ecological-therapeutic projects that articulate appetite as an animating force driving the constant emergence of body and world that we experience as life. Therapies intervene in appetites—building attachments to flavors and to plants—in order to influence the open-ended unfolding of humans and environments through each other. The co-constitutional relationship is often obscured in biomedical formulations of the body proper as a medical object, which is shaped by, but distinct from, its environment. *Dawa lishe* is one example of Tanzanians enriching our vocabulary with words that describe how disconnection, isolation, and attenuation of both ecological and social relations arise as pain, disease, and debility. *Dawa lishe* producers use the friction between diverse forms of knowledge to pry open a conceptual space in which to conceive of remedies that function by reworking the relationship between eating and being-in-the-world through appetites. By cultivating appetite as a mode of attending to the forces that nourish alternative forms of lushness, TRMEGA and EdenMark position their remedies to intervene not only in how people eat to be healthy but also in ideas of what constitutes health and healing and how efficacy is articulated.

Forgotten Flavors

Alex Uroki was the first to tell me that people have "forgotten" (*kusahau*) how to eat. We were sitting behind EdenMark Nutritive Supplies pharmacy and clinic in Bomang'ombe (Kilimanjaro). I had initially reached out to Uroki after the national registrar for traditional medicine held him up to me as a model of what traditional medicine in Tanzania could be, of what the government hoped to support through its registration and development of traditional medicine. I quickly came to understand, however, that Uroki was not a healer in any traditional sense. While Uroki is Mchagga, born and raised and currently living in the Kilimanjaro area, the "forgetting" he lamented did not reflect a nostalgia for the Chagga foods of his childhood[7] or a celebration of "African medicine."[8] In his investigation and formulation of EdenMark's plant-based therapies, he was not epistemologically committed to an earlier time when something was known, or to knowledge that is now lost. The "forgetting" he described to me as orienting his work explicitly disrupted the processes of colonial unknowing that inhere in the recognition and registration of traditional healers as a site for modern governance (see also chapter 3). For Uroki, "forgetting" glosses a loss of palate. He argues that people not only no longer know how to eat but, even more fundamentally, do not have the capacity to be attracted to the flavors that heal. "Modern" food, with its sugar, salt, and oil, has made more complex and harsher flavors both unfamiliar and undesirable. Bodies have forgotten their attachments to edgier tastes.[9] His remedies encourage the body to remember, and they strive to repair this damaged relationship to flavor. Many of the therapeutic foods and medicinal herbs that line the shelves inside his Bomang'ombe pharmacy (as well as those of the smaller pharmacy that he had recently been successful enough to open in the more populous town of Arusha, an hour's drive west) challenge the palate. In so doing, they locate healing in building more complex appetites that nourish healing relations.

We first spoke of the orienting drive of appetites in relation to dandelion greens and their bitterness.[10] At that time, Uroki was quite obsessed with dandelion greens. His large furnace-driven forced-air plant dryer at EdenMark Nutritive Supplies' production facility was filled with them. He sold them in his pharmacies in Bomang'ombe and Arusha, as well as on consignment in small shops in Moshi town. The dehydrated leaves were packaged whole to be added to soups and vegetables or ground into a powder to be taken as medicine. He was also experimenting with dandelion as one of the foundational ingredients in what later came to be a very popular kidney formula. Uroki argued that people would agree to consume these bitter greens as a medicine prescribed in dosages (one teaspoon three times a day mixed in a cup of warm water), even if they would refuse to consume them as food during a meal. He hypothesized, however, that, by taking this medicine (and others like it), people would learn to tolerate bitterness, remind their palate of the sharper flavors it is capable of savoring, and build an appetite for these plants. Eventually, he predicted, many would also come to wonder what the plant is. They would come to realize that they can collect this and other common plants themselves.

Learning to eat dandelion greens as a medicine would ideally catalyze the development of a more extensive relationship with the plant, in which the person would learn to identify, gather, and prepare it. Uroki points out both that food calls one into different relations with the world, and dandelion greens as a food have the potential to encourage a person to notice the world differently. Cultivating a palate for their bitterness, he suggests, leads a person not only to discern the complexities of dandelions' flavor but also to identify its distinctive green leaves more effectively among the many "uncultivated" plants with which it grows. Therefore, such a cultivated palate may even invite foraging. Appetite animates the forms of attention according to which people build attachments. According to this understanding, losing a desire for bitterness allows the attachments that cultivate this desire to wither.

Uroki explicitly framed his effort to commercialize therapeutic herbal medicine as being aimed at helping to change which foods people consume and which flavors they desire. In these interventions into the palate, Uroki apprehended both flavor as an effect of relations and appetites for flavors as worthwhile sites of intervention—more than mere dimensions of subjective experience. I find Judith Farquhar's work on the power of flavor and the efficacy of appetites in Chinese medicine helpful in accounting for Uroki's work in Tanzania. Farquhar acknowledges the challenge of articulating flavor's capacity to catalyze physical change. "Participants in Westernized cosmopolitan culture," she writes, "are firmly under the sway of an idealist and visually biased aesthetics of the senses: we know that flavors are in the domain of pleasure, which must be epiphenomenal to bodily transformation such as illness or therapy."[11] To understand healing in China, as she argues, or, as I would argue, in Tanzania, it is necessary to provincialize these "aesthetics of the senses." Uroki's efforts offer a space to think an alternative political phenomenology of the eating body. As he suggests, learning to be attracted to and moved by dandelions' bitterness is one way to learn—one remedy through which to cultivate the capacity—to be attracted to and moved by different land relations. Flavors are not metaphors. For Uroki, they are a way of sensing that which moves us, and as such, they are a way of apprehending the forces that compose the body. Interventions into the palate engage flavor to cultivate connections to people, plants, and place through both taste and memory. They drive what foods we desire, shape what we consume, influence what we grow or gather, compel relations of exchange, and thereby organize the ways we dwell.

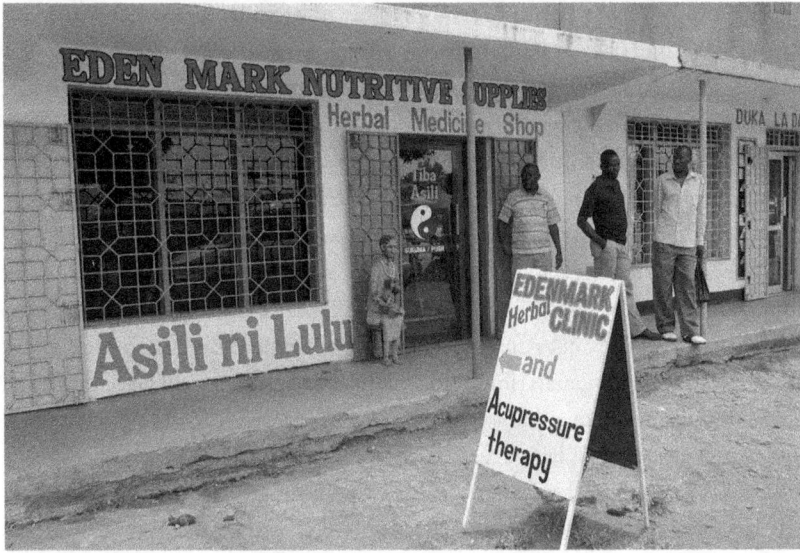

FIGURE 2.1 The primary clinic and pharmacy of EdenMark Nutritive Supplies, which is owned and managed by Alex Uroki, Bomang'ombe, Tanzania. Photo by author, 2014.

## Knowledge at the Crossroads

Uroki's primary pharmacy and clinic lies at the major crossing that defines Bomang'ombe today. EdenMark's bright blue façade calls out from the center of a strip of shops tucked behind a row of roadside vegetable stands. The shouts of scouts from the *daladalas* (minibuses) stopped across the street animate the clinic on hot days when the door is propped open. Bomang'ombe shifted centers a few decades ago, pulled by this highway and the promise of access to bigger markets, contact with more diverse communities, and exposure to a wider array of ideas and goods. The smooth paved road carries cars, trucks, buses, and motorcycles west to the city of Arusha, up to the border of Kenya, and beyond, as well as east, around the southern edge of Mount Kilimanjaro, through the town of Moshi and eventually down to the coast, to the Indian Ocean, with its own expansive history of trade and travel. This new Bomang'ombe is shaped by the throughness of contemporary capital, the mobility of labor, and the movement of commodities. The old Bomang'ombe is not far away, however. Turning south off the highway, any number of small dirt roads lead through older residential areas and toward the flats. Here the earth at times cracks from dryness and at other times is flooded with water from heavy rains pouring off the mountain's slopes.

The old Bomang'ombe remembers that the loud, dusty, exhaust-filled throughness of its newer self is only one among many currents that have pulled people, animals, plants, and things through this space. It harbors the old winds of north–south migrations. The smaller road heading north climbs up the western slope of Mount Kilimanjaro through the settled farms and fields of those who have dwelled at times for generations on these fertile slopes, then through the town of Sanya Juu, and arrives at the large-scale, foreign-owned commercial farms that ring the national park's border and grow green beans and coffee for export. Some Maasai still try to move their cattle through this area. Then, the smaller road continues down the mountain, south of Bomang'ombe, where Pare have settled for generations and where the privatization of large swaths of land is less of an obstacle, and village market days are lively, as animals, harvests, and things gather and change hands.

Uroki's learning and his work—the way he thinks about and enacts his own relations with plants and healing—have emerged at the intersection of forces shaping Bomang'ombe and the broader world in which it sits. Part of Uroki's genius lies in how his therapies open up lines of movement, possibilities of throughness, and new attachments.[12] The plants on his pharmacy shelves pull their users through times, spaces, and relations—through the tensions of the contemporary world, the realities of ancestral presence, and the pain of dispossessions felt but rendered invisible. It took me time to see the effect of this attention to movement and the energies and desires that animate it, rather than the properties of objects. In the previous chapter, I thought with TRMEGA's compost about knowledge—or at least growing knowledgeable—as indexing a process of witnessing, a sensitivity to change, a desire to respond, and a commitment to remaining in relation, rather than knowledge as mastery of an object. In this chapter, I am thinking with EdenMark's appetites to apprehend agency as an emergent, relational quality formed in the heterotemporal flows that shape people, plants, and place, rather than as the discrete property of a subject. Apprehending agency as movement is critical to accounting for the constantly shifting sites of interconnection that are experienced as lushness. *Dawa lishe* locates the power to heal in movements that invite a pooling and dispersing of the forces of life and liveliness, generating nourishing (inter)connections. Harming, in contrast, is a greedy harnessing—the overuse and overconsumption—of the forces of life and liveliness that leave barrenness in its wake.

Reflecting on the Maasai visiting EdenMark pushed me to account for healing as a process of creatively increasing connections and nourishing relations that stem from an embodied understanding of movement as the steady state. When I started spending my days at EdenMark, I wondered at their arrival.

Always men. Never alone. Often purchasing a range of Uroki's therapies, for an amount that many others never could have afforded. Maasai knowledge of therapeutic plants is iconic among Tanzanians. Hearing of my research into therapeutic plants, many Tanzanians recommended that I work with the Maasai. They assured me that all Maasai know medicines, and they described Maasai as being intimately aware of plants' therapeutic qualities. Fathers teach sons the names and uses of plants in the landscapes through which they move. Mothers teach daughters as they migrate. Knowing medicines is a way of moving through space, and the Maasai are valorized (and persecuted) for being people who move. They move with their cattle in daily, as well as annual, migrations. In responding to the pressures of the postcolonial economy, some have capitalized on this knowledge. Throughout the country, in even the smallest markets, Maasai men, blanketed with the legitimacy of their red and purple *shuka*, and women, in their dark indigo wraps and elaborate beadwork, sell medicines. So, I wondered, Why would they turn to this Chagga man's commodified medicines when they surely knew these plants and where to collect them? The surprise did not disappear with my creeping shame at the banality of the question. Rather, it shifted as I developed a broader understanding of the reasons that many Tanzanians are turning to this "new" sort of herbalism and the kinds of potency that Uroki's plant-based remedies offer. Uroki's expertise lay in his skill at orienting to the diversity of these heterotemporal flows—establishing a space to gather, work with, and direct them—and offering people the possibility of creatively orienting toward their own becoming.

African healers have long exploited the power and vulnerability of the crossroads. They have approached crossroads as liminal spaces where spirits might gather, witches might bury medicine, and the living and the dead might overlap. Accounting for such geographies of healing draws attention to forms of knowledge and ways of knowing that have been conceived in constant movement rather than stabilized objects, in improvisational practice rather than epistemological commitments, and in folding timescapes rather than linear timelines. This history resonates in Uroki's approach and enables us to see the ways in which he offers the crossroads at which his clinic sits as a place from which to reimagine the spaces and times of healing. The interconnections and transformations possible at the crossroads of different kinds of knowledge, forms of embodiment, histories of body-land relations, and traditions of healing articulate appetite as a desiring force that can alter body-land relations. Uroki's remedies at EdenMark and the advice through which they are dispersed mobilize appetite as a point of therapeutic intervention and theoretical provocation. Uroki and EdenMark's remedies both embody these crossroads.[13]

EdenMark's clinic and pharmacies are registered as traditional healing premises. Uroki has formally registered himself and those who work for him with the government as traditional healers, yet his herbs come from many places. His curiosity about alternative therapies started well before this current registration system was in place. His passionate search for effective therapeutic plants and his preoccupation with all sorts of wellness regimes began when his wife developed cancer. Together, they sought a range of ways to address her pain and debility. Later, when Uroki decided to start a business, he made forays back to his natal village high on the slopes of Mount Kilimanjaro. He described to me how he would exchange a pound of meat or sugar for the opportunity to sit with elders who knew the therapeutic potential of certain plants. Now, as the regional leader of a traditional healers' association, Uroki reaches out to other healers in order to help them register with the government and to think about how they produce and package their herbals. In return, he asks them about their therapies and learns more about the therapeutic value of the plants they use. He is also a voracious reader of both pieces online (ranging from herbalists' blogs to Harvard's Integrative Medicine research and Cornell's Complementary and Alternative Medicine research) and books he purchases in Dar es Salaam, in Nairobi, or from itinerant booksellers who come through Kilimanjaro (most of these are published in the United States). He has developed networks throughout the country through which to buy therapeutic plants from growers in different corners of Tanzania. He has taught himself acupressure and commissioned carpenters to make instruments to assist him in massage. He has also tapped into the Indian community in Tanzania and attends workshops on pranic healing and develops friendships that teach him about other ways of addressing illness and debility.

Uroki's eclecticness—his movement back and forth among African, Indian, Chinese, and American practices—engenders the crossroads as he refuses to pledge allegiance to the geopolitical units that were drawn up in the "age of empire" and that are still critical to tradition's role as an object of modern governance. So much work has shown that these units are sustained through their reinforcement in political formations, institutional boundary-making, cultural claims, and knowledge formations.[14] Interest in the possibilities of traditional medicine both inside and outside of Africa emerges in part as a reaction to insistences that biomedicine is "Western." Claire Wendland's beautiful and compelling ethnography of Malawian medical students, *A Heart for the Work*, pushes back against this ideological claim and argues that Africans have long been part of the development of biomedicine. She captures how Malawian doctors and doctors-to-be are reworking the relationship between medicine and politics through their lived experience.[15] By complicating biomedicine's dominant identity as "Western," she acknowledges the ways that its colonial history is alive today but also refuses the binary it engenders between the Western as cosmopolitan or global and the non-Western as traditional or local. The problem, she demonstrates, is that when biomedicine's ongoing link with racial capitalism is reduced to a simple binary between the Western and the non-Western, biomedicine can never be local. Uroki would add that its other(s) can also never be global.

Wendland and Uroki, I suggest, are pushing against the same problem from different sides: Historical and political arguments that define the therapeutic landscape through a sharp binary between biomedicine and traditional medicine exclude both distinctive formulations of biomedicine in Africa as "global" (or, more accurately, as equivalent to a formulation of biomedicine in the United States) and distinctive formulations of non-biomedical practices as *more than* local. Wendland and her Malawian biomedical colleagues then help render visible the epistemic and ontological implications of Uroki's eclecticism. While other healers leverage their localness, working to turn it into an asset, a form of credibility, despite the consequences of such subjectification, Uroki refuses this strategy. Rather, he seeks to undo the binary itself by cultivating myriad connections and holding space for their creative interconnection.

Undoing this binary, however, does not mean turning away the state's efforts to define traditional healing. It means instead that while the state office that registered him as a traditional healer—and has sought his assistance in registering others—works to strategically separate food from medicine, indigenous therapies from foreign therapies, and plant-based medicines from other practices, Uroki works to craft intersections. Since Uroki does not require epistemological commensurability, he is not obliged to uphold the hierarchy through which ontological singularity is generated. Rather, he is comfortable working in several registers simultaneously. Uroki enthusiastically mobilizes multiple knowledges in the service of shaping experience and intervening in states of being. His remedies disrupt efforts to purify the traditional and the modern, the local and the global, the natural and the cultural, the African and the Western.[16] Healing in his clinic is, in fact, a process of offering a range of languages and materials through which the experience of pain, debility, and misfortune might be engaged and transformed. Uroki cultivates diverse assemblages that can stand for a moment and enable action. That is, he draws things and practices previously divided by modernity's aspirations into proximity with one another. He remains undeterred by assertions that such gatherings are illogical, mythical, and even impossible—and he experiments. He looks for effect.

His intentional work at the intersection of so many traditions and registers of healing creates the friction that animates his creativity. He uses the frictions between forms of knowledge to pry open a conceptual space in which to attend to physical complaints while working to disrupt the violence that medical cures and public health have been used to justify. Through his eclecticism, Uroki dismantles the binary between biomedicine and traditional medicine, showing its ideological character and illustrating its ethnographic falsehood. In the process, he redefines appetite in order to direct attention to the energies that compel the open-ended unfolding of bodies and environments through each other—and to make these dynamics actionable. This fluidity is often obscured in biomedical formulations of the body as a medical object, which is rendered material in its separation from the social and historical and discrete in its distinction from the environment. *Dawa lishe* enriches the conceptual apparatus available to describe the social and material embeddedness of suffering, pain, disease, debility, and misfortune in order to better intervene in these experiences.

## Political Phenomenology of the Eating Body

To fully understand the provocation of *dawa lishe*, it is necessary to recognize it as a response within a longer trajectory of historical changes. Only then can we begin to see how cultivating an appetite for plants that pulls people into other economies of taste acts as a tactic to unsettle the relations between eating and being that structure conceptions of power and modes of governance. Hannah Landecker's pivotal work, which elucidates the changes in scientific understandings of nutrition and the shifting techniques of governing bodies in relation to food from the nineteenth to the twenty-first centuries, is helpful in gaining analytical leverage.[17] She starts by describing how consumption energized politics as well as bodies in the nineteenth century. At the time, arguments that adequate food was required to fuel laboring bodies shaped industry's programs to maximize productivity alongside the demands of workers' movements in the early twentieth century. She then tracks a shift in the focus of both scientific and political debates from an industrial metabolism that had been concerned with the sufficiency of consumption to a postindustrial metabolism that was concerned with efficiency, regulation, and risk.[18] Industrial metabolism's generative metaphor of the factory, animated by the times and tempos of energetics, gave way to postindustrial metabolism's iterative interface, animated by the times and tempos of information. Caught in the strong pull of the knowledge economy, the nutritional sciences of the twenty-first century have come to operationalize nourishment as the digestion of biochemical "informa-

FIGURE 2.2 John Ogondiek (*left*) and Victor Wiketye (*right*) from the National Institute of Medical Research's Ngongongare Research Station on Traditional Medicine talk with Alex Uroki (*middle*) and examine a solar dryer at the EdenMark production facility, Bomang'ombe, Tanzania. Photo by author, 2015.

tion" and food as a site of exposure.[19] Landecker does not represent this paradigm shift as one from less to more knowledge, or from bad to good science; rather, her analysis highlights how diet and metabolism offer historically and materially grounded vocabularies that reflect the ways that people see themselves in the world and that enable people to describe the options they see for being in their own bodies in a given historical moment. These vocabularies also reflect the ways that people see others in the world and the options they see for them being in their bodies.[20]

The scientific legacies that Landecker recounts naturalize versions of the body and forms of bodiliness that are severed from social and ecological relations by racial capitalism. "There is the uncomfortable sense," she writes, "that not only has the equation between eating and being ceased to function, [but also] it itself has had a role in producing contemporary ills."[21] Caloric logics, for instance, often still guide humanitarian interventions into severe malnutrition and famine in Africa, while biochemical logics increasingly shape understandings of the coexistence of malnutrition and obesity.[22] Both draw attention away from the complex social, political, economic, and ecological dynamics that shape these conditions and thereby also away from broad structural solutions to them. The plant(ing) remedies of TRMEGA and EdenMark simultaneously take up and divert the concerns of the field by offering alternative experiences of bodiliness through remedies that cultivate attachments with previously alienated land through the love of certain plants. These remedies work by expanding people's capacity to grow (with and through) plants as well as by thickening social relations that nourish through the sharing of composting techniques, seeds, and new recipes.

Industrialization and so-called postindustrialism have taken different shapes in Africa than in North America. East Africans have long experienced the equations of being and eating as they are formulated through the industrial logics of conversion, as efforts to position their bodies and land as sites of extraction. Histories of exploration, colonialism, and development describe the shifting processes through which people, land, animals, minerals, and now even biodiversity itself are cast as natural resources to be exploited first for the growth of the empire and then for the strength of the nation. In Tanzania, the work of obfuscating the value that is created from the conversion of raw materials into manufactured goods is also the work of ignoring the violence that shapes everyday life. The ways that petropolitics, land grabs, mining, industrial agriculture, and waste (of both extractive and knowledge-based industries) remake national landscapes are palpable. Tanzanians and others on the continent live in, through, and around the seemingly insatiable desire to consume that which drives the self-devouring growth of capitalist (re)production.[23] Consequently, the options that Tanzanians see for being in bodies in this historical moment are shaped by their position in circuits of global capital, their continued proximity to sites of massive extraction, and a turn toward the knowledge economy. *Dawa lishe* is born of these complexities and also seeks to address the recurring injury and relentless exhaustion of bodies and lands that they generate. It names efforts to expand the vocabulary of the nutritional sciences in order to render visible the social, political, and ecological relations that compose postcolonial being in the world.[24]

African therapeutics offer a rich vocabulary for conceptualizing the relationships between eating and power, as well as for distinguishing between acts of feeding and fortifying that are healing and those that are harmful. From rumors of witchcraft that sucks the life force out of people to create zombies willing to labor endlessly to cartoons accusing leaders of eating citizens' wealth, people subtly and not so subtly call out the propensity for those in positions of power to accumulate wealth by consuming others' strength and vitality for their personal gain.[25] In his now-classic historical and sociological study of politics in Africa, *The State in Africa*, Jean-François Bayart describes the significance of eating and consuming things to African conceptions of power as "the politics of the belly."[26] As with speakers of other Bantu languages, Tanzanians reflect on eating through the concepts of *uganga* and *uchawi* and their richly diverse historical connections to sovereignty (for further discussions of these concepts, see the Introduction and chapter 3). Eating and feeding are sites of healing (*uganga*) when they articulate relations of care and good governance,[27] and they are sites of harming (*uchawi*) when they nourish the forces of destruction and drain the forces that generate life.

A wide range of African therapeutic practices carefully structure rituals of eating in order to reconstitute the relations that generate more liveliness and ensure fertility. Food is offered during *ngoma*—large, all-night gatherings that involve drumming and calling on or calling out those who are possessing the afflicted. Similarly, eating—and eating together with kin and community—is central to many more modest healing rituals as well.[28] Kitchen hearths and cooking pots are central features in ceremonies to prepare medicine, remove curses, and combat bewitchment.[29] Bundles of therapeutic roots and barks are stuffed inside a freshly slaughtered chicken and boiled or simmered in a gravy. Medicinal herbs are added to soups or sauces. The effect may be, as phytochemists tell us, that the substances held responsible for a plant's therapeutic qualities are rendered effective when dissolved in fat, but it may also be that food shared from one plate with certain significant others generates the kind of closeness that promotes health and wellness. Yet, the efficacy of ritual meals cannot be reduced to parallel stories of nutrients (a vocabulary to address the biological body) or of care within households (a host of social relations supporting the biological body).

Thinking with *dawa lishe* is thinking in and from Tanzania, where rich histories of African therapeutics have approached the materiality of the body as always already simultaneously biological, social, and ecological. The temporalities and spatialities of the body enacted in these practices gesture to imaginative horizons not easily visible within biomedicine. The social therapeutic projects of TRMEGA and EdenMark innovate within worlds in which familiar vocabularies articulate the body as the very substance of social relations with both living and dead, human and nonhuman.[30] This opens up ways of thinking about knowledge and knowing, as well as efficacy. The tradition of the "medicine bag," for example, offers a lexicon for describing the healer's "body" as the materialization of relations and describing "knowledge" as a process of being in intimate relation (accommodating, feeding, coaxing, and hosting). It illustrates how African therapeutics enacts a metaphysics of relation, as captured in experiences of cohabitation, rather than a metaphysics of purity, as captured in experiences of eradication.

In many parts of sub-Saharan Africa, healers describe being forced to accept the ancestral "medicine bag." Refusal makes life untenable; this is often experienced as a life-threatening illness. The power of the medicine bag offers one way of capturing the process of subjugating oneself to the demands and desires of forces born of the past as one becomes a healer. The relations established with these forces will continue to be central in a healer's practice as they treat others.[31] Therefore, acceptance of the medicine bag cannot be reduced to a scholarly process of learning to identify, prepare, and prescribe particular plants. As you will hear healers affirm in chapter 3, one does not learn to heal at school. Accepting the medicine bag means learning to accommodate, appease, and satiate that which is pressing or climbing on one. Thus, hospitality, not mastery, animates the healer's expertise and bodily capacity. As the "medicine bag" reasserts itself throughout the generations, it dramatically highlights the lived experience of the body as "more than one but less than many."[32] It gives form to the ways that healers apprehend corporal materiality as cohabitation.

The medicine bag also describes how conceptualizing cohabitation as corporal materiality supports conceptualizing epistemic practice as hosting. Whether or not an individual "believes" in "traditional medicine," the lexicon inherited through histories of healing enables appetites to be apprehended as a nexus of relations and a site of mediation among human and nonhuman others. Public and private stories of ancestors, spirits, and devilish others make demands on the living, illustrate eating as a practice of meeting and engaging.[33] They capture the experience of being in a body, or bodiliness, as an effect of the coordination and sequencing of myriad encounters. Such an articulation of corporality in African therapeutics challenges biomedicine's political and economic commitments to the body as the ground for a unified subject that acquires knowledge, makes rational decisions, and holds a right to sovereignty over their corporeal being.[34] The stories of food and politics that Landecker describes shaping nutritional science have pressed on bodies in Tanzania since at least the late nineteenth century, but not only. Tanzanians also have a long history of thinking about power through and as eating, as well as a history of acting on the social, political, familial, and environmental relations shaping bodies through appetites. These diverse vocabularies are at play as people attend to the politics and realities of this historical moment and strive to generate new ways to be in bodies and attend to body-land relations.

The producers of therapeutic foods and herbal medicines that capture my attention in this book draw on these generative frictions. Their social-ecological projects, glossed as *dawa lishe*, recognize therapeutic efficacy as an effect of good relations, not only as control over biological facts. In so doing, they render visible the forms of power that are reproduced through biological conceptions of the body, public health notions of community, and environmental health's articulation of land, and they open up space for innovation.

As practitioners work at the intersection of these diverse local and global legacies, working through the conceptual and practical ruins of postcolonial healing, they take eating and feeding as grounds for theoretical inspiration and ethical reckoning.[35] The histories of African therapeutics—when freed from being boxed in by being labeled as "culture" in public health—give rise to the possibility of apprehending appetites as animating forces that weave bodies from and into the world, as the very "substance of endurance."[36] EdenMark and TRMEGA apprehend appetites as collective, even at times ecological, compulsions. The way that their plant(ing) remedies work with appetite elicits what Tironi and Rodriguez-Giralt refer to as an "ecological mode of attention."[37] As with Uroki's dandelion greens above, many of EdenMark's and TRMEGA's remedies are imagined as an invitation to cultivate new forms of palatability, to cultivate one's desires in ways that promise expanded connections—to people, places, and plants. Through these remedies appetite, as an object of therapeutic intervention, shifts: it is less a characteristic or quality of a biological body than it is the desiring force that bodies forth enduring relations. If bodies cannot be taken as a priori—stable (no less standardized) material facts that precede the moment of therapy—then a remedy's potency is not (only) a measure of its effect on the body proper but a measure of its effective choreography of relations (social, ecological, ancestral) enacting the corporal.[38] The efficacy of conceptualizing appetite in this way lies in its potential to engage individual desires while still socializing and politicizing disease, distress, and debility.

Uroki, Mama Nguya, and others in this book are attuned to relations between people and plants and the ways that taste and desire inhere in body and landscape (form and context). They are cultivating a vocabulary through which to recognize the prolonged systematic harm that is woven throughout particular practices of care, and they elucidate connections between medical treatment and social abandonment that are systematically obfuscated through biomedical regimes of knowledge. Their plant(ing) remedies offer a way to apprehend a dispersed body extending across the land through its incorporations and excorporations, its compositions and decompositions, its histories and labors. In so doing, these remedies engender a new therapeutic focus, perhaps best described by indigenous feminist scholars as "body-land," and they engage appetite as the forces through which this simultaneously corporal and ecological form of being emerges.[39] *Dawa lishe* seeks to draw together efforts to experiment with the efficacy of appetite as a therapeutic, social, and political project. It gives language to the potential of such a reorientation of healing toward body-lands, to incite the collective work of cultivating corporalities that might survive the toxicity that is everyday modern life.

### (Non)knowledge at the Crossroads

Uroki reads nutritional research as well as more popular discussions and alternative medicine posts online. His clinic bookshelves hold books that outline the nutritional value of various fruits and vegetables, as well as the links between specific micronutrients and claims about their healing effects. Uroki encourages his staff to study these books, too. On slow afternoons, I would find Romana, who managed the EdenMark clinic in Bomang'ombe, taking notes as she read the books he had left out on the desk. Many of the labels on Eden-Mark's remedies list nutritional information gleaned from a variety of sources, including scientific articles and the popular press.

Uroki does not ignore or resist nutritional research, but he is not under the illusion that a simple "lack of knowledge" is the reason that people are ill or that "information" will remedy their complaints. While Romana and her colleague Jenipha, who Uroki hired to staff his Bomang'ombe pharmacy, did discuss with clients what constitutes a good diet and often recommended that they eat more green leafy vegetables or fresh fruit, such "education" never defined their encounters. That is, Romana and Jenipha did not imagine that the problem was ignorance and that (more or better) information would redirect appetites. Therefore, they did not mobilize knowledge as the solution to ill health, weakness, fatigue, or decline. Rather, they offered plant-based therapies to shift the generative power of appetite to, in turn, shape the desires that animated the body in everyday life.

Over the years that I observed Romana and Jenipha's work, I noticed that both were particularly attentive to the ways that desire folded itself around particular foods and herbal medicines. For instance, as Romana and Jenipha reminded men of the importance of eating fruit, they acknowledged that many considered fresh fruit "children's food." Therefore, they would offer the *wazee* (elders) who came with complaints of fatigue, or those whom they assessed as lacking vitamins or minerals, baobab fruit (*ubuyu*) dried and ground into a powder so that it could be mixed in a glass of water and taken as a medicinal juice. In offering Uroki's *ubuyu* powder, Romana and Jenipha asserted not only that this baobab is therapeutic but also that fruits are appropriate for adult men. They used this alternate form (dried powder) to coax an opening to cultivate an appetite for fruit, while also rearticulating *wazee*-ness. By engaging the men's appetites as the animating force through which their bodies and subjectivities are materialized, Romana and Jenipha render actionable a notion of healing that includes eating the "right food."

In EdenMark, data describing the nutritional value of particular plants catalyzes conversations about eating. This became particularly clear in observing Romana's exchanges in response to the quantum magnetic resonance analyzer's (QMRA) computer-generated assertions that a patient lacked specific micronutrients and the questions about eating that then followed.[40] The QMRA is a small machine that is plugged into a laptop computer and used as an individual diagnostic aid. Most of those that I saw in Tanzania appeared to be made in Korea. It is not clear when these machines entered Tanzania, but by 2010, they could be found in alternative medicine clinics, in the backs of small shops, and in "pop-up" ventures. Their diagnostic capacity in biomedical terms is limited at best. There may be some general population-based algorithms that shape the reports generated. However, given the lack of information about how the machine is programmed and any oversight or system of accountability, medical professionals both inside and outside of Tanzania suggest that the results are dubious. *Dawa lishe* producers participate in broader public debates about the QMRA and other "alternative health" technologies circulating in Tanzania. Those who encountered the QMRA in clinics they visited also debated these technologies—some were skeptical, some curious. All evaluated it less according to any objective truths it revealed and more in terms of how it sparked conversation between clinic staff and their clients. Romana used it to develop relationships with those who came to EdenMark and to support clients in expanding and strengthening the connections in their lives more broadly.

In observing the QMRA's use in several herbal clinics in the area, I found that, while staff's understanding of the evidence the QMRA generated varied, their procedures were the same. The person running the test entered the *mgonjwa*'s name, age, address, height, and weight. The person being tested was asked to remove all metal from their bodies—watches, jewelry, phones, coins in their pockets—before being handed two metal rods to hold. These rods were attached by wires to the QMRA, which was, in turn, attached to the laptop. The "run" command was clicked, and for several minutes, the laptop screen showed the silhouette of a human body being scanned, a heart pumping, and charts forming as images whirred, beeped, and flashed. Eventually, a long health report appeared. This assessment listed, among other things, the levels of particular micronutrients available in the person's body, including vitamins C, B, and A and minerals such as zinc and calcium.

Romana, who usually ran the QMRA tests at EdenMark, was uniquely skilled in leveraging the results to develop a broader understanding of the conditions of the person's life and the relationships that sustained it. In her hands, the QMRA did not mimic a biomedical test. No clear line was drawn from deficiency through biological function to physical complaint. Rather, Romana mobilized the report in a subtle, winding, compassionate effort to elicit from the *mgonjwa* an articulation of the relations that constituted their strengths and weaknesses. Romana was sensitive to the fact that probing the availability, circulation, and consumption of food can feel intrusive. Questions about eating expose intimate relations. Who could procure food and who could not? Or who is able to prepare food and when? Why might meals not be cooked regularly? Why did a particular person not get enough food in a household? Did others? Answers to such questions render visible the power dynamics in households through both mundane and more dramatic moments. The QMRA opened a space to broach intimate topics to which offense could easily be taken.

Romana offered patients the chance to interpret the results with her. She and the person tested (and frequently kin who accompanied them) worked together to incorporate the test results into a history of events and relations salient to their life. But if they failed, they might simply turn away from the results as unhelpful or superfluous, or they would dismiss the report, declaring the results inaccurate. In EdenMark, the process of valuing the results of the QMRA required experimenting with the ways that quantitative results can be used to unfurl qualitative experience. The report produced by this strange machine emerged as an object of speculation around which Romana opened a space in which to think together with her clients about the forms of power that pressed on them.

These conversations about the dynamics of eating initiated conversations about the political, social, economic, and material dynamics that shaped their ability to eat well. This process did not center a lack of nutritional knowledge as the primary problem or the acquisition of correct or sufficient knowledge as the primary solution; rather, Romana engaged those who came to her in crafting a space in which the complex social, material, and ethical relations structuring their immediate experience of being in a body might be glimpsed. Her collaborative style cultivated an ecological mode of attention in order to re-story what it means to eat well in postcolonial Tanzania.

Eating and Pain

One hot afternoon, when much of Bomang'ombe, including the clinic, was without power and the blades on the fan sat unhelpfully still, I turned to the logbook where Romana dutifully recorded the names of all patients who had been assessed with the QMRA. Next to each person's name, Romana indicated the formulas she recommended to them. I took many turns through this log-book on such quiet afternoons, asking Romana questions about people, their conditions, and her recommended therapies. Some days, when the clinic was slow, we would try the cell phone numbers they had provided to see if we could follow up. I had noted that there seemed to be a preponderance of patients listed as having an issue with *vidonda vya tumbo* (sores or wounds of the stomach). Curious, I counted the number of those assessed with the QMRA over the previous seven months to whom Romana had prescribed EdenMark's peptic ulcer formula. I found that she had prescribed it to well over half (64 percent) of them (74 percent of female patients and 54 percent of male patients assessed were then prescribed peptic ulcer medicine). These percentages do not assert any sort of prevalence of peptic ulcers in the population. They do not even tell us how many people bought EdenMark's peptic ulcer formula. They do, however, reflect that people came into EdenMark in significant numbers, complaining that it hurt to eat.

People paid TZS10,000 (about USD5) to be assessed by the QMRA. Sometimes, that was all their money, so they took the assessment and the list of recommended treatments home to discuss with others who might help them. Sometimes people decided to buy only a few of the treatments recommended. Others who knew their cash was limited chose not to spend the TZS10,000 on the QMRA. They described their symptoms to Romana and Jenipha, who would then recommend treatments based on their complaints and history. Others came to the pharmacy desk seeking treatment for someone at home with a *shida ya tumboni* (problem of the stomach). Jenipha would sell them the peptic ulcer medicine over the counter to take back to their *mgonjwa*. Romana stressed to me that, because only those assessed by the QMRA appeared in the logbook, the percentages I had calculated were an underrepresentation of sales. They did confirm, however, the sense that Romana, Jenipha, and I all shared that, in Uroki's clinic, one of the most common diagnoses was peptic ulcers.

How pain, injury, and debility are storied is important because different stories inspire different responses. Healing, after all, is fundamentally interventionist. How might this widespread experience of a burning sensation in the stomach that complicates eating by making it painful be storied if eating is that which nourishes body-lands? What sort of response does the story generate, and what remedies does it make possible? I am particularly interested in EdenMark's skill at mobilizing diverse vocabularies to surface how this widespread, chronic pain when eating speaks to the experience of being in a body in Tanzania in this historical moment.

In Tanzania, complaints of pain that interrupt nourishment are not only evidence of a hole in the stomach lining, a buildup of acid, or even the proliferation of the *Helicobacter pylori* bacteria. While such explanations are at times incorporated into illness narratives, those who come into EdenMark with such stomach pain are motivated less by evidence of the truth of their wounds than by the search for the truth of a moment that generates so many bodies that struggle to digest what they need to live and grow. Uroki and his staff strive to respond to this call. They use the term "peptic ulcers" to draw connections with biomedical symptomatology even as they trouble its fidelity to the epistemological boundaries of clinical knowledge. The experience of stomach pain in EdenMark is taken as evidence both of a particular person's physical condition and of a historical moment in which the very possibility of nourishment is fundamentally challenging. Diagnosing ulcers leads Romana and Jenipha to reflect with those who come to the clinic on the conditions that make eating regularly increasingly difficult, given the available food. In this way, a specific diagnosis harnesses biomedical vocabulary to draw attention to the fact that experiencing pain when eating is a problem. The process of diagnosing locates the alarming nature of this pain in the erosion of (individual and societal) relationships that sustain life and liveliness and acknowledges that doing nothing threatens to further isolate one from the generative forces of commensality.

Romana and Jenipha asserted that irregular eating habits are a widespread problem due to economic and social realities that required people to work outside their homes, leaving no one at home to cook multiple meals a day. They were not alone in this analysis. I heard similar assertions inside *dawa lishe* clinics and herbal shops, as well as inside neighbors' and friends' homes. Romana and Jenipha themselves exemplified those who worked many hours outside the house in offices, clinics, schools, and market stalls. When I was observing at EdenMark, I often contributed by buying Romana, Jenipha, and myself lunch at the canteen next door: rice and beans and gravy, or bananas and meat. We entered through the back, directly into the kitchen, and the mama who ran the place took good care of us. She also sold fried bread in the mornings and poured steaming cups of sweet, milky tea throughout the day. Such small canteens dotted commercial areas, and similar street stalls could be found around bus stations and markets. Those with money bought breakfast or lunch when they were out. This frequently meant food cooked with more oil and salt than at home and meals with more meat. Not everyone could afford to eat out every day, however, and even those on a salary hesitated to eat such "heavy" food routinely. Tanzanians located their accounts of the increasing impossibility of eating well and eating frequently enough in macro-level changes: Not only did both men and women work for wages outside of the house, but elders were less likely to live with their adult children, and younger children were more likely to be at school.

The rise of the "modern" family has changed the shape of households in Tanzania. The effect was gradual, and others have traced these changes in the family and its rhythms through colonization, missionization, the imposition of taxes, the need for wage labor, rising urbanization, population growth, land scarcity, and shifts in inheritance patterns. In Kilimanjaro, where EdenMark sits, land has become particularly expensive and relatively unavailable. Parents can rarely divide their holdings among their children in a way that would allow all (or any) to provide for their own families solely from the land. Many Tanzanians describe how the negative effects of these changes have been exacerbated by neoliberal reforms and structural adjustments. Even so, these trends began before the 1980s, and they are wrapped up in visions of development that many still hold dear. Peptic ulcers index this double-bind: a problem with dwelling; with the relations that constitute a homestead, a family, a lineage; and land in the contemporary moment. The diagnosis of peptic ulcers leads Tanzanians to plot their stomach pain within the lived history of these social and material transformations. Stories that account for social, ecological, and bodily wounding open a space to make historical sense of the experiential paradox that consuming nourishment hurts.

On another afternoon when the clinic was slow, we decided to look through the QMRA logbook once again. It was the season for weeding, a time when all farmers needed to be in their fields to ensure that their new shoots had the space and light they needed to flourish. As many explained to me, health concerns had to wait until the plants' futures were more secure. It seemed like a good time to follow up on my impression that *ubuyu* juice was frequently prescribed together with EdenMark's peptic ulcer formula. As I turned the pages of the logbook, I counted the times that *ubuyu* powder was recommended. I found that, of the 472 people Romana recorded having assessed in the previous seven months (from just before the short rains in September 2013 through most of the long rains in April 2014), she had prescribed *ubuyu* powder to almost a quarter of them (113).

Romana offered *ubuyu* powder not only to elders who had no appetite for fresh fruit but also to people who struggled to eat regularly. She gifted it to Nuru, a young man with HIV/AIDS, for instance, when he complained of painful muscle cramps during one of his visits to Uroki's clinic. Romana and Jenipha were careful to warn him that he would still need to eat in addition to drinking the juice. If he drank the juice, however, Romana suggested that he would eventually start to feel his muscles relax. Romana shared her own experience of drinking this *ubuyu* juice to alleviate cramps. Before she started working at EdenMark, she told us, she went through a period in which the muscles in her legs would cramp painfully at night. In the morning, she would have a hard time walking until she stretched. Someone gave her baobab powder to mix with water and drink as juice. She used it for a "long time," after which she found that she not only began walking more easily but also that her cholesterol had gone down. If taken as a discrete encounter, the exchange might be seen as suggesting that *ubuyu* powder was a cure for Nuru's cramping muscles. Yet while Romana certainly hoped that the juice would help decrease his pain, she offered him the *ubuyu* powder as a way of mediating the relations through which Nuru's life is possible and increasing the attachments from which he can gain strength.

Nuru lives with his grandparents nearby. As he navigates life with HIV/AIDS, he has found strength in the two women at EdenMark. He drops by frequently to talk, complain, joke, get advice, and be buoyed by their moral support. Both Romana and Jenipha had developed a deep affection for this charismatic young man, who seemed determined to make a life in the face of incredible adversity. Through their increasingly familiar conversations, they came to know that eating regularly was not easy for him. His grandmother did not cook three meals a day. While Romana and Jenipha made critical comments about this among themselves, Nuru did not seem resentful. He appreciated that his grandmother had taken him in, and he recognized that she was elderly with her own burdens and work. Nuru admitted, however, that he was in a bind. Whether or not his grandmother had said it explicitly, Nuru knew that she would take offense if he—as a grown male—attempted to cook meals in her kitchen. Jenipha and Romana agreed that she would perceive his use of the kitchen as an accusation that she did not adequately care for him. It might also indicate to others that, intentionally or not, she was failing to act in a way that promoted her household's vitality and liveliness. Yet, Nuru did not have the money to discreetly supplement his grandmother's meals with food from a canteen or restaurant. Therefore, even though the antiretroviral drugs he was receiving from the clinic made him particularly hungry, he was not able to satiate this hunger at home. The *ubuyu* powder Romana and Jenipha gifted Nuru was less a cure-all than a response to a complex situation in which they assessed that Nuru could not eat enough or well. Because it could be easily mixed with water and drunk outside of the house, it offered a way of eating that supported Nuru's relationship with his grandmother, reducing the impact of her inability to cook multiple meals and making what she could do feel more sufficient. It also further cultivated an attachment between Nuru, Romana, and Jenipha.

In Tanzania, children frequently move between and across households to live with grandparents or aunts and uncles, which extends care and support through lineages. Older children who are living with working aunts with very young or no children of their own provide them with companionship and help. Parents might send a child to live with their grandparents as part of their effort to provide support for elders. I remember sitting with my friend Mariamu in 2000 as she wept after relinquishing her youngest and last child to an aunt for three months so that the child would know her family and so that Mariamu would not grow too close to and overprotective of her only daughter. Such traditions of movement composed and decomposed households in the flow of a lineage's reproductive energies.

In the 1990s and early 2000s, however, AIDS stretched the resources of many and changed the landscape of familial care.[41] Some have highlighted an "orphan crisis" in sub-Saharan Africa, as a growing number of children struggled without kin who could take them in after their parents had died of AIDS.[42] Others have examined the complex dynamics of child-headed households as older children strove to care for their younger siblings.[43] Still others have noted that increased burdens have fallen on grandparents, as they were left to care for grandchildren—as in Nuru's case.[44]

While the ulcers that line the throats and stomachs of people with AIDS are not the same as peptic ulcers, the questions of eating and pain resonated in the experience of both. *Dawa lishe* calls for remedies to address the depleted, undernourished bodies generated in these crises, where patterns of dwelling that support liveliness and growth have been ruptured, leaving homes in ruins and the sick abandoned.

Romana talked for many weeks about taking me with her to bring Eden-Mark remedies to Sarah, a woman bedridden with HIV/AIDS. It was never clear to me if Romana knew Sarah from school or from when she sold vegetables in the market or both. They were not close friends, but at one time, they had lived their lives in close proximity. Since then, their paths had periodically crossed. On the way to Sarah's house, Romana told me that she first learned of Sarah's condition when a woman came to her, saying that her neighbor was very ill and needed help. At that time, Romana had only a vague sense that she might know the woman who was ill, but she followed the concerned neighbor. Romana recognized Sarah immediately, despite her frail condition. Sarah was confined to her bed, unable to speak or sit on her own, let alone stand, walk, cook, or care for her two sons with whom she lived.

There have been a few moments in my years of friendship and work with Romana when I have seen her particularly moved and called to respond and to act well beyond the responsibilities of her job at EdenMark. With Sarah, with Nuru, and with me. Romana described being shaken when she first saw Sarah. Remembering that she had part of a canister of EdenMark's Immune Booster formula left over at home, Romana ran home to fetch the remaining herbal powder. The next day, Romana told Uroki about Sarah, explaining her precarious state, her incapacity, and her aloneness. Uroki gifted Sarah another canister of Immune Booster as well as some African Potato and ArthRid. Romana brought them to her, with instructions on how to prepare them, how often to take them, and told Sarah to stay on her antiretrovirals in the process. It was after this that she began talking about taking me to meet Sarah.

Several weeks later, during planting season, when that clinic was typically slow because the planting prevented people from attending to all but their bodies' most urgent demands, Romana decided it was time to bring another canister of the Immune Booster formula. When I arrived at the clinic, she gathered up the medicines and insisted I climb back into my car. She directed me to drive away from EdenMark and the commercial center of Bomang'ombe, which has built up along the paved Moshi-Arusha road. We soon started winding through the dirt roads that reached toward the old Bomang'ombe. Slowly, the sandy, twisting roads grew narrower and narrower, until they were often barely distinguishable from foot paths. When we pulled up to a small, L-shaped structure comprising several rooms, each with a door facing out toward a barren courtyard, Romana announced we had arrived. Sarah and her two sons rented three of these rooms for sleeping, living, storing, and preparing food. They cooked on a charcoal stove outside, or at times, for protection from sun, rain, or wind, under the narrow overhang of the corrugated iron roof.

When Romana and I arrived, Sarah was sitting up in bed. Her thin body had sunken into the concave center of her foam mattress, and her blankets had knotted into the nest of someone confined to this space day and night. Her legs were stiff, not bending easily at her command, but she was eager to show us that she could stand up, with a little help from Romana. Her arms moved awkwardly, and her speech slurred, but her eyes were large and expressive. They held such fear. She was trembling. Often in a quiet pause as we sat with her, her eyes would start to tear. Romana would tell her that she would be fine, that she was much better, not to worry, to pray to God. Sarah insisted that the medicine that Romana brought was making all the difference. She told us emphatically that she takes the medicine from the hospital twice a day, morning and night, and she takes the medicine from Romana three times a day—morning, afternoon, and night. Inquiring gently, we learned that she had taken these medicines this particular morning with only cold water, no food, not even tea. Romana grew animated, insisting that the weakness that remained was from lack of food, that Sarah must eat.

Romana asked about the two sons, Simon and Zawadi, assessing that they were the last of Sarah's daily support. Zawadi, the younger of the two, was in school. He was eight years old, and he had been diagnosed with HIV in infancy. Zawadi's father was in Arusha. Sarah's comments seemed to imply that he was also sick. Under her breath, only to me, Romana remarked sarcastically that his biggest contribution to Zawadi's care was to tell him that he did not actually have HIV and that he should stop taking his medicine. Zawadi's father, Sarah told us, had not come to see his son at Christmas. Romana speculated that he could not face his own fear of death and illness, or his responsibility for a child. She suggested that his abandonment was a denial. The government treatment center where Zawadi received his antiretroviral medication had enrolled him in a Korean program for children with AIDS, ensuring that he received not only his medicines but also other support, including food, school fees, and other essentials. Simon, the older of Sarah's two sons, was seventeen years old and not HIV positive. Simon's father had left Sarah long ago. On the day of our visit, we learned that Simon was no longer permitted to attend school because he did not have the proper shoes. Black shoes with black shoelaces were required to study. As this thin, lanky teen was growing fast, he had outgrown his shoes, but there was no money to buy a new pair. Still wanting to study, he had shown up for school in flip-flops and been sent home.

Despite not being in school, Simon was nowhere to be seen. His mother swore that he had just left. Romana went to look for him to no avail. We waited. We told his mother that we would take him to buy shoes. She insisted that he was around; he would be back. We talked a bit more. Romana asked the neighbors if they had seen Simon that morning. Just as we were about to leave, a tall, thin teenager with a pleasant smile walked in the front door. He had a gentle hand-shake and a kind voice. As we sat down together, Romana quickly launched into lecturing him about caring for his mother, about the need for him to stay close to her. What if she fell, if she needed something? If he went so far away, how was she going to call him? How would he hear? Cell phones might be said to be ubiquitous in Tanzania, but there were certainly not two in this household, and perhaps even the one old flip phone did not have enough prepaid shillings in it to so much as send a text. Romana scolded him for not cooking *uji* (a common porridge made from cornmeal) for his mother that morning.

During this scolding, Sarah was shaking. She could not look at any of us. Tears welled up in her eyes. Simon, who had entered the room gentle, seemed to grow harder as he looked at his mother withering away on the bed that they all shared. Her weakness, her inability to care for him or for herself, seemed to anger him. Romana reminded him that there were many years before Zawadi was born, during which she cared only for Simon, without having another child. She tried to impress upon him how much his mother loved him and how she had worked to care for him, and she asserted that now he must care for her. We told him that we would take him to the market to get new shoes if he wanted to return to school. His face lightened a little. Romana took the opportunity to insist that he make *uji* for his mother each morning before going to school. She should eat something with her morning medicine, and the rest should be put in a thermos, tucked under a little stool near her bed so that she could reach it during the day when no one would be around.

We took Simon to the market. There were a few moments when he wore a half-smile as we wandered through the stalls looking for a pair of black dress shoes that fit him. When we found a pair, Romana began negotiating the price. She asked the saleswoman to have pity. She told her and the woman selling next to her the story of Simon's mother and her condition. The saleswoman growled, "This is business! There are no donations in business. I give my *sadaka* (offering) at church." Simon's unexpected gift of shoes shifted quickly from feeling like a moment in which he was being cared for to one in which he was being pitied. Tears welled up in his eyes. After we bought the shoes, he tucked them under his arm. The smile did not return. We did not give him a ride back home as I had expected we would. Romana said goodbye, and he walked off slowly through the market, shoulders hunched as if he had been beaten down.

As Romana and I walked back to the car, she fumed about the saleswoman who kept repeating that *sadaka* are for church, that her work in the market is a business. "So many tears," Romana sighed. Considering Simon's anger and humiliation, his tears, I wondered out loud if it had been necessary to share his mother's story. "It is necessary," Romana insisted, "that people know that you have a problem. If you hide you cannot get any help." I had wanted to hide. I suspect that Simon had, too. Yet, in the face of my awkwardness and heartbreak, Romana remained resolute: Healing is not about eliminating vulnerabilities; rather, it is a process of holding, sharing, and distributing them in efforts to foster vitality. Trying to deny vulnerability would deny the very lines of connection through which one might be nourished. The harm here was not that Simon needed help but that the saleswoman would not recognize the ties that bound her livelihood to his liveliness.[45]

When we got back to EdenMark, Nuru was in the parking lot. He looked good. There was a big smile on his face. He seemed to have gained weight. We all entered the clinic together to find Jenipha behind the pharmacy counter and one of Romana's children waiting, ready with tales of school and teachers. But Romana's mind seemed to wander. She turned to me at one point and asked if it might be good for Nuru to talk to Simon, to give him some advice: How, she wondered, could she facilitate the introduction? Shaken by the precarity and pain of Sarah's small family, Romana sought to foster more connection. One has to be willing to share one's problems, weaknesses, and struggles in order to seek new ways to become embedded in a rich set of supportive relations. The other option is to become more and more isolated. Food for the mother requires shoes for the son. Healing requires intervening to increase the relations that articulate the necessary coexistence of vulnerability and vitality.

Romana was struggling against the limits of possibility within EdenMark's project. Uroki was also building a business, even if he did not respond with the same harshness as the woman in the market. Romana encouraged Uroki to gift Sarah medicines. He not only agreed but also supported Romana going to spend a workday with Sarah. These ruptures of EdenMark's routine made visible EdenMark's search for a way to intervene that went beyond sitting in the clinic, collecting histories, running QMRA tests, and prescribing herbals. Romana strove to shift household patterns and bolster relationship dynamics that might continue to strengthen Sarah even after our visit ended. She engaged healing as a process and grappled with how to reconcile this with the fact that EdenMark offers a product. She pushed against the limits of the pharmaceutical imagination according to which herbal products are governed.

HIV/AIDS has generated particularly dynamic debates inside and outside the clinic about the more-than-biological properties of effective treatment. In examining efforts to extend antiretroviral therapy (ART) in Africa, Vinh-Kim Nguyen coined the term "experimentality" to describe styles of governance and techniques of self that foster the social conditions and collective rhythms critical to antiretrovirals' efficacy.[46] Nguyen, who is both a medical doctor and an anthropologist, describes clinical researchers' awareness that the efficacy of antiretroviral drugs requires that the material realities of social life reflect the laboratory conditions in which the drugs demonstrate effects.[47] Key to the significance of Nguyen's argument is that HIV/AIDS programming provides evidence that biomedicine understands the efficacy of pharmaceuticals as inextricable from the social, material infrastructure in which they operate. Health development projects critical to moving a trial drug from the laboratory to the community address poverty, rurality, or gender oppression as obstacles that lie outside of pharmaceutical science proper and yet are critical to demonstrating the trial drug's efficacy in the world. The individual biological body as the site of illness, and therefore the efficacy of treatments for that illness, is maintained through the separation of professional spheres of practice and regimes of knowledge: Pharmaceutical science remains accountable to the material reality of the biological body, and public health intervenes separately in the cultures that prevent their treatments from working. Nguyen's work helps me to understand how power works by formulating social relations, agricultural landscapes, economic pressures, and religious traditions as context.

The precarity of life in the time of AIDS, however, has exposed the fiction that environments (cultural or ecological) are merely a backdrop for bodies. The obfuscation of the social and ecological collectives through which ill bodies are constituted and maintained raises complex ethical questions. Romana was shaken by these questions. As am I. Romana sought to engage ART as social and chemical, infrastructural and biological. She sought to increase the density of the relations into which they entered. Her outreach, and the herbal products she mobilized, strove to increase the connections through which Sarah lived. This included the social and biological conditions in which her antiretroviral drugs might work. And yet, Romana felt the insufficiency of modes of healing tied to business models fundamentally centered on products.

I see EdenMark's rearticulation of appetite in the service of cultivating an ecological mode of attention as an evocative gesture. It reveals a desire to reconceptualize therapeutic potency in ways that recognize that environments and bodies (context and form) emerge together through the labors of everyday life. With the dandelion greens above, Uroki hints that learning to welcome the bitterness of foods long forgotten would ideally lead people to eat differently. When he envisions people collecting their own dandelion greens when they are in season, he points to an efficacy of appetite that resides beyond the commodity form. Perhaps even to the process of dislodging the commodity form. The gesture remains aspirational, however. Insofar as the herbal medicine and the shoes were offered as an exception—whether as gift or as donation, with compassion or pity—they reinforce the limits of response and the shape of obligations in the economies of contemporary medicine. Uroki's business model, after all, has him tracking the sales of his canisters of dandelion greens rather than those who are "eating differently." When Romana saw that Sarah needed to eat, she offered nourishing medicines; however, she never commented on the barren courtyard. She intervened in the relations critical to the efficacy of the herbal remedy by attempting to attend to the tension between Sarah and her son Simon over the shoes. This was, however, a personal effort, not a systematic part of EdenMark's project. Unlike those in TRMEGA, Romana did not have access to a way of cultivating a space of relations that would address the more profound traces of structural violence and attenuation of social connections.[48]

## Eating in the Time of AIDS II

I met Jane Satiel Mwalyego in 2014, when she was a radiant, strong, fifty-one-year-old with a talent for motivational speaking and a passion for organizing. Her energy, however, belied the complicated story of her path to the breezy stone porch of the National Institute of Medical Research in Ngongongare, where my friends Victor and John and I first spoke with her. Jane was the third-born child in her natal family. Her parents passed away when she was in Standard 4 (perhaps about nine years old). "We were children raising ourselves," she remembers, "living in a risky environment" (*watoto mazingira hatarishi*, a phrase used now by NGOs that translates to "vulnerable children"). Somehow, Jane managed to stay in school for three years after her parents' deaths. She dropped out of school after completing Standard 7. She married and had four children of her own. Her husband was a security officer. When her elder brother died of AIDS in the 1990s, she and her husband took in his three children. His death marked the beginning of the devastation this disease has wrought in Jane's family. Over time, AIDS would claim the lives of five of Jane's seven brothers and sisters, as well as her husband.

When her husband's death in 2000 left their household without an income, Jane started small-scale trading at the market in an attempt to scrape together enough to support the seven children for whom she was responsible. Two years later, however, she herself fell ill. Without a husband, father, or mother, and with no house left to her, she was driven to begging. She decided to return to her natal village in the Mbeya region. In her father's house, she found her brothers and sisters married with their own children. The second-born in Jane's natal family, her remaining older sibling, died while Jane herself was so sick that she was confined to her bed. Jane's condition at this time was much like Sarah's, as described above. Jane remembers the mattress she was lying on as she missed her older brother's funeral. He left behind a three-year-old and a pregnant wife. This sister-in-law would give birth only shortly before she herself died of AIDS. "Children surrounded me there," Jane recollects. During this time of grief, a younger cousin visited and found Jane on the mattress drifting in and out of consciousness. Moved by her desperate condition, he decided to take her back to Arusha with him for treatment. In Jane's telling, he scolded Jane's younger brothers and sisters before leaving for Arusha, saying, "Idiots, you are killing my sister here, while you sit drinking alcohol. You know nothing.

I will take her now. You will see." He left Jane's children to be cared for by her younger brothers and sisters. He enrolled her in a clinic near him, gave her a place to stay, and provided her with sufficient food. What she remembers is that each time she was given food, she thought of her children.

Slowly, Jane grew stronger. She started being able to go out and earn a little money washing clothes for others. Yet she remained wracked with anxiety. She worried about her children and wondered if they were eating. She learned that they had been left to fend for themselves in Mbeya. The older ones had started to leave. One of the children of her eldest brother went to Dar es Salaam to become a shuttle bus conductor. Another one was taken by a woman in Iringa who needed help caring for her children. Her own firstborn son also eventually went to Dar es Salaam. The cousin who had taken her to Arusha for treatment brought a pastor over and told him that he was now handing Jane over to him. The cousin expressed his desire to build a house for Jane and her children in Arusha. He did not want her to go back to Mbeya, as all of the kin there who might have supported her had died. Jane remembers that, on the very day in 2007 that he handed her over to the pastor, he returned to his office in the mining center and died. His passing left her terrified by her sense of loneliness and vulnerability.

The pastor kept his word to Jane's cousin. He helped Jane rent a house and collect her children, as well as those orphaned by her siblings' deaths. Healthier and motivated by her children's presence, Jane started waking up earlier to wash clothes and then taking the TZS5000 (approximately $2) that she would earn for this work to the market to purchase carrots in bulk. After bundling the carrots into smaller units, she would resell them. By the time evening fell, she would have earned TZS10,000 to TZS15,000. In the morning, she would wash clothes again. She used the money she made to supplement the basic supplies that the pastor gave to her: cooking oil, flour, beans, and children's clothes. After a year, she was known in the market. Others who worked there advised her to leave aside the washing and to come earlier in the morning. She started buying a TZS30,000 bag of carrots on credit in the morning and selling the smaller units throughout the day. In the evening, she would repay the TZS30,000 and keep whatever profit she had made.

Jane's friends in the market connected her with other widows who made their living through small-scale trading in the market. Jane told them, "I live with HIV, and I have a strong desire to talk." In 2008, she formed a small widows' association. The women elected her chairperson. The members contributed modest dues that they could invest in members cyclically. Within the year, Jane received a TZS300,000 loan from the association to help pay her daughter's tuition fees at a teachers' training college. Furthermore, after discovering that Jane tended to arrive at the market later than others in the association because she lived so far away, a woman offered Jane and her children a home that she had bought in Usariver but did not intend to live in immediately. Jane took her up on the offer and moved closer to the market where she worked. By this time, her niece had passed Standard 7 and started secondary school.

Then, in April of that year, shortly after moving, her eldest son died in a car accident in Dar es Salaam. She buried him there. She was devastated. Grief dampened her will to live. She remembers feeling demoralized, having worked so hard through pain and exhaustion to support her children. "If they, too, were dying, then what was the point in living?" she wondered. Jane stopped working at the market. She also stopped taking her antiretrovirals (ARVs). She remembers, "I did not see any reason for continuing to take the medicine. I took medicine so that I could help my children, but now the children were dying." Her friends from the market came to the house and asked if she was taking her medicine. Jane would lie and say she was, but they could see that the medicine bottle had not been touched. "The doctor had told me that if I stopped taking the medicine, I would die. So, I realized this was a way to die. I stopped swallowing the medicine." Within two weeks, Jane grew confused and suicidal. She remembers the compulsion she would feel, when a car drove by the house, to throw herself in front of it. Her daughter returned from teachers' college. Jane felt her daughter's appeals call on her heart (*ananitia moyo*), "So, mother now you want to die because brother has died? So let us all die! Should I stop my studies and we just stay here? What about your nieces and nephews who[m] you are raising? What are you going to do with them?" Her daughter's words shook her. Jane resumed taking the medicine, but her body did not accept the drugs as easily as they had before. She could be constipated for a week at a time. She had no appetite. At the clinic, the doctor told her, "Eee, if you are careless, you will die. You will leave your children. You know that the medicine is poison. You are needed [by your children]. When you start [ART], it is for your whole life." Jane locates her decline in the death of her son, profound in both its singular grief and in the attenuation of the life force that embodied her family's ongoingness. She marked the turning point in her daughter's return and the call back to the relations that remained.

Jane met Mama Nguya in 2010, two years after her son had died. She was still weak and depleted. She went to the market each day, although she was only able to sit at the table in front of her small piles of carrots and sleep. Gesturing to Mama Nguya behind her, Jane remembers this first encounter in great detail.

When I met this woman, I explained that I was so-and-so and I use medicine. She told me, "I am a retired officer. There is nothing I can do to help." Yet as we were talking, she could see that I really needed help. Although she repeated, "I have retired. I don't have anything that I can do for you," she added, "I do hope to help people like you [in the future]. I am planning to start an office. But what can we do now?" I admitted that I did not know. I believed her that she had nothing. Before we parted, she gave me *mlonge*. She told me at the time that, "These leaves are good. From today onward, you will be using this [*mlonge*]." She asked what I generally eat at home. She advised me to change the time of *ugali* that I eat, urging me to eat a traditional diet (*vyakula vya asili*). She told me to love eating pumpkins and sweet potatoes. These things are not expensive. She encouraged me to eat guava when it was available. "When you get money do not buy bread or Blueband [a processed margarine], but try to eat original, natural, traditional things."

For both women, this was a turning point—in their lives, in the development of TRMEGA, and in their work in the world.

Jane not only followed this advice but also started working together with Mama Nguya to open her office. Unfortunately, after a few short months, Mama Nguya's father fell ill, and she had to travel to her natal home in Kagera, in northwestern Tanzania, to help nurse him. Jane remembers, "*Mama* left me in the office. She reminded me to eat greens and *mlonge*, and what to do to stay strong. And she said, 'When God wishes, I will be back.'" Nguya stayed in the village for three months. During this time, her father passed away, and she helped to bury him. When she returned to Usariver, Jane had changed. Mama Nguya exclaims, "She was beautiful!" and Jane teases, "She didn't remember me! The first day she returned [to her home in Usariver], I went to give her my condolences. She thought that I was one of her neighbors!"

Mama Nguya also found that Jane was no longer alone. Even before Mama Nguya left to care for her father, Jane had started doing what she does: gathering people. She felt compelled to share what she was learning. "I didn't keep it in ... I went to churches, villages, and various groups. When she [Mama Nguya] returned home, she found we were many and healthy. She was surprised and she asked me, 'What happened?' I told her, 'I don't know.'" Drawing others around her, cultivating a community in which she could find strength, was part of Jane's healing. Much is missed in public health rhetoric if we hear Jane's story as a linear account of "disseminating knowledge." For Jane, healing (*uganga*) encompasses acts of extending, nourishing, strengthening. Lower viral loads may be an important indicator of healing and health in the clinic, but *uganga* rests in the nourishment of the life force that leads to social connection and generativity.

Every six months, Jane faithfully returns to the local clinic for a check-up. Since meeting Mama Nguya, the doctor and nurses have seen her dramatic improvement. They frequently question her about her diet, and she tells them how she is managing it. While the clinic typically keeps individual health records, she requested that she be able to take her results back to Mama Nguya each time so that, together, they could track any changes. Both women attribute her improvement to eating well, to *dawa lishe*. Most important to Jane, however, is the idea that her increased strength and wellness might be used to motivate and benefit others. To encourage them to nourish their bodies, to eat *mlonge*, and to shift their relationship with the environments they live in, all in an effort to live with AIDS and to live in bodies filled with ARVs. "Many of us have children who are HIV positive," Jane mused one afternoon, "and I want them to see that they can be strong; that they can live a good life. Even my own children, they say, 'If today I will be found to be HIV positive, I will be OK.' We inspire courage in them."

Jane's dynamism cannot be denied. To be in her presence is to be swept up in her energy. Her story is full of the social and institutional formations of AIDS development from the 1990s and early 2000s with its emphasis on the support of kin, the role of the church, confessional narratives animating the creation of identity groups, and the benefits of microfinance associations. She understands her life through the fragments of investments around women's empowerment, health development, and income generation. The vocabulary and rhythms of her story are shaped by narratives of progress. She offers her account as evidence of her ability to take advantage of—to be worthy of—national and international health investments. Therefore, the turn at the climax of the story, where *mlonge* invites her into healing as land relations and disrupts these narratives of progress by opening them to critiques of capital, is all the more powerful. Jane lives out the aspirational gestures Uroki makes when he describes his herbal remedies as stimulating appetites that draw people into denser webs of relation with both people and plants.

In her work with TRMEGA, Jane uses her story to narrate the social and biological wounding of poverty, AIDS, and social abandonment. She not only suggests that the wounding others have experienced can become the precondition for their healing, but also, through her outreach work, she argues that such social harms can be mobilized as the precondition for social transformation. Her efforts resonate with those Laurence Ralph describes in his ethnographic work on AIDS, addiction, and violence on the South Side of Chicago. Ralph insists that we need to think more deeply about "the relationship between forms of bodily injury and forms of social injury." In "What Wounds Enable," he calls on us "to deepen our understanding of the myriad forms that injury can take and the myriad manifestations that those injuries can have."[49] *Dawa lishe*, I argue, does this. It strives to address myriad forms of injury—not just disease. Jane is open about living with AIDS. For her, this openness is critical. As she stated emphatically to me, "I am a testimony." Her personal narrative, as a component of her connection to TRMEGA and of TRMEGA's international funding, might be seen as evidence that supports arguments of biological or therapeutic citizenship. Indeed, it is not evidence against this idea. Jane is savvy and therefore aware that, after the neoliberalization of health care in Africa, claims to rights, care, and protection are increasingly made through nongovernmental organizations to an international community of sorts. Yet, this analytical turn misses the therapeutic and political openings that Jane and Mama Nguya strive to create through TRMEGA. Jane's testimony is also a way of living into the space that Romana insisted was essential for Simon in the market: Sharing vulnerabilities, extending toward others, invites connection and thickens the web of relations that support life and liveliness. *Dawa lishe* is a proposition, a lexicon for increasing social relations (with humans and nonhumans) and nourishing collective capacity in response to illness and injuries, and a set of experiments that locates healing in social and physical transformation. In taking up this proposition, TRMEGA and others disrupt the forms of regulation, categorization, and knowledge that shape the mechanisms of inclusion and exclusion that are central to modern health governance. Their work exposes points of friction generated by valuing healing situated in social-ecological collectives, and pose the right to navigate these frictions as a right to reinvent the terms of modern (therapeutic) sovereignty.

## Conclusion

*Dawa lishe* offers a version of lushness that describes dense social and ecological relations as the space for response and healing. Plant(ing) remedies not only re-spatialize healing but also re-temporalize it. Efficacy does not rest only on catalyzing a finite moment of cure. Claims to efficaciousness are evaluations of a remedy's ability to collaborate in an unfolding of bodily strength and capacity through social, ecological, and therapeutic responses.

EdenMark and TRMEGA apprehend appetite as the force that generates the relationships through which life and liveliness are either cultivated or undermined. The palate becomes a place for thoughtful intervention and a site for ethical reflection. Which flavors animate the relationships that cultivate the forms of lushness that are healing, and which flavors animate the relationships that cultivate the forms of lushness that are harming?

Through the loss of a taste for bitterness, Uroki offers one concrete example of how changes in economy and appetite have resulted in people forgetting how to eat. The sweet and salty flavors favored by global capital economies have overwhelmed palates, he argues, and in so doing, they have attenuated the capacity to appreciate other flavors. Seduced by flavors produced in the webs of global capital, people become attached to the destructive relations that are undermining the liveliness of their bodies and lands. When Uroki formulates dandelion greens as remedy, he speaks of not only the plant's phytochemical properties but also the positive impacts of collaborating with the plant to cultivate the palates of those who take his medicines. Developing a taste for their bitterness is one way of deepening a taste for the ways of living and eating that will reshape social, economic, and ecological relations. Remember, he noted that if people commit to eating the canister of greens as medicine, then after a while, they are likely to realize that the plant grows wild near them. When the greens are available outside their front door, or on the road as they return from work, they will not want to pay for the canister of dried greens; instead, they will begin to collect the greens on their own. This remedy thus works in part by attuning the senses and inviting one into more multifaceted and complex relations with the world around them.

Uroki's attention to flavor recognizes it as a property that can draw one closer to the world, and he gestures to the ways that it might trouble the commodity form of his remedies. He primarily engages with plants, however, as a natural resource. Dried and packaged, EdenMark's herbal formulas cannot reproduce themselves, even if they do, as Romana demonstrates, nourish and intensify the social relations that might reproduce them. Mama Nguya and Jane join Uroki and Romana in attending to flavor. They too see flavor as a stimulant for appetites that animate body-land relations. TRMEGA's herbal formulas strive not only to attune users to the world around them but also to move the world through users as they draw them into projects extending gardens.

The appetites that both EdenMark's and TRMEGA's remedies aim to increase are for flavors that have been erased by development. They aim to stimulate a hunger for the entanglements these forgotten flavors bring. This is a hunger different from the hunger of poverty, which is structured by enclosures of land as that have been delimited as property, or of the deprivation that follows in the wake of drought or torrential rains that violently wash away a newly planted field, both of which are growing more common due to climate change. It is a hunger different from that produced by antiretroviral drugs, which can be so ferocious that people report stopping their ARVs rather than being driven to eat more than their share from communal plates and feeling as if they are taking food from their children's mouths. And it is different from the hunger of diabetes or hypertension or other chronic diseases that deplete bodies at increasing rates in Tanzania. The appetites cultivated by EdenMark and TRMEGA—such as learning to love pumpkin—nourish those living in the wake of slow violence. These cravings work against economic, political, and therapeutic regimes that trap African bodies and lands in binary logics that articulate their worth in terms of their service as a resource for national/capital development. Their efficacy rests instead in their ability to draw bodies into alternative modes of consumption and production that foster people's attunement to deeply relational ways of being.

EdenMark's and TRMEGA's plant remedies coax bodies into new relations; tasting rekindles desire and opens senses; learning to apprehend harsher flavors invites more careful attention to the world. Uroki suggests that people might come to see healing plants in the landscape as they move from place to place. He hopes that they will develop the capacity to respond to illness or weakness with what is at hand. Yet, Romana, who sells and gifts EdenMark products, illustrates the limits of these prepared herbal formulas when they focus strictly on individual biological injury in the midst of social and ecological precarity. By contrast, Mama Nguya unfolds her formulas across networks that strive to seed dense patches of green, assisting members with gathering plants around them. In this way, her remedies gently begin to reconnect people and plants to each other and to the land from which they have been alienated. The herbal formulas—as well as the larger social projects of which they are part—challenge the forms of knowledge that authorize therapeutic efficacy and elevate interventions into the palate as a ground for politics.

In *As We Have Always Done*, Michi Saagiig Nishnaabeg scholar and artist Leanne Betasamosake Simpson teaches that the opposite of dispossession is not possession but rather attachment.[50] While the specific forms of colonial dispossession in Africa and North America are different, the struggle to center ways of dwelling that generate more life opens a space for thinking together. The work of refusing the forms of disposability that inhere in the economization of life, land, and labor in settler-colonial and postcolonial worlds suggests a possible trajectory along which to build solidarity. *Dawa lishe* names a way of exploring new attachments for a more just future: one that draws on traditions of collectively evaluating, naming, and intervening in unjust accumulation (*uchawi*), one that builds ongoingness through ever-cycling processes of composition and decomposition, one that orients toward a *politics of lushness* (that is, the entanglements, accommodations, and nourishment that generate more life).

As Uroki articulated with his dandelion greens at the beginning of this chapter and as both Romana and Jane illustrated, appetite is a way of drawing people toward these attachments. *Dawa lishe* dismantles and reinvents the notion of appetite, resisting the ways that it is commonly taken up in public health—as an individual quality or preference, a side effect of disease or treatment, or a context through which to support pharmaceutical efficacy. Instead, appetites are the desires that drive relations and energize response. Appetites animate response(ability) through acts of ingestion and digestion. Eating pulls the whole world into and through oneself. The process is transformative, building "body" by consuming plants and releasing "land" through bodily excrements that turn into soil. A focus on eating unsettles representations of bodies, helping us to understand them as entities not bounded by skin. Instead, this focus directs attention to the ways that bodies take form through processes of composition, decomposition, and recomposition.[51] As such, they are spread over space and time. Uroki and Nguya, Romana and Jane, cultivate appetite as the material and conceptual focus for a notion of health and healing that addresses such a temporally and spatially dispersed body. They position appetites as social, cultural, and physical forces that move bodies through the world and the world through bodies.

Appetite as the force through which body and land are brought into being offers a vocabulary that undermines epistemic dependency on reified objects and the scales that inhere in them. In opening a space for discussion and debate about complex interscalar relations, appetite renders visible connections between agriculture and chronic disease, land relations and health. This version of appetite helps to formulate questions that are impossible to ask within nutritional sciences or the medical disciplines that draw on them. Taking appetite as the focus of therapeutic interventions opens the analytical horizon and allows Tanzanians to ask, "How might people come to hunger for foods and crave flavors that restore, repair, and reinvent nourishing body-land relations?" "How might attending to a damaged relation with flavor intervene in the ways that the expansion of industrial agriculture is converging with the rising threat of chronic disease?"

While Uroki, Mama Nguya, and others listen and often concur with analyses of the hunger, poverty, and barrenness generated by the increasing enclosure of land as property and speak eloquently of the insufficiency of remedies that are limited to working on discrete biological bodies, their interest is not in the production of critique. Instead, *dawa lishe* points to work that operates in a different social tense: the speculative future. The remedies rekindle ways of knowing and being that might reconfigure the political order and cultivate relationships that might body forth people and plants otherwise—relationships that support possible alternative futures. The times and places of eating bodies refuse to be contained within the liberal subject and its discrete lifeline. They also decenter a dominant version of bodily sovereignty that locates rights in an individual bounded by their skin. What sort of alternative sovereignty might be articulated in these continual incorporations and excorporations of the eating body? Which practices generate the temporalities that are needed to attune to these movements, and which analytical modes or storytelling practices render visible struggles over possible forms of sovereignty? In reinventing appetite as a site of body-land healing, EdenMark and TRMEGA set the stage for a more expansive articulation of reproductive justice.

3

# Registers of
# Knowledge

Mama Nguya first offered *dawa lishe* in a speculative gesture, as a way of naming nascent efforts to build common cause with others who also refuse the epistemological and ontological commitments shaping modern notions of health and healing. *Medicines That Feed Us* explores *dawa lishe* as more than a clever name for these projects. In our work together, I saw Mama Nguya's phrase as an invitation to tell a different story and as an incitement to theory that is rooted in an analysis of how these projects trouble the settlements made in national and international efforts to develop "traditional medicine" as a modern category of knowledge and practice. While Alex Uroki collaborates with independent healers in his research for new herbal remedies and Mama Nguya imports medicinal plants from her home in Kagera, the work of EdenMark and TRMEGA exceeds the definitions of traditional medicine. The challenge arises when Uroki's and Mama Nguya's desire to circulate their plant-based formulas requires them to apply to state regulatory agencies for certification. They have cultivated a subtle dexterity that enables them to move their products across regulatory tracks as food, drug, or traditional medicine. By working across these fault lines, they refuse to have their plant(ing) remedies circumscribed by any singular method or mode of storying efficacy—via nutrients, active ingredients, or indigenous knowledge. These strategic maneuvers trouble the state's efforts to capture any products that slip between the disciplining structures of medicine and agriculture within traditional medicine.

Because my effort to account for *dawa lishe* must also account for its complex relationship with traditional medicine, this chapter takes a detour through this history with a focus on the institutionalization of the category of "traditional medicine" in Tanzania. I show how national and international efforts to position traditional medicine as a solution to problems of knowledge and politics consolidated forms of unknowing that rendered colonial dispossession and ongoing extraction invisible, making the persistent depletions and chronic injury that followed in their wake inactionable. *Dawa lishe* calls on us to recognize traditional medicine as a product of such practices of "colonial unknowing" as they are enacted through scientific, legal, and bureaucratic practices. It is an invitation to work collectively to rekindle the relations that will enable both seeing and healing the wounds of dispossession. In these ways, *dawa lishe* has potentially radical implications.

Traditional medicine, as a modern category of knowledge and practice, emerged over the course of the late nineteenth century and twentieth century. It is rooted in the colonial desire to intervene in precolonial practices that address the flourishing of bodies and ecologies in order to assert control over labor and land. East Africans experienced colonial dispossession both as a threat to political sovereignty and as a violent injury to body-land liveliness. Struggles for control over land were also struggles for control over the forms of lushness that defined precolonial versions of health. Colonial dispossession therefore provoked responses from healers that the colonial state experienced as resistance. Colonial administrators reacted by targeting healers directly and implementing policies that systematically dismantled the precolonial relations between peoples, ecologies, and things on which healers' ways of knowing and power to act were based.

German (1880–1918) and British (1919–61) administrations' struggles for control over colonial subjects and lands in Tanganyika became in part a struggle with healers and a problem with healing. As the field of science and technology studies has shown, the age of empire sought solutions to problems of politics in conjunction with problems of knowledge, or at least problems of politics spurred on the creation of epistemological and ontological practices that obscured dispossession. As colonial boundaries delimiting legitimate forms of knowledge were naturalized, healers were separated from their basis of power, bodies were separated from lands, plants were separated from ancestors, and social relations were separated from physical being. Colonial science, law, and economy rearticulated the nature of harm. Ways of knowing that understood healing as good land relations were rendered primitive, magical, or unintelligible.

The British instituted legal distinctions between "witchcraft" and "native medicines" in order to divide what administrators saw as troublesome social relations from useful natural resources. Native medicine, as a sphere appropriate for scientific investigation and economic development, was a precursor to the postcolonial category of traditional medicine. The historical arc from colonial native medicines to postcolonial traditional medicine is one trajectory along which to examine how anthropology, botany, and chemistry grew through their participation in colonial projects and the naturalization of African dispossession. Particularly important to the argument in this chapter is that the new subfields of ethnobotany, phytochemistry, and ethnopharmacology refined the terms in which strategic forgetting, disruption, and obfuscation of the relations that animated African therapeutics structured participation in modern knowledge and politics. The history of traditional medicine as a modern category of knowledge and practice starts with this broader reformulation of the relations of science and capital to nature. Empire not only remade African landscapes through crude resource extraction but also transformed them through the circulation, study, and movement of plants and people.

Conceptions of reason and reasonableness were shaped by these scientific disciplines, and these definitions were further codified by colonial clinics and courts as part of the state's everyday workings. In the process, colonial clinics and courts institutionalized the displacement of forms of justice founded on practices designed to eliminate evil and replaced them with forms of justice founded on evaluations of individual bodily integrity, scientific efficacy, and legal rights.[1] David Arnold uses the term "the imperial pharmakon" to describe the specific configuration of science, law, and economy through which colonial administrators managed the constitutive ambivalence of forms of healing that were also threats to empire.[2] Postcolonial leaders have continued to mobilize the language and logics of the imperial pharmakon to shape research and drive policy in order to promote traditional medicine in the service of nation-building. Understanding traditional medicine as an imperial project reframes the story of colonialism and postcolonialism in Tanzania by foregrounding the ways in which the state mobilizes the imperial pharmakon to govern bodies and plants, to structure the economization of labor and land, and to circumscribe possibilities for justice. The resulting story enables an analysis of *dawa lishe* as a call to (re)kindle alternative plant relations and decenter the relationship between harming, healing, and memory solidified in the modern category of "traditional medicine."

International and national policymakers, scientific institutions, and complementary medicine programs echo ethnobotanical methods when they position "traditional healers" as custodians of ancient knowledge—minds that must be mined before they die. Such claims not only are anachronistic but also obscure how institutionalizing traditional medicine has naturalized the historicity of colonial objects of knowledge. Ahistorical assertions that contemporary practices are living pieces of an indigenous past to be capitalized on—a national resource or a global heritage—facilitate techniques of unknowing that make it difficult to speak of alternative forms of expertise, objects of social power, and notions of justice.[3] In this chapter, I examine the relations between the processes of un/knowing that forged traditional medicine in Tanzania and the terms, objects, and institutions through which struggles for justice have been imagined (and limited). By developing research into traditional medicine, the newly independent government in Tanzania hoped to find substitutes for expensive European and American pharmaceuticals and to bypass unfair patent laws that protected monopolies on their production. Justice would be seen in stocking an expanding network of clinics and dispensaries. Such modernization continued to divide bodies from land, however, and limited the vocabulary in which dispossession could be addressed to that of political economy.

Decolonizing our therapeutic knowledge requires assessing the forms of expertise, kinds of evidence, and techniques of institutionalization that have formulated the contemporary category of traditional medicine and naturalized particular answers to the questions about the therapeutic value of plants. It also requires attending to the ways that African healing practice and therapeutic innovation repeatedly exceed the answers to these questions of knowledge as they are embedded in modernist projects.[4] The state's recognition of traditional medicine has never fully captured the wide range of practices that strive to catalyze growth, fullness, maturation, extension, strength, and fertility of people and land. Each chapter of this book traces innovative experiments in Tanzania that unsettle the binary of traditional and modern medicine and offer a creative space to rethink the challenges of politics, ecology, and health faced today.[5] Recognizing how the imperial pharmakon continues to embed techniques of unknowing in official registers of traditional medicine enables us to see how Tanzanians are generating therapeutic value on/through/with plants in ways that explicitly refuse the terms of (dis)possession that shape liberal notions of justice.

This chapter examines the new registration system for healers, their medicines, and their premises that was rolled out in Tanzania in 2011. I consider the conditions of possibility for, and the limit points of, this registration system and then turn to look at how healers are grappling with the restrictions and fictions that inhere in it. The historical narrative that unfolds throughout this chapter frames the challenge that the newly appointed registrar of traditional medicine faced when we sat down together in Dar es Salaam in 2016. I was curious about how remedies, such as those I was following out of EdenMark and TRMEGA, were categorized in this new system. With a sigh, she stated, "They are difficult to categorize." The solution she suggested requires that their development and commercialization commit them to one legal category and bureaucratic pathway: drugs, food, or traditional medicine. Modernization, as Latour has taught us, demands such purification and renders impurity as primitiveness. Experiencing *dawa lishe*'s taxonomic difficulty as resistance—a refusal to be fixed ontologically for the sake of governance—rather than insufficient development or a lack of authenticity requires a political reading of the commitments embedded in the categories that are central to the state, their forms of institutionalization, and their modes of enforcement. This chapter is dedicated to that reading, to describing the historical conditions and affective landscapes in which *dawa lishe* may be apprehended as potential, rather than as a problem.

## Conditions of Possibility

In 2002, the Tanzanian legislature ratified the Traditional and Alternative Medicines Act and, with it, enacted its call for a new registration system. At the time, however, it was not clear how the call would be answered. Fragments of other registration systems complicated the landscape. Over the decades, as colonial and postcolonial states strove to control healers and healing, other registration systems had been tried. As one effort gave way to another, the names of healers and plants were rolled over, using one uneven system as the basis for the next. Furthermore, discrete efforts to tap healers or their plants as resources of a country's economic development had resulted in a variety of programs evidenced by the certificates of various sorts on the walls of healers' homes, from trainings in hospitals to short courses with nongovernmental organizations to memberships in a local or national organization.

An unexpected series of events, however, catalyzed the total suspension of all healing practices in Tanzania in 2009. The radical move to make healing illegal created a clean slate of sorts for the state to later initiate the new registration that the 2002 legislative act had called for. The timing of this new registration system's rollout was also shaped by a second event whose story reveals the complicated entanglements into which this registration system entered. This first section describes these events—first, the violence against people with albinism, and then a wildly popular cure-all called the "cup of life." These events bookended the sudden suspension of healing in Tanzania in 2009. Although neither surfaces in official narratives of the registration system, ethnographic research makes it possible to see them not only as part of the context in which the registration system was implemented but also as active markers of the limits against which the registration system was defined. Thus, insisting that they are part of the story of registration opens up an analytical space in which to grapple with the boundaries of the modern category of traditional medicine by illuminating the tensions and fears that shape its current incarnation and haunt those tasked with implementing it. I tell each story in some detail in order to concretely illustrate both the intricate ways in which the colonial past lives on in the present and the fictions and fissures, resistances and refusals, that shape healing today.

### Suspending All Healing and the Attacks on People with Albinism

In 2006, as the body parts of people with albinism were coming to be valued as powerful components in medicines that ensure the accumulation of power, wealth, and position, a wave of murders of people with albinism began in Tanzania.[6] Men, women, and children died as they were dismembered—most often their attackers hacking off an arm or leg, at times genitals and hair. Many more were injured or debilitated, and all people with albinism learned to arrange their lives around the fear of this violence. Yet prejudice against people with albinism in Tanzania is much older than the brutal attacks that have marked the past two decades. Elders recount stories of people with albinism being ghosts—holding an ambiguous and dangerous position between the living and the dead, the embodied and the disembodied. Accusations have been leveled against new mothers that their albino newborn is evidence of an affair with a white man or that they have been taken by the spirit of a European colonist.[7] Albino skin has long been a site of deeply contested notions over the capacities and vulnerabilities that structure bodies and worlds. However, as Zihada Msembo, a sixty-year-old woman and a leader in the Tanzanian Albino Society, noted, "When I was growing up, it was not like this. It was just stigma, but not people coming to cut up bodies."[8] Only after the turn of the millennium did prejudice against people with albinism begin to be played out through another powerful frame—the brutal, disorienting, and transformative power of circulating body parts, especially in areas of the country where local economies feed global extraction (i.e., mining, commercial fishing, etc.).[9] This correlation corroborated a link between the violent extraction of body parts as a resource for medicine and the demand for these medicines by people desiring to grow rich through the unrestrained exploitation of natural resources.

In January 2009, as the violent attacks continued, the prime minister announced the immediate suspension of all licenses held by healers and a prohibition on all healing practices. The prime minister's decision caught off guard those in the Office of Traditional Medicine in the Ministry of Health, the Institute of Traditional Medicine at the Muhimbili Health Sciences Center, and the Department of Traditional Medicine at the National Institute of Medical Research. This decision to suspend all healers followed colonial patterns of framing healing as witchcraft. In so doing, it struck out against the state's scientific and bureaucratic project to develop traditional medicine. The state-supported research institutions found that, overnight, the healers with whom they had been working were unregistered—overnight, it had become illegal to work with them. The prime minister's impulsive declaration appeared to be a political response to the racial overtones of stories increasingly circulating in the American and European press about the "harvesting" of body parts from people with albinism. The gruesome tales re-energized colonial imagery of a dark, violent, superstitious Africa just at the time that Tanzania was attracting attention from the United States as a relatively peaceful country strategically located in East Africa with increasingly "business-friendly" economic policies and the promise of oil extraction. Then-President Jakaya Kikwete referred to these attacks against people with albinism as not just "disgusting" but also a "big embarrassment for the nation."[10] By unregistering healers and suspending healing, the Tanzanian government hoped to preserve US confidence in current and future investment. The governmental response, however, did little to stem the violence.[11]

Slowly those invested in developing the state's legal and bureaucratic framework around traditional medicine came to realize that this unexpected and blunt suspension of all healing created a unique opportunity. A year and a half later, in 2011, when the government rolled out its new, more elaborate and more stringent registration system, it was the sole gateway to a legitimate healing practice in the country. Previous registrations were invalidated, and all healers were required to comply with the demands of the system that included registering healers, their medicines, and their premises. While the erasure of previous registrations might promise a more even landscape in which to implement this new three-tiered system, the rollout in practice privileged some regions of the country over others. The lists provided by the national registrar of traditional medicine in 2016 show that over half of the approximately three thousand healers registered in the first three years of the new system's rollout were registered in the Lake Zone.[12] The disproportionately high level of registration in the area of northern Tanzania around Lake Victoria—an area that borders Kenya, Uganda, Rwanda, Burundi, and the Democratic Republic of the Congo—correlated with a higher prevalence of the brutal attacks against people with albinism in this area. In addition, the Lake Zone experienced more intensive state interventions in response to this violence.[13]

The quick success of the rollout of the registration system in this one zone of Tanzania may indicate healers' desire for government protections and, as a result, a greater willingness of healers to be registered and/or the state's desire to show it is doing something and, as a result, a greater effort by government officials to register healers there. Given the violence, healers everywhere, but presumably more intensively in the Lake Zone, feared being swept into the state's efforts to address "witchcraft" and targeted unjustly. As I will discuss in more detail below, colonial ordinances against witchcraft rolled over into the postcolonial state legal statutes and carried with them the history of colonial accusations of witchcraft being used against healers who were seen as a threat to colonial governance. Perhaps the government's significant success in registering healers in the Lake Zone indexes healers' hopes that registration would legitimate their practices, support their healing work, and protect them from any targeting that they might otherwise experience as a result of the government's efforts to address the violence against people with albinism.

The geographical distribution of registered healers may also reflect a different approach by Traditional Medicine Coordinators in areas where the bureaucratic management of healers is under intense scrutiny. The desire to be able to identify and locate people quickly in the wake of another violent incident prioritizes the implementation of registration in the service of surveillance and control. Given the disproportionate representation of registered traditional healers in the Lake Zone, the primary goal of registrations in this area has an outsized influence on the status of the overall state project. Coordinators recognized by the Baraza la Tiba Asili na Tiba Mbadala (Council of Traditional Healing and Alternative Healing) as particularly successful promote strategies for registering healers in their areas, shaped by concerns that they will be called on to control those propagating violence rather than other concerns that might shape a registration system. They are not, for instance, guided as strongly by concerns over how to surface debates on the boundaries of contemporary medicine; how to develop collaborations that challenge modern divisions between medicine and agriculture, body and land; or how to stimulate progressive thinking about concepts of property. They do not see their work—as do those in the projects drawn together as *dawa lishe*—as reinventing relations between medical science, law, and economy, to address the chronic depletion of social, biological, and ecological liveliness. Instead, even as the violence had waned, the states' struggle for control continues to narrow the notions of justice and healing articulated through the legal and political technologies that govern traditional medicine.

### Rolling Out the New Registration System and the Cup of Life

The first line of a 2011 joint technical report from the National Institute of Medical Research and the Institute of Traditional Medicine began, "Something big and quite unusual has been happening since August 2010 in Samunge village in Loliondo, Arusha."[14] As Tanzania suspended all (non-biomedical) healing, a "miracle" cure-all distributed by a retired Lutheran pastor had become wildly popular. The drama swept up medical professionals, government scientists, churches, and the Tanzania People's Defense Force, as well as the general public. The "cup of life," as some referred to it, reached peak popularity the same year that the government rolled out the new registration system described above.

The story, as Rev. Ambilikie Mwasapila tells it, however, began with a dream in 1991, two years after he committed to the North Central Diocese of Evangelical Lutheran Church to serve Loliondo, a small parish in the northwestern corner of the Ngorongoro Conservation Area. In Mwasapila's dream, God told him to collect a particular root to treat a woman said to have HIV/AIDS. While the call felt compelling, he hesitated. He was just settling into Loliondo, a place many other pastoral staff found challenging due to the lack of roads, scarce electricity, and thin cell phone coverage.[15] The challenges of the poor infrastructure did not shield his new life or those of his partitioners from regional tensions, national changes, or global pandemics. Mwasapila's dream came amid pressures both on local land relations as the government's restrictions in the wildlife area exacerbated long-standing tensions between the agricultural Songo and the pastoral Maasai, and on social worlds as HIV/AIDS weakened bodies and devastated kin structures. These dynamics undermined the ecological and social basis of health in Loliondo in 1991, just as the last socialist Minister of Health stepped down and the structural adjustment policies demanded by the International Monetary Fund (IMF) began to be implemented more thoroughly and consistently. In the midst of all of this, he waited for another sign.

It was a decade later, when Mwasapila retired from his parish and began thinking about moving, that God spoke to him again and revealed that Mwasapila had more work in Loliondo. He listened and did not move. Several times over the next few years, God came to him, directing where he should live and bringing suggestive visions of future work. In 2009, God showed Mwasapila where to collect the root he had first envisioned in 1991 and again told him to administer it to a woman in the village who was rumored to be living with HIV/AIDS. This time, Mwasapila obeyed. Sometime later, God revealed to him that the woman had been cured and suggested that, in his hands, the root could be used to treat not only HIV/AIDS but also a range of other chronic diseases.

News of Mwasapila's root spread quickly. By February 2011, the Ngorongoro Conservation Area Authority reported that tens of thousands of people had drunk Mwasapila's medicine.[16] People from Tanzania, as well as other countries in the region, including Kenya, Uganda, Burundi, and the Democratic Republic of the Congo, continued to arrive in cars and minibuses and were willing to wait in line for up to five days. The wealthiest began arriving by helicopter. Government ministers visited the retired pastor and drank his medicinal tea.[17] Even after reports of people dying in line began to circulate, the sick and the hopeful continued to come. News stories showed relatives pouring this "cup of life" into their loved ones through nasal gastric tubes. Mwasapila came to be known as the *Babu wa Loliondo* (grandfather of Loliondo). As the crowd continued to swamp the small village of Samunge, the church built latrines to manage the deteriorating hygienic conditions, and Tanzanian troops were brought in to maintain order. Vodacom and Airtel built cell towers, gifted Mwasapila cell phones, and provided his helpers with brightly colored T-shirts featuring their companies' logos. The doctors with whom I work in Kilimanjaro said that not only patients but also many of their colleagues went to Loliondo. Reflecting on that time now, Tanzanians describe going to Loliondo as "tourists" as often as they report going as *wagonjwa* (people who are ill). Something unusual was happening, and it was not to be missed!

The story of Babu captured international attention, with reports of this retired pastor's cure-all therapy airing on the BBC and CNN, as well as posted in the *Guardian*, the *New York Times*, the *Washington Post*, the *International Financial Times*, and a range of African newspapers. Even more than this international coverage, rumors circulated that the 30-km line of waiting people could be seen in Google Earth satellite images. These indexed Tanzanians' collective sense that the gathering on the plain of Loliondo was "big." Even planetary.

At the height of his popularity, only a few months after the new 2011 registration system was launched, Ministry of Health officials hand-delivered registration forms to Babu. Although they worked diligently to convince him to register, Babu remained unpersuaded. He adamantly declared that he was a pastor, a channel for God's will, a mere servant, not a healer. The cup of life, he insisted, was an act of God and only therapeutic as long as God continued to work through his hands. He reportedly ladled the medicinal tea into every cup personally, even at the height of the rush, when reports estimated that he served over one thousand cups a day. He welcomed government scientists and invited them to take samples of the root back to their laboratories, but he assured them that they would find nothing. The power of the medicine was not in the active ingredients, he asserted, but rather in the way that God worked through him to transform the plant into a therapy. Only the tea made by him and consumed in Loliondo would be effective.

The National Institute of Medical Research's Traditional Medicine Research Unit, together with the Institute of Traditional Medicine in the Muhimbili, did investigate the root in Babu's cup. They identified it as *Carissa edulis* (known as *mugariga* in Kiswahili) and reported antidiabetic, anticonvulsant, cardiotonic, hepatoprotective, and antioxidant activity. They recommended more studies. The scientists at the National Institute of Medical Research (NIMR) Ngongongare Research Station also attributed a range of therapeutic effects to *mugariga*. They argued that laboratory studies should be conducted while simultaneously engaging in initiatives to catalyze more immediate commercialization in the form of teas or other herbal products.

Yet, for Babu, all of this was beside the point. His refusal to be a discrete agent, a willful subject, or an intentional actor for the Tanzanian government— marked by his insistence that he was a channel for God's will—was also a refusal to participate in the making of the subjects and objects that registration strove to generate (e.g., healers and medicines). It was a refusal to mobilize his work in the service of efforts to formulate a Tanzanian or African "traditional medicine" that articulated efficacy through science and expertise through bureaucratic recognition. By refusing to register as a healer or acknowledge the attribution of laboratory results, Babu refused to mobilize his labor, or the labor of the plants in his remedy, as resources for the nation. In this encounter, he leans into modernist distinctions between religion and science, insisting that they are distinct categories of practice and incommensurable forms of knowing and knowledge. The parallel truth claims of the pastor and the scientist formulated the separateness of the realms in which they each work.

By withholding his participation, Babu reveals the registration system's role as a technology of inclusion and exclusion, of making known as well as making unknown and unknowable. That is, the state designed the new registration system to refine the subjects and objects of legitimate political and economic participation. The registration process lays out a pathway to becoming the kind of healer and having the kind of relationship with plants that opens one up to being coordinated by legal, economic, bureaucratic, and scientific practices. It turns one and one's plants into resources. The extraordinary popularity of Babu's cup of life, especially in the wake of other healing being pushed underground by the abrupt suspension of healing between 2009 and 2011, focused the public's attention on the state's request to include Babu and his refusal. This test of the registration system posed complicated questions: Which forms of difference can be digested within contemporary traditional medicine? Which remain indigestible?

Understanding both the violence against people with albinism and the popularity of the cup of life as critical to a history of the 2011 registration system insists on remembering that the subjects and objects recognized through registration are not ahistorical realities now protected by national law. They are, instead, part of a history of interventions that dismantle precolonial worlds by redefining healing and the forms of sovereignty that it generates. An insistence on remembering that which is rendered unknowable forfeits the elegance and ease of a tightly bound narrative that is working within the state's epistemological and ontological commitments. It opts to embrace the messy realities and their impossible tellings through lexicons of witchcraft, magic, and myth, as well as spirits, ancestors, and gods. It insists on including those practices that articulate healers and plants as more than resources for development.

The stakes are high, for it is the partial connections, misrecognitions, and mistranslations generated when techniques of unknowing are experienced as incomplete (palpably insufficient and fully unsatisfying) that catalyze new possibilities for innovation, survival, and resistance. In widening our analytical lens—in starting with these two events that bookended the suspension of healing and cleared the rolls in preparation for the rollout of the new national registration system—we can wonder with Tanzanian healers about the possibility and desirability of participating in the production of (representational) knowledge and politics, and we can see the radical work of those who imagine something else. We can see the challenge that *dawa lishe* poses: its rearticulation of body and land through new temporalities and alternative possessions.

The registration system, however, is only the latest turn in a much longer story. To articulate the double-bind that generates *dawa lishe* as both a concrete set of social, ecological, and therapeutic practices and a speculative proposition, it is necessary to turn to the nineteenth and twentieth centuries for a history of the colonial interventions into precolonial therapeutics in the region that redefined healing and, with it, the forms of sovereignty and notions of justice it enabled.

## Colonial Interventions in African Therapeutics

In precolonial Eastern and Central Africa, practices that shaped investments into reproduction, lineage, growth, maturity, and continuance constituted *uganga*, or what is often glossed in English translations as "healing." While *uganga* existed in relationship with political and social power, the specifics of the relationship between sovereign entities and *waganga* (healers) varied widely across a given territory and over time. In some areas, chiefs themselves held esoteric healing knowledge. In other areas, people with this knowledge worked within the king's court, and in others still, healers engaged in less established and therefore less predictable relations with political leaders, sometimes supporting, sometimes challenging, and sometimes redirecting their power.[18] *Waganga* influenced issues as wide-ranging as where communities should settle, when rains would come, when planting or harvesting should begin, and how kinship relations should be managed. To do so, they mobilized plants, minerals, animals, words, and a range of objects in efforts to intervene in affairs that affected the coherence of groups, the organization of social networks, and the capacities of bodies. The tension between *uganga* (healing, acts that thicken connections, catalyze growth, stimulate continuation, and invigorate bodies) and *uchawi* (harming, acts that attenuate life, thin relations, cut off continuance, and deplete capacities) framed questions of justice, techniques of governance, and styles of resistance. Those who explicitly endeavored to navigate these tensions for others were *waganga*, healers—or, in Steven Feierman's important repositioning of our analytic, "peasant intellectuals."[19] Yet, the morality of their expertise was not fixed. As the Kiswahili proverb intones: those who know how to heal know how to kill. References to others as *waganga* or *wachawi* contributed to unfolding public assessments of the justness of their actions—that is, the ends for which they used their knowledge and abilities.[20] Did they generate more life, longer lineages, and meaningful lushness, or did they attenuate life, shorten lineages, and facilitate overly specific growth that undermines lushness, facilitates lopsided accumulation, and limits future response?

Colonial laws, medical services, and infrastructures intervened within the complex frame of *uganga* and *uchawi* by disciplining both the experts and the knowledge that guided ethical relations among the vitalities of precolonial people and things, bodies, and elements.[21] In colonial Tanganyika, first German (1880–1919) and then British (1919–61) scientists, government officials, doctors, priests, and nuns worked to translate the diverse array of local practices, technologies, threats, and entities as African versions of colonial spheres of expertise (e.g., as African medicine, African religion, African governance).[22]

The "imperial pharmakon"—as David Arnold refers to the eighteenth- and nineteenth-century colonial formulation of the tensions between poison and remedy—shaped the processes that created such ontological equivalence, facilitated the naturalization of distinctions between matter and spirit, and organized difference through race.[23] "Native medicine" and "witchcraft" emerged as legal categories as well as institutional foci through these processes of translation.[24] As colonial distinctions between types of practices, "native medicine" and "witchcraft" did not trace existing local ways of knowing or acting faithfully. Instead, they mistranslated and displaced *uganga* and *uchawi*, thereby reordering relations between healing and governance through colonial configurations of knowledge. As categories of knowledge as well as law, "native medicine" and "witchcraft" enabled a diverse array of precolonial healing practices to be sorted according to their imagined usefulness to colonial projects.

Native medicine came to index materials, experts, histories, and practices that were thought to have potential scientific and economic value for colonial states. It defined the focus of colonial research on plants and became the precursor for the postcolonial category of traditional medicine, with its goal of developing *materia medica* pharmacologically. In colonial Tanganyika, as elsewhere in the British colonies, research into native medicines mobilized modern notions of toxicity and its relationship with remedy to intervene in precolonial poison cultures.[25] Naturalists and scientists who were interested in substances that Africans used for medicine, hunting, fishing, and agriculture established collections useful for research by sorting them according to the "active principles" of poisons and therapeutics.[26] Colonial science and law rendered precolonial relations with plants and other potent powers (ancestors, spirits, jealousies) illegible to the state and developed a modern notion of toxicity to redefine efficacy in the service of empire.[27] Botanical histories crafted linear accounts that retroactively established the consistent, identifiable existence of specific plants over time and their intrinsic therapeutic capacities in the service of colonial (and later nationalist) agendas.[28] In relation to these new botanical entities, *waganga* were characterized as custodians of traditional knowledge, primitive herbalists, or charlatans. Their role as political actors, regardless of which category colonial logics put them into, was sidelined.[29]

Yet, the British remained all too aware of the political nature of African therapeutics. The memory of Maji Maji, a widespread resistance movement against earlier German colonial rule that was coordinated by healers through the circulation of the therapeutic water (Kiswahili, *maji* means water), endured. Fearing healers' power to organize opposition to their colonial administration, they introduced a witchcraft ordinance into each of their colonial territories. Even as the ordinances were found to be confusing in practice, colonial officers argued to uphold them in order to have legal recourse to deal with *waganga* whose therapies threatened colonial rule.[30] Provincial and district officers grappled most directly with the difficulties of implementing imperial policy on the ground. When the government required healers to request permission to practice in each new province or district they entered, for instance, letters from provincial and district officers granted this permission. Each letter stated clearly that the identified healer had assured their office that they were not entering in order to practice "witchcraft."[31] In this way, colonial officers endeavored to control healers' movements. Their tactics had immediate effects on healers and longer-term implications on what constituted healing and healing expertise.

Each revision to the original 1922 Witchcraft Ordinance continued to divide African therapeutics into practices to be disciplined by law (as witchcraft) and practices to be disciplined by science (as native medicines). Legal statutes cast witchcraft as fraud and witches, as well as the healers who fought them through divination and ordeal, as charlatans. In doing so, they maintained the possibility that herbalism was a protoscience and that herbalists were custodians of knowledge that had been obtained over the centuries through trial and error. Legal statutes and policy initiatives established and defended the epistemological and ontological terms through which the colonial governmental, scientific, and religious organizations remade the landscape of African therapeutics (in some cases quite literally, as exemplified by the elaborate botanical gardens at the Amani Research Station). This effort to separate African therapeutics into two distinct and governable categories—native medicine and witchcraft—redefined healing, thereby disrupting existing precolonial understandings of justice and dismantling the institutions that maintained them.[32]

Particularly relevant to the broader argument of this book is that these policies concerning native medicine disarticulated knower and known, rendering knowledge a question of mastery over materials rather than a question of the ethics of relations. This move is exemplified in the way that the British mobilized the term *mganga* as a professional title for a person with technical knowledge of plants' therapeutic capacities, rather than as a term through which the public asserted their evaluation of the ends to which a person worked (that is, evaluations of whether their actions were in the service of healing or of harming). The colonial state limited its attention to plant relations that could be engaged as resource relations. It focused ideas of therapeutic value around plants that could be imagined as contributing to modern technoscience expertise as inspiration for drugs, foods, pesticides, or the like. This radical constriction of the engagements that mattered required an unknowing of plant relations within other spheres, such as kinship; that is, it obfuscated the relations that fostered the co-emergence of lineage and land central to the conceptions of ongoingness in African therapeutics. These tactics of colonial unknowing not only dismantled precolonial relations but also did so in a way that rendered dispossession invisible. The colonial past lives on in the present configurations of traditional medicine as a postcolonial heir to "native medicine."

As African independence took the shape of the modern state, many legal statutes (including a witchcraft ordinance in many former Anglophone colonies) rolled over from colonial law and established the legal infrastructure for the newly independent republics. At the same time, national health care services based on scientific medicine (and expanding colonial and missionary infrastructure) sustained an interest in plants as primitive medicines, that is, as not-yet-elucidated bundles of active chemical compounds. All that could not be imagined as knowable through the sciences of toxicology and pharmacology were deemed irrational or superstitious and experienced as unruly, as witchcraft. Taking colonial efforts to separate witchcraft and herbalism as critical points in the genealogy of postcolonial traditional medicine historicizes the forms of expertise, kinds of knowledge, and techniques of governance that elaborate the contemporary category of knowledge and practice. In so doing, it renders visible the particular notion of toxicity and its relationship with remedy and memory authorized through the state's research and promotion of traditional medicine as a product of specific historical configurations of science, law, and economy. This story of modern traditional medicine denaturalizes now well-institutionalized answers to the questions: What are plants? Who can know them? And what counts as knowing and knowledge?

The centrality of phytochemistry in efforts to modernize traditional medicine in Tanzania has its roots in the colonial efforts to separate African therapeutics into witchcraft and herbalism. Unpacking this historical legacy is critical to accounting for social-ecological-therapeutic projects that refuse to be confined by the epistemological and ontological commitments of the imperial pharmakon embedded in the postcolonial policies, institutional forms, legal codes, and scientific work articulating traditional medicine. *Dawa lishe* experiments within the fissures and friction of traditional medicine in order to render chronic depletion and persistent injury visible as an effect of alienation from land and labor. Plant(ing) remedies draw people closer to the land through the extension of gardens and the cultivation of appetites. Producers expand the concept of nourishment (*lishe*) as an effort to reclaim an ethical vocabulary with which to hold medicine (*dawa*) accountable to relations that promote the flourishing of body-lands. Accounting for this complex relationship between *dawa lishe* and traditional medicine requires recognizing that historical struggles over how to know, discipline, control, and mobilize plants in Tanzania were fundamentally struggles over how to articulate *uganga* and *uchawi*, healing and harming.

The tensions that define this history and the friction in the questions, when raised, surfaced in the accounts of Babu and the cup of life, as above. Babu's position as a pastor and his ability to mobilize strong Christian rhetoric worked to preempt any accusations of witchcraft. It was also the platform from which his refusal to register was more fundamentally a refusal to position himself as knowing subject (healer) with mastery over a material object (the *mugariga* tree's therapeutic properties). Babu welcomed the state scientists and gave them his formula to take to the laboratory. The friction did not result as much from any active resistance to being subjected to the power of the state as it did from being subjected to the power of God. Babu argued that he and his cup were, in fact, physical manifestations of, or channels for, God's power. In this way, he challenged the state's efforts to locate agency in him and efficacy in the plant. Without a subject who knows who could be owner, and without an object that can be known and therefore owned, the relations Babu depicted troubled efforts to create property out of the therapeutic qualities of the formula. Yet nothing Babu did could prevent the state's assertion of dominion over a different history that they used to explain the root Babu was boiling for his medicinal tea. *Mugariga* reemerged in the scientific reports as *Carissa edulis* and as a plant regularly used by Maasai when cooking meat. The telling of this alternative narrative retroactively gave the plant an ethnobotanical history. Journalists and anthropologists followed suit, referring to *mugariga* as a "well known traditional medicine."[33] This "retrobotanizing" joined broader efforts to transform therapeutic relations with plants into a national resource for development in an effort to manage the constitutive ambivalence of healing practices that are also a threat to state control.[34] The next section details the history of the postcolonial Tanzanian government's efforts as they were embodied by those who traveled to Loliondo to negotiate with Babu and went back to the lab with his roots.

## Institutionalizing "Traditional Medicine"

Like the colonial category of "native medicines" that preceded it, the postcolonial concept of "traditional medicine" articulates a relationship between expertise, knowledge, and property that is driven by economic logics. The articulation of these logics in postcolonial legal statutes, medical research, and public health initiatives frames the land relations relevant to health care as those that exploit natural resources for national health. From this perspective, the three-tiered registration system, which was called into being by the 2002 Traditional and Alternative Medicines Act and rolled out in 2011, is only the most recent in a series of efforts over the past sixty years to control healing by institutionalizing traditional medicine. This system's efficacy, therefore, is not tied to the faithfulness with which it reflects the multiplicity of modes of healing on the ground or recognizes the ways that historical continuities with precolonial practices shape diverse notions of healing. Rather, like the registration systems before it, this new system aims to intervene in the complicated relations through which healers, plants, and place are seen to emerge in ethnographic work. It aims to establish more elaborate legal subjects (healers) and to render a priori their difference in kind from the objects they mobilize (medicines) and the contexts in which they work (premises). This epistemic and ontic work of the registration system refines earlier ways of representing the labor of healers and plants available to the state as resource and articulating place as background to human agency rather than as a dynamic of human–nonhuman relations. In this section, I trace the history of traditional medicine's institutionalization to show how it has delimited the sphere of political possibility for healers and censored alternative ways of both knowing through relation and being-with through place. I offer this history as critical to the broader genealogies in this book harnessed to account for how *dawa lishe* pushes against the limits of modern traditional medicine, even as it draws on African therapeutic practices in striving to articulate healing as, and through, lushness.

After gaining independence in 1961, the country's socialist government prioritized building a national health care service. By the 1970s, it was clear that, even while spending a substantial percentage of its foreign currency reserves on pharmaceuticals, Tanzania could not properly stock the new hospitals, health centers, and dispensaries that it was building together with its citizens. The political will to investigate African traditional medicine grew in large part out of frustration with the cost of European and American pharmaceuticals and with the patents that protected monopolies on their production. Traditional medicine and, more specifically, research into therapeutic plants promised a path toward the sort of "self-reliance" central to Nyerere's vision of African socialism, as put forward in the Arusha Declaration.[35] In this context, the newly independent government began experimenting with bureaucratic technologies that would render healers and their plants available to the state as resources for scientific development.

Initially, healers retained the uncertain legal status they had had under German and British colonial administrations, continuing to be neither explicitly protected nor prohibited. The colonial Witchcraft Ordinance, with only slight modifications, rolled over into national law at independence and remained available to officials striving to control healers deemed a threat to the state or disruptive to the social order.[36] The Medical Practitioners and Dentists Ordinance referenced traditional medicine and traditional healers as exceptions to broad guidelines for medical practice; however, it did not attempt to articulate their rights or responsibilities. The Ministry of Culture and Sport assumed responsibility for the registration of healers. In the early 1970s, the state tried to organize traditional healers through a professional association, Umoja Waganga wa Tanzania (UWATA). Yet, the deep divisions among healers that had developed under the colonial government's strategic and often violent efforts to separate healers from their social bases of power undermined this first postcolonial effort to organize the healers. Jealousies and rivalries intensified. Group dynamics were further complicated by the postcolonial state's privileging of plant matter as central to the generation of (scientific) knowledge and (economic) value, as well as the desire to decenter the power of ecological and social dynamics, including those with ancestors and spirits. Fed by distrust, infighting among healers grew, and cooperation quickly dissolved, as healers jockeyed for control in this new organization. Within a year, the president of Tanzania banned UWATA.

Given the difficulty of organizing healers via professional associations, the government turned its energy toward investing in research to control the *materia medica* used by healers. Energized by a Pan-Africanist vision, President Julius Nyerere proposed scientific research as a first step in developing modern medicines that could supply the network of clinics that formed Tanzania's own national health care system. He imagined that the resulting medications could also be sold to Tanzania's newly independent African neighbors, thereby stimulating the regional economy in ways that would contribute to greater sufficiency and growing sovereignty. In 1974, Nyerere officially opened the Taasisi ya Dawa za Asili (Institute of Traditional Medicine) in Muhimbili Medical Center, the primary teaching and research hospital in the country. This department was explicitly established to investigate plant, animal, and mineral substances used by healers in Tanzania in order to identify leads for medical research that would facilitate the construction of a national pharmaceutical industry. Therefore, the methodological strategies enacted in the Taasisi examined the pharmacological effects of substances that might contribute to the development of modern medicines. Scientists were not focused on nutritional benefits that might articulate the value of foods, and there was no frame from which to consider the land relations mediated by the substances they examined.

Even after the disbanding of UWATA, then, the state continued to view healers as custodians of traditional knowledge that might be mined for ethnobotanical findings that could justify investments in laboratory studies. In addition, with the ratification of the Alma-Ata Declaration in 1978 and the extreme shortage of staff in Tanzania's new system of dispensaries and clinics, "traditional practitioners" were identified as a local untapped resource who could be mobilized to help develop primary health care.[37] This invigorated a focus on healers, and their medicines created a (brief) window in which medical anthropology played an integral part in organizing state research into traditional medicine in Tanzania. Margaret Deval, who had been an administrator at the University of Dar es Salaam, received government support to study for a PhD in medical anthropology.[38] Upon graduation, she became the first medical anthropologist at the Taasisi. She worked closely with the botanist Rogasian Mahunnah to conduct studies that recorded healers' knowledge of plants and therapeutic preparations in the service of the state's later laboratory investigations.[39]

In 1979, only five years after the Taasisi at Muhimbili opened, an act of parliament formed the NIMR and tasked it with promoting and carrying out medical research, including "research into various aspects of local traditional medical practice for the purposes of facilitating the development and application of herbal medicine."[40] For several decades, the NIMR ceded most active research on traditional medicine to the Taasisi. This research was focused on making traditional medicine a national (and natural) resource.

Insofar as traditional medicine is a modern category of knowledge and practice that is rendered intelligible through its institutionalization running parallel to that of biomedicine, it is inseparable from the history of the modern state's economy. During the 1980s, Tanzania experienced increasing international pressure to move away from its socialist policies, including its nationalization of many industries, price controls, and villagization.[41] Fluctuations in the international economy, a war with Uganda, significant drought in the mid-1970s and again in the mid-1980s, and the burdens of building a modern state after colonialism had combined to overwhelm these efforts at "self-reliance." Nyerere attempted to mediate the growing pressure on his government through a National Economic Survival Program in 1981 and a Structural Adjustment Program in 1983. Yet, amid growing economic concerns at home and intense international pressure animated by the ideologies of the Cold War, he stood down as president in the 1985 general election. His successor, Ali Hassan Mwingi (1985–95), quickly embraced a program of economic liberalization.

This new economic environment further consolidated the focus on therapeutic plants as a central material through which to negotiate the shifting alignments of science, law, and capital. In 1989, the mandate to register traditional healers shifted from the Ministry of Culture to the Ministry of Health. The task fell first on the small Office of Traditional Medicine in the Ministry of Health. In 1991, as the last socialist Minister of Health in Tanzania stepped down, the Taasisi at Muhimbili was upgraded from a research center to a research institute (the Institute of Traditional Medicine, or ITM). ITM has remained dedicated to the scientific investigation of traditional medicine.[42]

In 1992, as the country grappled with the growing consequences of structural adjustment policies and the realities of neoliberal economics, the government convened an advisory committee on traditional medicine. This committee was tasked with designing institutional structures that would facilitate the disciplining of both human and plant labor so that they might be mobilized more effectively as resources for the state. NIMR staff participated in this advisory committee, as well as those from the Ministry of Forestry, the Ministry of Health, ITM, and others. Professor Mahunnah, whose ethnobotanical work shaped the early years of the Taasisi, chaired this effort. While the intensity of the committee's work fluctuated over the next decade, its conclusions about how to manage healers and plants as resources for national development came to shape the content of the Traditional and Alternative Medicines Act, which the Tanzanian parliament passed in 2002.

The 2002 act shows that, as the state's interest in therapeutic plants expanded, so too did its need to regulate institutions capitalizing on plants, both as leads for pharmaceutical development and as products in a growing global herbal industry. International initiatives like those facilitated by the World Health Organization's (WHO) Program on Traditional Medicine energized efforts to develop this act and its counterparts elsewhere on the continent.[43] The specific wording of Tanzania's 2002 act also built on a model law that the Organization of African Unity (now known as the African Union) had developed. With its ratification, Tanzania found itself a leader in formulating legislation that addressed traditional medicine on the continent.[44] Professor Mahunnah, as a professor of botany and the head of the Taasisi, as well as chair of the governmental advisory committee on traditional medicine, was invited to international meetings to describe how Tanzania had developed and successfully moved the act through parliament. During this time, he transitioned from chairing the governmental advisory committee to chairing the Baraza la Tiba Asili na Tiba Mbadala (Council of Traditional Healing and Alternative Healing), the implementing organ that Part II of the 2002 act had called into being. From this position within the Ministry of Health, he has continued to work internationally to develop intellectual property law in Africa as it relates to promoting and protecting African medicine. This work strives to align science, law, and economy in ways that will manage the constitutive ambivalence of healing in the service of national and regional development.

Since 2005, the Baraza, as it is widely known, has assumed responsibility for monitoring, regulating, promoting, and supporting traditional medicine. The Baraza was explicitly mandated to implement the registration of both traditional and alternative healers, health delivery facilities, and "raw" traditional medicines (Part IV of the act). Creating the infrastructure for this new, more involved registration system that regulates healers, their medicines, and their premises, as well as beginning the process of registering, has been a primary focus of state efforts concerning traditional medicine over the past fifteen years. It strives to capture practices that cannot be regulated as drugs or foods by the Tanzania Food and Drugs Authority (TFDA) and render them legible through the logics of the imperial pharmakon, even if as raw materials, natural resources, and indigenous knowledge. To this end, between 2005 and 2010, teams of scientists, healers, and government employees brought together other "stakeholders" in a series of meetings to develop the details of a new registration system. The act mandated a registration system that mirrored the form of certification and control that existed for biomedicine, including mechanisms to register practitioners, their premises, and their medicines in succession. This bureaucratic rationalization solidifies the modernist divides between knower and known, subject and object, physical and the moral or ethical, form and context, and body and land that the plant(ing) remedies highlighted in this book unsettle.

In addition to the Baraza, the 2002 act compelled NIMR to create a Traditional Medicine Research Unit. In 2008, twenty-nine years after NIMR was established, it opened two sites dedicated to traditional medicine. The largest, in Dar es Salaam (known as "Mabibo" after the neighborhood where it is located), is staffed primarily by chemists and pharmacists. It focuses its work on developing prototype herbal products.[45] The second site, the Ngongongare Research Station, in northern Tanzania, was acquired by NIMR from an Italian development organization the same year. The scientists assigned to this research station inherited relationships with five "traditional healers."

While competitive struggles with the Mabibo lab limited Ngongongare's phytochemical investigations, researchers there have been proactive in experimenting with relations between healers and scientists, as well as the sorts of entanglements and obligations that might lead to knowledge.[46] John Ogondiek and Victor Wiketye, as I describe in more detail below, developed uniquely collaborative relations through which the limit points of traditional medicine, as articulated in the 2002 act, could be felt. They cultivated a space in which the incommensurabilities between so-called modern and traditional healing could be held jointly by researchers and healers, insisted on engagements that recognized the inextricability of healers and their medicines, and remained in ongoing conversation with healers over what sort of knowledge was needed and what ways of knowing were ethical. However, this work remained relatively invisible.

The ITM, the Baraza, and NIMR all work in conjunction with other regulatory offices. Science—including the identification of therapeutic plants, the elucidation of chemical compounds, and the formulation and standardization of remedies—motivates governmental policies, bureaucratic procedures, and legal precedents about traditional medicine. Intellectual property protections can only be sought for registered medicines. Both NIMR and ITM seek out the TFDA to prepare for registering their herbal products as a first step before seeking patents for their formulas and trademarks for their products. In turn, their work pushes the TFDA and other regulatory bodies to develop the necessary procedures and protocols to address herbal products in Tanzania. The TFDA coordinates with the Tanzania Bureau of Standards (TBS) and the Business Registrations and Licensing Agency (BRELA). The former develops standards and certifies products as consistent with those standards, while the latter facilitates intellectual property applications and mediates the registration of patents, trademarks, and geographical indicators. The interlocking work of these institutions shapes relations among knowledge, policy, law, and economy. Together, they have strengthened the gatekeeping role of experimental science in legitimating traditional medicine in postcolonial Tanzania by concentrating the adjudication of plants' therapeutic value in the laboratory. In this way, plants remain the primary site of drug exploration, laboratories continue to play an authoritative role in producing scientific and economic knowledge about plants, and intellectual property regimes increasingly circumscribe debates about justice in postcolonial Tanzania. Land relations that engage matter as its relations and articulate knowing in (inter)action are rendered invisible as science, economy, and law collaborate to confer legitimacy only to those forms of efficacy located inside plants' biological and chemical capacities and knowledge inside the heads of scientists who elucidate these plants' internal capacities.

The new registration system might best be understood as a bureaucratic and legal technology that is designed to connect and coordinate the work of these institutions by rendering healers and plants legible and discerning which relationships and practices are legitimate.[47] Its power lies in its potential as an apparatus through which "persons and things can alter from themselves,"[48] as will be shown even more vividly in the descriptions below of efforts to implement the registration system. Plants become *materia medica*.[49] Healers become custodians of knowledge, which may be shaped into a professional position. The places in which plants are collected recede into the background, and clinics become merely context. These transformations consolidate a relationship between expertise, knowledge, and property that was initiated by the mobilization of the imperial pharmakon and is embedded in modern notions of toxicity. Articulating power as inhering in these transformations reveals how institutionalizing traditional medicine secures the social, economic, and institutional bases of healing that the Tanzanian state promotes. The next section traces the details of how these relations are established bureaucratically. *Dawa lishe* strives to name efforts to unsettle the investments in knowledge, expertise, and property that the state has sought to render axiomatic. This section and the next establish the context in which EdenMark, TRMEGA, and other innovative projects craft a vocabulary to apprehend the ongoing impact of colonial dispossession through chronic illness, frame justice as the dismantling of forces of depletion, exhaustion, and exploitation, and attend to healing as practices to (re)configure relations through lushness.

## Scaling the Traditional with the Modern

The first phase of the government's efforts to roll out the 2011 registration system focused on the registration of healers and, to a lesser extent, their assistants, allowing the issue of the healers' premises and medicines to sit in the background. As discussed above, the 2002 Traditional and Alternative Medicines Act tasked the full-time registrar of the Baraza with implementing a system to approve and record healers and their premises. The legislature did not budget for additional support staff, and one person could not register all healers and premises in Tanzania themselves. The national registrar, therefore, enlisted provincial and district-level Ministry of Health staff to appoint Traditional Medicine Coordinators. The infrastructure of traditional medicine is built on the existing institutional and administrative scaffolding of the national health care system. In this way, the bureaucracy for traditional medicine not only mirrors that of biomedicine but also parasitizes it.

Existing biomedical staff took up or were assigned the role of Traditional Medicine Coordinator at each level—district, provincial, and regional. Most often, the responsibilities of Traditional Medicine Coordinators were added to those of the social welfare officers within the Ministry of Health system. When I first reached out to the provincial offices in Kilimanjaro in 2013, however, the provincial coordinator was a pediatrician whose workload at the district hospital already exceeded her resources. She did not experience the additional role as an honor. Instead, she complained of the devastating consequences of children arriving at the hospital "too late" after they had been attended to with herbs and by healers. In these complaints, healers appear as obstacles to effective biomedical care. They serve as evidence for her argument that the Traditional Medicine Coordinator's responsibility is to control and restrict healers' work. The only positive role she envisions for healers is as community members who are well positioned to facilitate getting sick children to the hospital more quickly. Her approach exemplifies the bias that inheres in scaling traditional medicine's representation structure through existing biomedical staff. Already overstretched staff, with little or no interest in traditional medicine, and at times with explicit hostility toward healers and their treatments, were saddled with serving as a Traditional Medicine Coordinator at either the provincial or the district level.

Victor Wiketye and John Ogondiek, with whom I explored healers' uneven experiences of the 2011 registration system, discussed these issues at length with healers and each other. Victor and John had assumed the leadership of the NIMR Ngongongare Research Station, with its focus on phytochemistry and traditional medicine, in 2008 when it was passed to the Tanzanian government by the Italian NGO partnership that had originally built it to facilitate their own research into traditional medicine. Neither Victor, as director, nor John, as senior research officer, however, were trained as phytochemists. Victor had originally been hired by NIMR as a veterinarian to manage the animals used in clinical trials. John had been trained in Russia and Ukraine as a pharmacist. Still, they were creative and energetic and hoped that productive work would be possible in this northern research station. Both deeply curious, they immediately began cultivating their own relationships with the healers with whom their Italian predecessors had collaborated. In 2009, the prime minister's suspension of all traditional healing disrupted their work, rendering their relationships illegal at worst and informal at best. The suddenness of the announcement left them disoriented. Given their positions, however, the Ministry of Health pulled them into efforts to devise the new registration system, and they were on the front lines as the new system was rolled out. Both expended considerable professional and personal energy to support the healers working with the NIMR research station to register. Yet, two years after the registration system was in effect, none of the healers with whom they collaborated at NIMR had yet been able to successfully register.

In a creative attempt to draw attention to the problems of implementation, Victor and John hosted a demonstration table for one of the healers with whom they worked closely, Mama Fatuma, at the exhibition tent at the 2013 Annual Conference of the NIMR held that year in the Snowcrest Hotel in Arusha. They hoped that the irony of her presence at the professional conference under the banner of the NIMR, even as she was officially working unregistered (and therefore illegally), would highlight the need to better understand and attend to obstacles in the registration system's rollout. Mama Fatuma dried, ground, packaged, and labeled her medicines for the occasion. During the conference, participants frequented her attractive table, including the vice president of Tanzania who, after presiding over the conference's opening, was escorted around the exhibition tent by the director-general of the NIMR. Over the several days that the scientists and bureaucrats discussed national priorities for medical research, Mama Fatuma enjoyed brisk sales. While Victor and John appreciated the irony of the situation themselves, they feared that it was lost on the participants, who were simply excited to try her medicines to relieve their pains and attend to debilities.

In order to look more closely at these dynamics, Victor, John, and I explored how the registration system was implemented in Meru and Rombo, two districts in northern Tanzania (the first in Arusha and the second in Kilimanjaro).[50] We chose Meru, the district in Arusha where the Ngongongare Research Station was located and where Victor and John's efforts to collaborate with Mama Fatuma and other healers' efforts were foiled by their collective inability to complete the registration process. We saw from the national rolls that the stalled registrations that discouraged their work were not unique. Official registration numbers for Meru were particularly low. For contrast, we examined how the registration system was being implemented in Rombo, a district in Kilimanjaro. The national rolls showed an unusually high number of healers registered in Rombo, and the enthusiastic Traditional Medicine Coordinator's dynamic campaign had received national attention.

Between 2013 and 2014, Victor, John, and I jointly interviewed two district and two provincial Traditional Medicine Coordinators (the Meru and Rombo district-level officials and the Arusha and Kilimanjaro provincial-level officials), and we spoke with a range of officials in national headquarters. We also conducted eight extended, in-depth focus group–style conversations in Meru and Rombo. The groups ranged in size from four to fourteen people, and these conversations lasted from one and a half to three hours. The people who participated in our conversations were identified through a series of connections, including healers' associations, Traditional Medicine Coordinators' personal knowledge, our connection with the NIMR, and word of mouth. Not all were officially registered with the state as traditional healers. We were interested in these differences by district and in variation between individuals who had and had not registered.

We found that healers in both districts were thoughtfully weighing the potential risks and protections of registration. Both those who had (already) chosen to register and those who had not (yet) chosen to register suggested that the primary benefit was the protection that they imagined came with recognition by the state. Some healers linked registration and "being known" with "freedom" (*uhuru*). Recognition, they reasoned, would confer legitimacy. Healers mentioned that if they were "known" by the state, they could apply for loans at a bank, coordinate their treatments with doctors and health centers, and treat desperately ill people without being afraid of the consequences if anyone were to die in their care. They also anticipated that, as legal subjects, they would be entitled to legal protections. As one healer suggested, "Very sick people come to see us. One could die on our premises. If we are registered, the police will understand that we are healers. They will not accuse us of murder." The brutal attacks against people with albinism hung over this comment. Healers feared that the police would be suspicious of all healers, and courts would rush to decisions to make it look like the state was responding to the attacks. The violence then not only created the context in which the government had suspended all healing and bureaucratically established a "clean slate" across which to roll out the new registration system, but also it continues to shape relations between healers and the state.

The horizon of freedom is always determined by the social and material constraints of the position from which it is sought. Being known and legible, healers speculated, would help to protect them against the range of possible accusations that officials level against healers. The targeting of healers who are seen as challenging government control has a history that reaches back to German colonialism, was continued by the British, and continues as a tactic of the postcolonial Tanzanian state. Against this ongoing targeting, healers who decided to engage the state through registration did so to protect themselves from the state's violence. Yet they were also acutely aware that "being known" drew them into legal, bureaucratic, and therapeutic obligations that limited the scope of their practice and challenged how they could (legitimately) navigate relations between healing and harming.

## Registers and Refusals

Healers, in grappling with whether to register and what it means to register, highlight the ways in which the recognition of traditional human subjects (traditional healers) is tied to the recognition of modern nonhuman objects (plants, as they are conceived in botany and biochemistry). The colonial tactics that separated African healers from their social basis of power might also be said to have separated plants, ancestors, and other "things" from their social basis of power and dismantled the conditions of *dawa*'s efficacy.[51] Each move required the other, and each emerged over time in efforts that tacked back and forth, as institutional links between the recognition of healers and the recognition of plants developed. The latter emerges alongside the recognition of the chemical entities that structure proof of plants' safety and efficacy (or lack thereof). These forms of recognition, and the ways that the state structures the relationship between healers and plants, have been critical to eliciting traditional medicine's participation in representational forms of knowledge and politics. Insofar as the most recent effort to register healers, medicine, and premises is an ongoing effort to align (and institutionalize the alignment of) these techniques of recognition, it renders alternative forms of knowledge and politics intelligible only as religion, magic, fraud, or senseless violence. *Dawa lishe* names the efforts to cut a different path through this history.

Some healers who attended the focus group discussions that John, Victor, and I held—particularly those who lived and worked in Meru—explicitly "refused" the forms of subjection embedded in the demands of the registration documents. After a long, rather detailed conversation about the benefits (*faida*) of registration, Victor, John, and I realized that the healers were speaking hypothetically about the advantages they imagined could, in the best possible world, come from registration. Yet no one had registered. No one had tried to register. As the secretary of the healers' association explained the resistance that the registration forms generated among members, he said, "*Tulizikataa.*" In Swahili, this is most often translated as "we refused, rejected, or gave up on them," or occasionally, to capture the emotions, some might translate *Tulizikataa* as "we despised them." As a fellow member recalled, "When those forms came, they came by way of the secretary [of the local healers' association]. When the secretary phoned me, I went to read them. [And] I refused them."[52]

As the conversation continued, the healers slowly began to unfold the deeper epistemological and ontological issues that they saw the documents laying bare. As one *mzee* (elder)—who had first begun to apprentice under his grandfather's instruction in 1995 and, by the time of our interview in 2014, had become a well-established healer in his own right—deduced: "The government is not yet ready to recognize us. If they were willing to recognize us, then they wouldn't put in place such tough regulations, ones that would prevent us from being able to do our work." They saw the obstacles they experienced as embedded both in the content of the forms and in the process of engaging with the Traditional Medicine Coordinators. As another *mzee* asserted, those who sent the forms, took the fees, and evaluated and approved the application do not "mean [healers] well." One man, who described himself as specializing in treating stomach hernias, noted, "I myself still could not go to pick up these forms because these forms remain inadequate. The community around us does not appreciate traditional healers. Because we are living where others have some different beliefs, when some are told, 'This person is a doctor,' some see an *mchawi*, do they not?"

In one of our focus groups, after we had begun to build some rapport, an elder healer who had described himself as specializing in the treatment of male impotence, female infertility, and *tambazi* (literally "creeping" or, as he translates it himself, "rheumatism") turned to Victor and John. He appealed to their previously expressed interest in his and other healers' work and knowledge, even as he recognized Victor and John's role as scientists at the NIMR Ngongongare Research Station and, therefore, as government employees. He asked them, "When you read your regulations, which professional tools (*vitendea kazi*) do you see the traditional healer described as being able to use? The government has told us not to divine (*kupiga bao*). Now, how will we know the [state of the] patient? When a doctor is evaluating the patient in the hospital, she uses devices such as those she puts in her ears. She examines the [patient's] lungs. For a traditional healer, he also has instruments that he uses to assess the problems of the patient. Which tools do you see that the traditional healers are supposed to use?" This healer drew an equivalence between kinds of knowledge, comparing the ways that biomedical doctors know to his use of divination as technologies through which he comes to "know the patient." In this way, he framed the process of enrolling healers and plants in the state's version of traditional medicine as also a process of unknowing older practices of healing, their relations with the past, and their social-ecological basis of power.

Colonial and postcolonial debates over healing (*uganga*) and witchcraft (*uchawi*) have shaped contemporary practices of what the elder healer above called *kupiga bao*, as well as other methods of divination. When divination calls out specific individuals as the perpetrators of harm (witches, *mchawi*), it is seen as inciting violence and is therefore illegal. In the past half-century, healers in Tanzania have devised new ways of intervening in complaints without naming particular individuals as those who "sent" illness or misfortune. When legal debates reduce divination to the specificity of the accusations and strive to discern which speech should be governed as violence, however, they tend to miss the substance of the work of divination.

*Kupiga bao* or *ramili* and dreaming[53] describe a process and a space for engaging with ancestors and spirits—a relation that not only embodies a mode of coming to know but also constitutes knowledge. Knowing a patient, and knowing plants, is a coming or drawing together of the relations through which they are comprised. That is, coming to know through divination and dreaming is not a process of coming to possess information (to know *about* a patient or plants). Rather, divination and dreaming are paths by which healers and patients are drawn into immediate and palpable relationships with the forces that animate or threaten the patient's wellness and the plant potency. They work by opening a space in which healers might wrestle, coax, or cajole ancestors, spirits, and other forces that animated salient pasts and futures and where they might therefore transform the relations that cultivate both a patient's bodiliness and the ecological vitality, which makes such bodiliness possible. These are modes of knowing that might find resonance with what public health officials refer to as "social determinants of health," if we expanded the social in this phrase to include nonhumans. Divination invites understanding healing to include reparations for harm to lineage, community, and ecology, as well as injury to individual bodies. It provides a glimpse into the relations obfuscated by the imperial pharmakon's articulation of toxicity, remedy, and memory. As healers continue to argue for the value of divination—even in the circumscribed and vulnerable position currently permitted by the modern state—they point to alternative ways of apprehending the relationship between harming, healing, and the past-alive-in-the-present.

Another elder who had been practicing since 1974 patiently explained in response to John's questions that "some practices are necessary"; that is, they are defining of that which constitutes healing. Such essential practices include those extended in the passing of the medicine bag, the process through which that capacity to be a healer moves through the generations. When the state does not recognize these practices, registration is not a simple means of representation; rather, registration becomes an intervention into what is recognizable. This elder described feeling the state's effort to change healing bear down on him and his successors as he faced the registration forms.

> You know, doctor, this traditional healing is not a form of healing learned in school. No one, I think, among us here . . . nobody went to class to learn these traditional practices. Indeed, this is the meaning of calling it traditional healing (*uganga asili*). One inherited this knowledge from their grandfather, another inherited that from their home or lineage and still another was called by his father and told that "I am about to die now so I ask you to carry on for me this origin of mine (*asili*, meaning origin, tradition, source)." So, when the government enters this work, it

has to adequately address its people so that they want to collaborate. It is necessary that we do this and this and this. It cannot be that the government makes its regulations and comes to me and asks me just to fill in the forms even before I have understood what these efforts are all about: where they have started and where they are ending.

Healers understood the cunning of the forms. The forms are one mechanism that the state uses to articulate the trajectories through which healers might be defined as political actors, through which individuals are conferred the power to represent and to become representative.

While the Meru healers' refusals contrasted with Rombo healers' heavy silences in focus groups, the latter were not less likely to experience the epistemological and ontological gulf created by the state's distancing of divination and dreaming from traditional medicine. They did, however, approach this gulf differently, largely because of how the forms were mobilized in Rombo. The Traditional Medicine Coordinator in the district of Rombo wanted all who made or sold therapies to register: to list their names and addresses with the Ministry of Health and to be accountable to the state. He did not want what he saw as styles of healing or specialties to be an obstacle to registration. Instead, he privileged the ability to find any healer in his district if an incident arose in which they were accused of perpetrating harm. As a result, healers in Rombo were not encouraged to mention work in areas that could not be easily coordinated under the umbrella of traditional medicine. They were asked to list a few of the conditions they treated. An unspoken understanding emerged among those seeking to register: they knew to fill in the form with a few select details about their practice and leave others unstated. As a result, Rombo healers engaged with John, Victor, and me differently than Meru healers did. Whereas those from Meru urged us to become advocates for more subtle interpretations of *kupiga bao* or *ramli*, as well as for the active, agentive role of ancestral shades and *majini*, Rombo healers were more likely to deny such practices. We often felt that the conversations about divination and dreaming in formal group conversations in Rombo were stilted and rehearsed. While the Rombo Traditional Medicine Coordinator's approach enabled him to register healers in an area relatively efficiently, it did so by marginalizing relations with spirits and ancestors in state institutions. Yet they did not disappear in everyday life. Occasionally, healers would pull us aside afterward and reveal their use of divination or reliance on dreams. In fact, John received an ominous phone call from a healer one afternoon who had adamantly denied using *ramli* in the group the previous week. He claimed to have "seen" that something terrible was going to happen to John's sister (who lived quite a distance away in the middle of the country) if John did not enlist his help immediately! Such leaks and the uneasy laughs they elicited exposed the tensions catalyzed as the registration system sought to (further) dismantle relations between harming, healing, and the past-alive-in-the-present central to African therapeutics and to reinforce modern relations between toxicity, remedy, and memory.

Whether healers refused to register or whether they silently and strategically worked around the restrictions that inhere in registration, they were all well aware that the state sought to render impossible, mythical, fraudulent, and/or disruptive the intertemporal relationships that divination and dreaming potentiated. They recognize that the way to move through government processes is to explicitly articulate the centrality of plants. As a young Meru healer suggested, "For this reason, we should practice the ways of learning that link specific trees with the treatment of specific symptoms in a patient. There we will go along very well, and conflicts will come to an end and we will be appreciated and valued." In saying this, this healer identified the epistemological and ontological space ("there") where a plant's internal capacities address a patient's biological condition as a space of possible agreement. Learning to focus collaborations on botanical properties promises to avoid conflict and bring appreciation. Plants offered pathways for scientists to work. They served as boundary objects, transforming into raw materials to be converted into scientific knowledge even if the healer refused to fit themselves easily into the categories of the state and limit their role to that of a custodian of indigenous knowledge—as illustrated above as the NIMR scientists engaging Babu, his cup of life, and the *mugariga* root.

The three healers with whom Victor and John continued to work at the Ngongongare Research Station surfaced the links between their plant-based formulas and specific symptomatologies (if not also explicit biomedical diseases). All highlighted their work with plants when they approached Ministry of Health officials and identified plants as the objects around which they could sustain collaboration with their NIMR colleagues. Even if dreams and God figured in their practice, they knew to foreground their use of plants on the form, and they knew that sharing information about plants with Victor and John was the ground on which collective work was possible. You will remember that Mama Fatuma offered packaged and labeled herbal therapies at the NIMR conference described above. This was the effect of collaborations between Victor, John, and the healers with whom they worked that focused on cataloging plants, documenting their use, and correlating this use with patients' conditions. Together, they envisioned having a clinic on the grounds of the research station one day.

The development and promotion of traditional medicine as an institutionalized category of knowledge and practice is the product of the specific configuration of science, law, and economy through which the postcolonial state manages the constitutive ambivalence of the pharmakon, of *dawa*. The legacy of the imperial pharmakon as it has shaped modern notions of toxicity, as well as strategies for managing its harms, inheres in Tanzania's scientific organization, bureaucratic management, and legal codes. The state forges the grounds for the (political) agency of both healers and plants by dividing them from each other and from the context in which they co-labor. The 2011 registration aims to articulate even more clearly the subjects (healers), objects (medicines), and contexts (premises) of modern traditional medicine. Both healers and the Ministry of Health's Traditional Medicine Coordinators approach the epistemological and ontological gulf reinforced by registration in one of two ways. Some argue that practices illegible to science and law must be separated from those legitimating healers. Traditional Medicine Coordinators refuse to register healers they suspect of divining, and healers refuse to engage if they must hide their practice that centers connections with ancestors, spirits, or other unrecognized actors in order to register. Others recommend that the illegible practices be de-emphasized, suggesting that partial connections between healers and the state can begin a relationship through which (political) agency is imagined. This second strategy can be seen when Traditional Medicine Coordinators limit accepted answers on registration forms to healers' work with plants, and healers emphasize their role as herbalists in order to advocate for the right to practice. The effects of these strategies and their narrowing of the space and ways in which healers may act are evidenced in healers' laments that younger Tanzanians are less eager to "take up the medicine bag" than they once were; that is, they are less eager to learn and carry on their grandfather's or grandmother's *asili*. The elders not only complain that many youth decline to apprentice themselves and learn the plants, preparations, and applications but also bemoan that youth are not developing the kinds of relationships with ancestors, spirits, or disembodied others that shape such healing practices. Given that colonialization, missionization, and development have disassociated the "medicine bag" from regimes of power and governance, it is perhaps not surprising that many youth do not see this form of healing as a desirable pathway for building a future.

Yet this healing continues, and the state's efforts to incorporate and control these practices—through their efforts to separate and legitimize only herbalists, as well as cultivate popular conceptions of efficacy that are restricted to considering plants' internal chemical capacities—are always partial. Therefore, a second way in which healers and Ministry of Health officials navigate this gulf between healing and traditional medicine is by rendering invisible the potentially destabilizing histories incorporated into their practices. Some healers and Ministry officials—for example, those in Rombo, as described above—obscure relations that do not translate easily to, or coordinate easily with, scientific practice. In another instance, Mzee Elias read his ancestors' visitations in his dreams through his Christianity as the voice of God (similar to Babu above). Unsettling or disruptive relations are also concealed in the separation of the laboratory (as a space that generates expert knowledge) from its outsides (other spaces that generate everyday practical engagements). As I have written elsewhere, methodological strategies that complicate the simplifications necessary for formal phytochemical studies can result in ancestors that haunt laboratory hallways.[54] The techniques of unknowing relations with the past, ancestors, spirits, water, land, and nonhuman others that inhere in these strategies to articulate harming and healing structure dispossession, even as they render it invisible.

Healers and healing strain under the demands of the colonial legacies that have infiltrated research methodologies as well as the disciplines to which they have given rise and the institutions that support them. While moments of resistance, stories of challenges, and evidence of porosity can tell us about the workings of power, they do not in and of themselves engender new governmental futures. Traditional medicine is a difficult place from which to challenge the naturalization and universalization of biomedical notions of the body, time, efficacy, and property, because traditional medicine—as a modern category of knowledge and practice—is a product of these same histories. By contrast, *dawa lishe*, or "medicines that feed us," refuses the categories by which recognition has been ordered. Embedded in *dawa lishe* are critiques of modern medicine, industrial agriculture, and capitalism in its many forms—from colonialism through postindependence development. It does not rest at an account of dispossession, however. *Dawa lishe* seeks to reinvent possession and its links with the body, the environment, and social worlds. Simply expanding traditional medicine in an institutionalized form cannot realize *dawa lishe*'s potential. Producers and users engage with indigenous healing and therapeutic plants while simultaneously refusing to be defined—and therefore contained by—the indigenous or the traditional.

Some of those innovating the social-ecological-therapeutic projects that *dawa lishe* draws together register as healers; some do not. None see their work as traditional healing, even if they do talk of promoting traditional foods, seek out those they see as knowing traditional medicines, incorporate therapeutic plants in their herbals (and gardens), and promote their remedies as addressing health concerns. *Dawa lishe* is, in the national registrar's words, "difficult to categorize." This is because it seeks to escape the "cunning of recognition," as Povinelli calls the bureaucratic sleight of hand in which recognition hamstrings those not deemed "modern" or "cosmopolitan" by forcing them to enter into debates through their authentic performances of the traditional while simultaneously destroying the relations that support meaningful social and political power.[55]

## Toxicity and Remedy

This chapter recalls that traditional medicine was forged within a contentious history that displaced forms of justice founded on practices aimed at eliminating evil in favor of forms of justice based on evaluations of bodily integrity and legal rights.[56] Legal interventions into witchcraft, scientific interventions into native medicine, and bureaucratic interventions into local political structures all disrupted precolonial relations between people, plants, and places. Economic botany came to exemplify the emergence of plants as natural resources and defined by the usefulness of their biochemical characteristics, while intellectual property debates came to frame the sorts of fairness and kinds of value that could be claimed in contemporary Africa. These legacies of erasure continue to shape debates over the therapeutic value of plants and the terms through which questions of justice are posed in the defense, development, and promotion of traditional medicine.

The 2011 registration system for traditional healers, medicines, and premises called into being by the 2002 Traditional and Alternative Medicines Act defines the legal subjects and therapeutic objects of traditional medicine. Rolled out through the national health care service, this three-tiered system promotes forms of expertise, objects of knowledge, and contexts of knowing for modern traditional medicine that closely mirror those of biomedicine. Healers hold knowledge, medicines contain active ingredients, and premises facilitate the movement of bodies and the separation of pathogens. Registration, then, lays the groundwork for articulating differences that are critical to modern notions of harming and healing, toxicology and pharmacology, and murder and medicine. And it defines the obstacles. In 2018, I spoke with a newly minted MA in Traditional Medicine and Pharmaceutical Development from the Taasisi ya Dawa za Asili (Institute of Traditional Medicine) at Muhimbili who had just started working at the TFDA. He described his struggle to locate the sorts of toxicology reports and pharmacological research that he needed in order to register traditional medicines. While the failure of producers to list their ingredients on their therapies did at times make his work impossible, he was more frustrated with the dearth of studies on the ingredients that were listed. As a result, at the time of our conversation no traditional medicines from Tanzania had been registered with the TFDA. In addition, only eleven of the hundreds, if not thousands, of herbal products from elsewhere sold in Tanzania—from India, China, Yemen, Dubai, the United Arab Emirates, Kenya, and South Africa—had been registered with the TFDA.

Yet stopping here would miss the fact that Tanzania is witnessing an explosion of small-scale efforts to formulate, produce, and commercialize therapeutic foods and plant-based remedies. Some of these producers are (mis)translated through the process of registration as traditional healers; others reject this incorporation into the state. While some producers elude state regulatory organs, many commercializing therapeutic plants choose to register through the Small Industry Development Organization (SIDO) or as nonprofit, nongovernmental organizations. A few, like Alex Uroki, register as both a healer and a small business. Working across these two bureaucratic tracks for registration, these modern herbalists experiment with how therapeutic plants might be composed and decomposed, configured and reconfigured, in ways that sustain the flexibility required to continually escape being reduced to custodians of natural resources and managers of cultural assets. They seek a more dexterous and subtle position from which to respond to the ways that the past shapes the bodies and landscapes of the present.

*Dawa lishe* producers recognize these legacies as they take up plants as a critical nexus through which contemporary forms of power (science, law, and economy) are informed and therefore might also be transfigured. At times, they succeed in offering modes of healing and theorizing that resist the enclosures of nationally and internationally supported projects on traditional medicine. During my fieldwork, I learned to attune to the moments in which *dawa lishe* manages to interrupt pressures to conform to pharmacological techniques of managing toxicity and political techniques of managing the traditional: by troubling regulatory boundaries between foods and drugs. This unruliness, I argue, is stimulating Tanzanian experiments with alternative ways of generating therapeutic, ecological, and economic value and mediating their entanglements.

## Knowing Otherwise

Most of this book explores how movement across registers of knowledge and modes of attending to bodies and ecologies engenders an indeterminacy that opens up new ways of thinking about healing and sovereignty—new ways of enacting bodies and politics. This chapter, however, examines how these "registers of knowledge" came into being and identifies the techniques of enclosure that they facilitate. I locate the modern category of traditional medicine in a history of practices that have rendered plants medicinal. Colonial science first intervened in precolonial practices by reformulating the spaces in which plants' therapeutic value might be known. "Knowing practice" moved from hearth, home, and healer to laboratory, clinic, and pharmacy.[57] The postcolonial state, as evidenced through the most recent registration system, continues to refine these notions of who can know and what kind of things can be known by separating past and present, material and immaterial, subject and object, human and environment. These separations are central to conceptions of property and assertions of ownership that support modern forms of sovereignty. Yet, this version of knowledge systematically obscures the intimacy of plant–people relations. Readers may have noticed that it has been analytically difficult to stay centered on the intimacies of body and land in this chapter. The process of institutionalization is designed to render these intimacies "nonknowledge," in Agamben's sense of the word.[58] They can only be glimpsed in the stories. The broader argument of this book, then, might be seen as offering *dawa lishe* as a way of attuning to the possibility of such glimpses.

*Dawa lishe* recognizes solidarity among projects that cut across institutionalized registers of knowledge (food, drugs, traditional medicine), and it mobilizes each to pry open a space in the others; a space in which to think healing otherwise. Producers in small-scale entrepreneurial projects and nongovernmental organizations, through which their plant-based remedies extend, are using state scientific institutions and seeking legitimacy through both law and licensing. Their products tap regional appetites and engage global activist movements on permaculture, food sovereignty, women's empowerment, AIDS, disability, and more. Therapeutic foods and nutritious medicines also draw forward the forms, attentions, appetites, and rhythms of older modalities of African healing—plants that grandmothers and great-grandmothers used, formulas learned in visitations from ancestors, ritual tactics of address in public healing, gendered relations to particular plants and their planting. Producers and users engage in these diverse ways of knowing unevenly and creatively. They weave them together in ways that exploit fissures in the modern "order of things." They collapse categories of knowledge and mix methods of attending to the world in order to address the alienation of bodies from lands structuring colonial and postcolonial states and to surface the connections that biomedicine and modern agriculture obscure.

*Dawa lishe* troubles the modern category of traditional medicine by rekindling plant relations that exceed scientific logics and multiply forms of attachment.

The sense of solidarity it strives to recognize as a mode of healing lies in efforts to develop vocabularies and extend practices that draw people and land closer together. The experience of body-land being could be glimpsed in small gestures, such as one I had seen hundreds of times, but which only succeeded in capturing my imagination one afternoon in May 2015. I had gone with Mama Nguya and Jane to visit The Faraja Center, an organization founded and run by Martina Siara, which provides shelter and vocational training to young (under eighteen) single mothers and girls. We advised on the Center's large garden, and Mama Siara took the opportunity to have Mama Nguya and Jane speak to some of the girls there. Both women were, as always, charismatic and inspiring. The gesture that captured my attention, however, was humble and fleeting: one Mama Nguya never saw and Jane might well not remember. We were walking from the Faraja classroom through the small campus and around to the private garden of Mama Siara and her family. Our path took us along the side of a building where a narrow flower garden had been planted. Jane reached down and picked two leaves. The first was a weed that looked a bit like clover, but with broader, more triangular leaves. The second was an ornamental plant with purple and dark green leaves and small purple flowers. Jane casually plucked the leaves as she walked by and ate them. Later, when I asked her about this action, she laughed her easy laugh and told me that both were *dawa za tumbo*, medicine for the stomach. There was joy in discovering these little medicines unexpectedly in a flower garden, when one was on their way somewhere else. In twenty or so more yards, we arrived at our destination, the *mlonge* tree at Mama Siara's home. As Jane stood under this tall *mlonge* and joined Mama Nguya in advising how they might prune the tree both for fuller leafing and for ease of picking, she stretched up and grabbed a handful of *mlonge* leaves to eat. These mundane yet graceful gestures felt almost like greetings. Reaching up and taking in the very world she was moving through and the process of attuning or tuning herself to it.

A few days later, as my partner, daughter, and I were taking a particularly muddy Saturday walk in Machame (western Kilimanjaro, Uroki's home area), I saw the same clover-like plant and showed them that we could eat it. Three boys came bounding out of the bush to inform us that we should eat only the stems (Jane had happily eaten the leaves as well). They offered to guide us through the bush and show us something even better to eat. As we munched the slightly lemony leaves, we followed them down a narrow path until we came to a wild mulberry bush. All the children quickly switched to collecting berries. There is something very satisfying about knowing what one can eat and seeing the world as nourishing; the boys revelled in sharing that knowledge as much as they revelled in the sweetness of the berries. We all enjoyed not only being together, but also, collectively, taking the world into ourselves. Yet, these gestures of reaching, touching, plucking, smelling, and tasting are quotidian and subtle. It took a while for me to recognize them as remarkable and longer to recognize them as a material ground for knowledge.

For several years, each time I returned from Tanzania, I would stop in Croatia for a weeklong meeting of the Translating Vitalities group, a gathering of creative scholars, artists, and healing practitioners exploring the translations and transformation we experience as liveliness collaboratively. The landscape always jolted me a bit, as it was not like any I had known in North America or Africa. One afternoon, a friend and I were walking off the jetlag and reacquainting ourselves with the little town by the sea. Here, rosemary grows in unruly bushes five feet high along the road, and it edges people's yards and small backyard orchards. I would frequently reach out to pluck a piece of it absentmindedly as we talked and walked. I smelled it, tasted it, rubbed it on my skin. My friend stopped mid-sentence at one point and turned to me with a grin, asking why I was constantly reaching over the low stone walls into people's private gardens as we walked past and rubbing their plants on myself! It was that summer that I realized I carried back with me from my fieldwork a way of moving through and coming to relate to space through plants. Over the years, as I returned to Croatia and came to know these roads, yards, abandoned lots, and edges of the seaside, the range of plants I plucked expanded.

A year later, I was smiling. I had traveled to India with Mama Nguya, Rose, Dorcas, Uroki, and John. There is something electric about traveling with others to places with which none of you is familiar. Previously, we had only met when I traveled to Tanzania for fieldwork. My interlocutors took me in. Our relationships deepened and expanded with each visit, but in the context of them being the experts, me being the learner in Tanzania, and me returning to the United States to turn their expertise into my own through publications. This trip to India—an experiment initiated by an innovative administrator at Cornell University who wanted to develop structures to support less extractive, more just global partnerships—was a turning point. Since then, these friends and others have come to Cornell, we have met in Italy, and our plans for the places through which our relationship grows continue to expand. We have also developed collaborative projects, designed interventions, given public presentations, and co-published an essay together. But this story and my smile occurred on the first day that we were in Mysore, India, for the Engaged Cornell Healthy Foods, Healing Plants Partnership Convening.

We, the Tanzanian team, had finished breakfast together in the hostel and were walking up the road to the conference building to meet other teams from

India, Malawi, and the Dominican Republic. Our Tanzanian colleague, Dorcas Kibona, stepped off the road to examine the plants that grew along the shoulder, rubbing their leaves between her fingers, exclaiming with delight when she recognized one of them, and tasting a corner of a leaf she considered therapeutic. While the six of us ambled up the road chatting, these connections with the roadside flora felt like part of our early morning recognition of the new place in which we had woken up. Dorcas animated our excitement. I would not have thought of reflecting on this process of ingesting the landscape as an introduction to it—as taste and sight that motivated our meeting of this place—until Dorcas herself raised it later as we each introduced ourselves to the larger group. Dorcas's enthusiasm is contagious. She glows when she rises in front of a group; her charisma gathers a room of people together. Although the meeting would proceed mostly in English, each "team" included people who were not comfortable with this language. Dorcas introduced herself in Kiswahili, and our colleague, John Ogondiek, translated for her. She began by expressing her appreciation for being in Mysore and her hopes for collaboration. In doing so, she talked about walking from the hostel to the conference hall that morning and seeing plants that looked familiar. She noted that this part of India shares similarities with northern Tanzania. To elaborate her point or deepen the proof of it, she confessed that, on her way to the conference hall, she had picked some of these familiar-looking leaves, ones she knew to be medicinal, and had tasted them. John translated it as "and I chewed them." They taste the same, she concluded. There was a flash of surprise in the room. Some wore hesitant smiles. Confusion or shock passed across many of our Indian colleagues' faces. I felt that I could see them thinking, "Did she really say that she ate the dust-covered plants along the side of the road?" But Dorcas sped on with her introduction. The surprise faded, and I imagined that people assumed that they had misheard or that the translation was simply confusing.

During such visits, gestures that are so familiar they may seem unremarkable "at home" suddenly become notable events. By drawing attention to her own movement through the world, tasting, ingesting, and digesting, Dorcas not only offered an intimate recognition of geographic similarities but also demonstrated a mode of being in the world through which the body is continually made in/of place. This mode of being in the world, including the everyday gestures of becoming with plants, grounds ways of knowing. Knowledge is not a thing only mobilized in projects designed to change or transform. Knowing is a possible way of being-with that emerges from the actions through which one transforms and is transformed. Knowledge is these relations. With plants and other people. to know is to assemble, to collaborate in comings-together that shape the possibilities of endurance, healing, resistance, repair, and going on.

As we saw in chapter 1 ("Futures of Lushness") and chapter 2 ("Efficacy of Appetites"), *dawa lishe* works by kindling relations with plants that might reconstitute the body itself. Remedies do not stop at addressing a pathogen or an injured biological mechanism. They establish networks of relations—social and material—that bring forth that which has the capacity to be lively: individual humans, plants, homes, communities, and landscapes. Dorcas shows us that therapeutic knowledge, then, must be located in the bodily relations of the knower. It cannot be limited to a scientifically distanced articulation of the plant in itself. Dorcas was, after all, on her way to a workshop, where she was excited to build relations. She had applied for her first passport in anticipation of this trip. In this gathering organized by an American university with its professors and their partners, these relations were explicitly in the service of "knowledge." If knowledge is both social and material—always and inextricably—then it requires not only a meeting of minds but also the cultivation of bodily solidarity through co-laboring; that is, the collective work of people and plants as they have moved across Indian Ocean communities, reshaping landscapes in both East Africa and South Asia for centuries. Dorcas's gesture showed her openness to building bodily solidarity with the participants in the room: it argued that there is a solid place of familiarity from which to start our work together.

# 4

# Work of
# Time

In the *Ink of the Scholars*, the Senegalese philosopher Souleymane Bachir Diagne argues that time is not external to events, a context for living, or a long arc along which we might plot the development of various happenings. Time is not something out there to be managed; rather, time is the temporal insistence generated by engaging in action.[1] For Diagne, this insight is sparked by Engelbert Mveng as he reflects on the words of a "sorcerer" in Cameroon who told him that "man [*sic*] makes the time he needs." Diagne does not locate this provocation as a starting point to elucidate an "African" conception of time, rooted in something like "traditional medicine." Rather, he mobilizes Mveng and his sorcerer to inspire a re-reading of French philosopher Gaston Berger and the Hebrew prophets. He finds in them an argument that prospective vision—the capacity to foresee—is not a "seeing" into a future that awaits but rather an "acting" in the present. Diagne's work helps me to find in the innovative social-ecological-therapeutic projects of TRMEGA (Training, Research, Monitoring and Evaluation on Gender and AIDS) a search for the temporalities needed to address the durable bodily vulnerabilities and weaknesses glossed by biomedicine as chronic disease and treated with drugs for life. This chapter explores how TRMEGA's plant(ing) remedies engage biomedical therapies not only through chemical interactions but also through temporal interactions. Leaning into Diagne's insight, I approach *dawa lishe* as a (pre)position from which to ask: *What are the times needed to address healing in a toxic world*? What activities might generate the complex rhythms required to navigate the double-bind of modern lives dependent on the very toxicities that simultaneously undermine them? How might therapies do more than structure the consistent tempo of sustaining in the midst of chronic injury? Seeking answers to these questions trains attention on how plant(ing) remedies work with the friction between plant times, lifetimes, project times, and demographic times to generate affordances for interventions into the persistent depletion of bodies and land in postcolonial Tanzania.

Following TRMEGA's lead, I locate my account of the activities (re)shaping the time of the chronic in Tanzania in HIV/AIDS. "The chronic," as a temporal construction generated by epidemiological, clinical, and project work, offered a response to both problems of knowledge and problems of politics well before the rapid spread of HIV/AIDS in the 1980s became a global humanitarian emergency. I start this chapter with an analysis of the geopolitics of chronicity, figured through theories of the epidemiological transition, which suggest that the relative prevalence of infectious and noncommunicable diseases indexes the level of a country's development. As HIV/AIDS and its pharmaceutical treatments challenge this popular theory, it reveals chronicity itself as a public health technology for consolidating postcolonial temporal sensibilities. The advent of antiretroviral drugs transformed this fatal, infectious disease into a condition people could imagine living *with*. Acting-in-the-present to address HIV/AIDS generates a seeing-into-the-future (drugs for life) that offers newly dynamic formulations of time.

Yet, any individual experience of HIV/AIDS as a chronic condition depends on pharmaceutical access and adherence. The devastating impact of the virus not only motivated a global response but also rationalized interventions into intimate relations. The anthropologist and doctor Vinh-Kim Nguyen coined the term "experimentality" to account for the ways that NGOs, northern universities, hospitals, research institutes, churches, and pharmaceutical firms all strove to govern the lives of populations living with, or at risk for, HIV in the name of pharmaceutical efficacy. Life's rhythms need to reflect those of the lab for medications to produce predictable bodily responses, that is, to work. Mass

HIV treatment programs in Tanzania, as in other parts of Africa, restructured the clinic, the communities served, and individual lives. Rendering HIV/AIDS a chronic condition through pharmaceutical adherence required multiscalar (re)organization.

For many Tanzanians with and without an HIV/AIDS diagnosis, the scope of these interventions ties the experience of chronicity to debates over living through toxicity. Antiretroviral therapies (ARTs) organize collective reflection on the shared historical moment in which life (individual and communal) is only possible through the "toxicity" of such pharmaceutical regimes. The individual, collective, and institutional actions required to produce the experience of durability through adherence to these drugs structure a new relationship between time and toxicity. The making of the chronic through its relationship with the pharmakon continues to be reworked and refined as emerging disease specialists strive to elucidate links between HIV/AIDS, cancer, hypertension, and metabolic diseases. The entanglements of HIV/AIDS with the dramatic rise in the prevalence of noncommunicable diseases unsettle twentieth-century temporal sensibilities generated through theories of modernity rooted in infectious and noninfectious disease distribution. Yet, the dynamic activities making and remaking chronic time are in tension with experiential narratives.

In Tanzania, as elsewhere, pharmaceutical treatments rarely if ever secure a stable corporal experience indefinitely. Those living with a range of chronic diseases speak of the unpredictability of their day-to-day lives and anxiety about what the future will bring. After diagnosis, patients describe seeking the skills and creating the conditions to support living with the vagaries of their body. They focus on the creativity needed to meet the ongoing, yet always changing, challenges of fulfilling social expectations. Patient experiences reveal that therapeutic interventions for chronic disease do not conform to some version of Talcott Parsons's "sick role," in which sickness enables one to opt out of social and economic obligations for a discrete period of acute illness (what he calls sanctioned deviance).[2] Rather, a sense of wellness for those living with chronic disease surfaces through ongoing relations of care and accommodation. Chronic disease renders visible a search for temporalities to think and feel healing not as an achievement but as a quality of life.

When life with chronic disease becomes a struggle to adhere, wellness is forged in relation to structural obstacles to care. Public health studies in Tanzania and elsewhere in Africa identify cumbersome and expensive public transport, inadequate education, insufficient food, unstable living conditions, unsafe water, gender violence, poor health care services, and, more recently, climate change and environmental destruction as barriers to the consistent medical treatment that constitutes standard clinical care. Acknowledging lack and loss in social and material worlds outside of the clinic formulates problems that might be addressed through time-bound investments of humanitarian aid and/or development projects. Such a framing identifies the limited resources available to regularly adjust to physical and social challenges but sets beyond their scope full-throated critiques that articulate injury and illness as the embodiment of the dynamics of a historical moment—that is, instantiations of economies of abandonment, politics of disenfranchisement, and ecologies of self-devouring growth.

Efforts to articulate how bodies come to bear the attritional violence of extractive economies organized for capital accumulation have generated improvisational vocabularies. I find particularly helpful Lauren Berlant's offering of the term "slow death" to capture the persistent depletion that inheres when value is created through strategic abandonment, rather than defined by insistent attachment and strategic accommodation.[3] With this phrase, Berlant plays on the work of environmental historian Rob Nixon, who coined the term "slow violence" for those sources of deterioration that are so dispersed across time and space that they do not take shape as an event that mobilizes attention. Her work helps to articulate chronicity as the time of slow violence: "the physical wearing out of a population and the deterioration of people in that population that is very nearly a defining condition of their experience and historical existence."[4] The biomedical gaze often obscures these relentless depletions, narrowing the space of intervention to ever smaller universes inside the body and thereby naturalizing the forms of human and nonhuman relations that structure dispossession, barrenness, and exhaustion. *Dawa lishe*, however, seeks to elucidate these depletions by pointing us to social and therapeutic projects that seek to intervene in the forces draining human and nonhuman vitalities in postcolonial Tanzania.

Plant(ing) remedies strive to work against the social and material architectures of slow death that overwhelm any individual gesture of kindness and care, such as gifting a canister of herbal medicine for a woman wasting away in a barren courtyard and a pair of used shoes for her son growing up in the shadow of this mother's debility. In chapter 2, Simon's tears as he held his shoes and walked away from us in the market expose the frustration inherent in therapies that temporarily forestall death but do not generate more life. This limit point marked by herbal remedy as commodity—whether gifted or sold—is the starting point for developing *dawa lishe* as a practical and theoretical intervention. This chapter centers stories of plant(ing) remedies that work healing into the everyday spaces of life and lushness into the notion of home. By expanding the space of healing (to courtyards, gardens, and fields, for instance), TRMEGA members highlight the need to open healing beyond individual bodily well-being and survival to a broader vision of ongoingness.

I further develop the proposition of *dawa lishe* by describing TRMEGA's remedies as experiments with the times needed to articulate a meaningful notion of reproductive justice in the wake of AIDS. Plant(ing) remedies enroll bodies in rhythms that respond to but also expand beyond the tempos of pharmaceutical logics, public health demographics, and biomedical institutional coordination. They attend to the relational vitalities of humans and nonhumans. By working through and beyond the times of individual lives toward the continuance of lineage and land, TRMEGA, EdenMark, Dorkia Enterprises, and others offer remedies that work on the time(s) of the therapeutic. This experimental, temporal play cultivates modes of attention and forms of multispecies care that have been eclipsed in the clinic.

My fieldwork was in part a process of attuning to the rhythms and the temporalities of the forms of care I observed. As plant-based therapies move from the garden to the canister, as they are labeled and sold in shops, they work with the dynamics of economies and bodies familiar with managing bodily rhythms and capacities through pharmaceuticals. In this chapter, I draw on the reflections of TRMEGA members offered as they recount their experiences of using plant-based remedies, starting gardens, and contributing to the collective work of the organization. Their stories describe interventions that rework time at the intersection of the actions of pharmaceuticals and plants. TRMEGA's plant(ing) remedies choreograph temporal interruptions, layerings, and collusions. Their interventions stage engagements between people and plants, the actions that generate lifetimes and plant times. As members care for their own and others' bodies, they alter the times critical to the therapeutic. As Jane argued when explaining why her work through TRMEGA exceeds, without contradicting, the work of the clinic: Antiretrovirals (ARVs) repress the virus but do not nourish the body or the forces that give rise to it. They put the virus "to sleep," but they do not heal. Plant(ing) remedies, such as those of TRMEGA, generate new, hybrid rhythms along which being and health might unfold. The tempos and timescapes of plants focus attention on the dynamics of reproduction that pull bodies through land and land through bodies. The liveliness and lushness of gardens offer a different temporal experience through which to problematize the chronic and reimagine that which counts as healing.

The work of people and plants learning to nourish one another in these gardens disrupts the temporalities that render the slow violence of dispossession invisible. In so doing, it pries open the fault lines of modernity, undoing the normative conceptualizations of being and becoming in the postcolony. This is what makes *dawa lishe* more than another alternative or complementary therapy. Rather, it is a challenge to see both the immediate benefits and the long-term risks of defining chronicity as a quality that can be achieved through compliance with pharmaceutical regimes. It works across the distinctions between drugs and food, medicine and agriculture, to generate the times we need to heal both body and land.

## Unsettling the Geopolitics of Chronicity

In accounting for the ways that science, law, and economy forged temporalities that naturalized colonial dispossession, anthropologists have shown how narratives of development articulate the temporal horizons through which (racialized) hierarchies are maintained.[5] One particularly impactful example of how temporal frameworks in medicine not only described but actively engendered the (geo)political power dynamics is the theory of "epidemiological transition," which strove to interpret changing patterns of disease and mortality. This theory, articulated most prominently in Abdel Omran's 1971 article, "The Epidemiological Transition," mobilized the category of chronic disease to delineate levels of development and describe linear teleologies of progress. Omran was motivated by the fear that rising birth rates—what he and his contemporaries in the mid-twentieth century labeled the "population explosion"—would overburden the earth's "carrying capacity."[6] His epidemiological theory emerged from an urgent sense of the need to reduce global birth rates. He posited three evolutionary phases in demographic and health changes due to medical advances.[7] His origin point, the before-time in his analysis, was the "age of pestilence and famine," during which people experienced high rates of mortality, and population growth was negligible. Starting in the nineteenth century, Omran reasoned, people living in industrialized countries experienced a steady rise in standards of living, nutrition, and sanitation. These changes ushered in the "age of receding pandemics," for those in the so-called West, in which mortality rates declined and populations grew exponentially. The late nineteenth and early twentieth centuries brought the development of vaccines, antibiotics, and insulin. By the 1960s, when Omran was conceiving this theory, industrialized countries were seeing a decrease in the impact of infectious diseases on morbidity and mortality, as well as the emergence of chronic and degenerative diseases as the primary cause of morbidity and mortality. He hypothesized that this shift was the beginning of the "age of degenerative and man-made diseases," during which noncommunicable diseases increased, raising mortality rates and stabilizing population growth rates.

Omran's original hope was to motivate family planning and population control programs in developing countries in order to dampen the population growth anticipated as a side effect of their movement from "the age of pestilence and famine" to that of "receding pandemics." In the 1990s, as neoliberal economic policies spread through international development projects, his theory of "epidemiological transition" found new life as the dominant explanation for the rising prevalence of chronic diseases and their growing dominance in explanations of morbidity and mortality rates in "the West."[8] In both incarnations of Omran's theory, a specific articulation of changing patterns of morbidity and mortality in populations in industrialized countries is offered as an inevitable consequence, if not a goal, of development. In the second, chronic disease plays an explicit role in suturing biomedical logics into development rationales by tracing the arc of social evolution through the physical body.

On the ground, health practitioners navigate a much more complex burden of disease. Infectious diseases, including acute respiratory infections, diarrheal disease, malaria, tuberculosis, and AIDS, continue to plague populations, and new infectious diseases are emerging even as the prevalence of chronic diseases also continues to rise. Public health specialists have challenged the unidirectionality of this hypothesized transition.[9] The popularity of articulating health and disease patterns to describe the evolutionary character of demographic change among populations, however, remains in efforts to refine Omran's theory.[10] That is, chronic disease and its relation to acute infectious disease continue to serve as a site for rendering economic hierarchy and geopolitical power biological, even if the specifics become more complex.

HIV/AIDS, perhaps more thoroughly than any other disease, has upended the theory of an epidemiological transition.[11] Even as the impact of HIV/AIDS on mortality in Tanzania has decreased dramatically with widespread access to ARVs—dropping by 68 percent between 2009 and 2019 according to the Global Burden of Disease statistics and relinquishing its number one spot as the leading cause of death in Tanzania—HIV/AIDS remains the third most significant cause of death in the country.[12] This infectious disease still shapes health and health care in Africa.[13] The challenge that HIV/AIDS poses to even the most subtle versions of the epidemiological transition theory is not limited to (only) the continued high rates of infection. The HIV/AIDS epidemic also challenges assumptions, such as those embedded in Omran's writing, that chronic diseases are noncommunicable diseases.

Access to ART has enabled significant numbers of people with HIV/AIDS in Tanzania to lower their viral loads for significant periods of time. In interviews, many highlight the fact that as their strength returns, they are able to work, farm, cook, attend church or the mosque, care for their parents, and piece together the funds to send their children to school. In 2010, the World Health Organization (WHO) released recommendations for women receiving ART who were planning to become pregnant.[14] Previously, the emphasis on prevention through condom use or, in some programs, through abstinence precluded the possibility of health care providers counseling women with HIV/AIDS who were seeking to get pregnant on how to do so safely. The new protocols reflected adjustments that the WHO recommended, in light of evidence that women who had fully suppressed their viral load through the use of ART could be reasonably expected to give birth to an uninfected child. The effective suppression of viral loads made it conceivable that those living with HIV/AIDS might come to experience their condition as chronic rather than as fatal.[15]

Global inequalities of care for people with HIV/AIDS structure the actuality of people's experiences. Suggestions that "the age of AIDS" has given way to "the age of treatment" obscure the fact that many in Africa do not have meaningful access to treatment, and few find second- and third-line treatments available.[16] Furthermore, an insufficient and unreliable supply of ARTs, as well as inadequate distribution, complicates the possibility of mobilizing antiretroviral drugs as a means of preventing HIV. Only a few countries have enough drugs to use them in this way. Eileen Moyer and Anita Hardon argue that HIV/AIDS "remains exceptional in the age of treatment."[17] Others take up debates about the ability of ARTs to facilitate a return to the much-prized public health goal of a "normal life" in Africa or elsewhere.[18] The epidemic may have changed the meaning of chronicity, but not equally everywhere.

HIV/AIDS, as well as the ARTs that have emerged in response, have changed not only the complex social realities through which people are making lives and livelihoods but also the material realities of bodies. Critiques of the unidirectionality of theories of epidemiological transition cannot fully account for the entanglements of acute and chronic conditions. It is not only that the prevalence of communicable diseases continues to be high, even as the prevalence of noncommunicable diseases increases in developing countries, but also that strong categorical distinctions between infectious and noncommunicable diseases have obscured the dynamism of bodies with AIDS (and bodies on ARTs). Emerging disease specialists hypothesizing more fluid and complex interactions between microbial agents and noncommunicable diseases have established the relationship between infectious and chronic disease as a site of research.[19] These specialists point to associations such as that of *Helicobacter pylori* with peptic ulcers (see chapter 2) and gastric cancer. They also highlight studies that suggest possible infectious etiologies for cardiovascular disease and diabetes, conditions that are on the rise in Tanzania as they are throughout the continent.[20] The entanglements of HIV, hypertension, and metabolic conditions pose a particularly critical site of scientific investigation. In the wake of the AIDS epidemic, not only is the meaning of chronicity changing as it comes to be linked to pharmaceutical access and adherence, but also the very nature of chronic disease is being rethought.

It remains to be seen if this new research redefining the nature of chronic disease unsettles or reifies the geopolitics of chronicity. What is clear, however, is that AIDS continues to be the context in which the rising prevalence of chronic disease in Africa is understood both as a social fact and as a biological fact. The virus, as well as its pharmaceutical management regime, has reconstituted bodily realities as well as health care institutions to such an extent that public health specialists can no longer assume that even the most subtle version of the epidemiological transition articulates the demographic transitions of the late twentieth century. Researchers, clinicians, and broader publics are reflecting on the increasing prevalence of chronic diseases and wondering *what it is evidence of.*

## The Risk of the Chronic

As cutting-edge research into the relationship between HIV/AIDS, hypertension, and metabolic disorders unsettles this politics, it also opens a search for language to grasp the time of the chronic. The struggle this poses grew clearer to me one afternoon as I sat having lunch with an American biomedical doctor and chronic disease researcher from a top-tier medical school working in Tanzania. He turned to me at one point and asked, "*Is there no concept of chronic disease in African traditional medicine?*" Although his tone was congenial, and the lunch was convivial, the intellectual claustrophobia of the question's construction—what would it mean to answer yes or no?—conveyed his frustration. It also gave the question a rhetorical quality. Indeed, he did not wait long for a response but instead went on to reflect on the difficulty he has in convincing patients to adhere to the ongoing pharmaceutical treatments recommended for hypertension, diabetes, kidney disease, cancer, and AIDS. He lamented that some of his patients sought out alternative remedies for pain and illness in the face of these conditions before, during, and after trying clinical treatments. His chief concern revolved around how to motivate adherence—

how to foster his patients' commitment to managing their diagnosed conditions pharmacologically over the long haul. His challenge was greatest when convincing patients to take medicine even if they were not experiencing debilitating symptoms and, second, when supporting patients in continuing their medicines indefinitely even after they started feeling better, stronger, and more capable. He saw all stabilization of their symptoms as arising from medicine he prescribed, and he worried that his patients did not share this conviction.

By the time we met, my lunch companion had lived and worked in northern Tanzania for almost a decade and had no plans at that time of returning to the United States. He was raising his family in this mid-sized city, and unlike many expatriates, he spoke fluent Kiswahili. During his first years in the country, when he worked full-time in the medical wards of a tertiary hospital, he had witnessed the rapid rise of hypertension-related diseases. The growing challenges these and other chronic conditions posed to clinical care animated his turn toward research. He was sincere; his desire to be a good doctor and his distress over the obstacles he faced in providing standard evidence-based care were palpable. Long after our lunch, I continued to feel drawn to respond to the affective plea of his question, even if I was not willing to be bound by its assumptions. The negative construction—"Is there no . . ."—asserts that the "concept of chronic disease" is singular and uncontested. He did not invite a discussion of alternative ways people apprehend the attrition of strength and the dampening of energy levels—no less racing hearts, swollen limbs, shifts in appetite, blurry vision, or other symptoms—that he identifies as indications that a patient may have one of the chronic diseases on the rise in Tanzania. Rather, by mobilizing "traditional medicine" as a stand-in for local culture or, more broadly, "African-ness," he seems to wonder if the concept of "the chronic" is something that might be external to Africa, to Africans. The racialized divide between science and culture that haunts his question reflects decades of public health studies of the "epidemiological transition" that leverage "chronic disease" to rearticulate distinctions between the "West" and "Africa" in the service of a postcolonial social evolutionary theory. It points to a long history in which time and temporality are manifestations of power.

The affective plea of the doctor, however, is embedded in a history of risk. The notion that these chronic conditions put one perpetually "at risk" is central to the biomedical doctor's research and clinical work.[21] His plea is for his patients to see risk as he does. At the individual level, risk here is offered to the patient as the moment before the (potential) bodily harms linked to these conditions, as documented in epidemiological data and clinical experience. Hypertension, whether symptomatic or asymptomatic, puts one at risk for heart disease and stroke. Diabetes puts one at risk for nerve damage, ulcers, infections, heart and blood vessel disease, and kidney disease. Chronic kidney disease puts one at risk for acute renal failure and early cardiovascular disease. Leaving rheumatism unattended risks damage to a wide variety of other bodily systems, including skin, eyes, lungs, heart, and blood vessels. At another level, risk has come to function as an organizing logic in both medicine and public health. Health care practitioners assess how a patient's individual biological indicators compare to population-based data. In describing the patient's condition and recommending treatment regimes, they suggest the likelihood that an individual's condition will unfold similarly to the dominant trends in the population. Clinical assessments blur distinctions between prevention and treatment. Interventions strive to manage the risk that the condition will grow worse and have debilitating, even fatal, effects.[22] Efficacy is judged by a treatment's ability to lower the *probability* of acute damage by targeting the biological indicators that portend harm.

Not only the fluctuations of biological indicators but also their stabilization demand attention as clinical events that require intervention. Insofar as pharmaceuticals manage risk but do not eliminate disease, they complicate the ideology of cure.[23] The linear trajectory of diagnosis, treatment, and cure imagined for acute infections is not expected to ever be fully resolved within an individual instantiation of the chronic disease. Rather, drugs offer national health care systems as well as international development organizations a clinical solution to problems of population-based risk factors and disease prevalence. It is, therefore, not always evident when an individual is out of risk.[24] As a result, drug regimens for chronic disease are long; many, like ARTs, are imagined to be for life.[25] The times and tempos that structure adherence to these regimens—that stitch individual lives to medical interventions—have become a focus of clinical and public health work. Programmatic inconsistencies and institutional instabilities that disrupt the flow of treatment emerge as the grounds for professional complaint and public protest.[26] Thus, the rise of risk as an organizing logic has also accelerated the pharmaceuticalization of health—that is, the articulation of a right to pharmaceuticals as the right to health.

The possibility of lowering the burden of disease by intervening in individual bodies sidelines social-cultural and economic analyses and the forms of accountability that they might suggest. Social inequalities and environmental contamination are rendered "context," while patients' hesitation to take pharmaceuticals and queries about how to evaluate their effects over time come to be framed as barriers to public health that must be addressed.[27] Blame for acute events comes to be individualized as access to pharmaceuticals gives rise to the normative claims that people should be responsible for mediating their own risk.[28] Insofar as *dawa lishe* therapies seek to disrupt any stable focus on the individualization of risk, however, they might also be characterized as an intervention into the actuarial logics organizing biomedical diagnosis and treatment. In Tanzania, HIV/AIDS has generated social, ecological, and therapeutic projects striving to give language to the risks of dominant biomedical conceptualizations of chronicity and to experiment with responses that generate the times of healing.

## The Times of HIV/AIDS

A doctor in the Kagera region of western Tanzania diagnosed the first cases of AIDS in the country in 1983.[29] By the mid-1980s, the Tanzanian government, recognizing that the epidemic was widespread and growing, established the National AIDS Control Program within the Ministry of Health to coordinate surveillance and prevention efforts. During the first decade of the epidemic, some of the regions identified as most heavily affected were Kagera (Mama Nguya's home region), Dar es Salaam (where Uroki was living at the time), and Kilimanjaro (Dorcas's, Rose's, and Uroki's home region and where the two women lived at that time). Seroprevalence studies estimating particularly high HIV infection rates among people living in these areas drew intense international interest. The advent of antiretroviral therapy in 1995 fueled a sense of a possible future for communities devastated by the virus. Yet, a year later, as the antiretroviral triple-combination therapy had made it possible for Americans and Europeans to begin speaking of the possibility of living with AIDS rather than dying from AIDS, the therapy (and the imaginary) remained relatively inaccessible to most Tanzanians. Eight more years later, in 2004, the Tanzanian government established its National Care and Treatment Program to improve access to ARTs for people living with HIV/AIDS. After the implementation of this program, seroprevalence rates dropped from their high the year before, when 7.0 percent of adults were thought to be HIV positive, including 7.7 percent of women and 6.8 percent of men.[30] Both the consequences of the epidemic itself and the requirements of treatment have pressed Tanzanians to live in time differently.[31]

Identifying changes in HIV/AIDS programming over the past four decades helps to elucidate the kinds of temporal work demanded by global health aspirations to chronicity. Three distinct turns have characterized AIDS programming in Africa: pretreatment (the early 1980s to mid-1990s), the extensive infrastructural and programmatic developments to support testing and treatment (the mid-1990s to 2008), and the post-2008 withdrawal of funding and the narrowing of efforts around pharmaceutical distribution (2009 to the present).[32] The first phase of response during the early years of the AIDS epidemic was focused exclusively on epidemiological surveillance, public health awareness, disease prevention, and patient palliation. During this time, AIDS was equated with death—both social and physical. For health practitioners to reveal a diagnosis of HIV, no less for family members or community members to speculate about the cause of someone's sickness, was fraught.[33] For some, saying out loud that someone had AIDS was akin to wishing them dead, and such words had efficacy.[34] A dear friend who cared for her sister in late stages of the virus confessed the delicacy with which she and her other siblings addressed the issue of whether to wear gloves while tending to their sister's physical needs. This latex protection not only impeded sensation while cleaning, bathing, or tending to their sister, thereby creating distance, but also implied that she had HIV/AIDS. Even as the end drew near, such an implication was seen as a form of violence. In these early years, social scientists conducted a plethora of studies about the stigmatization of those with AIDS. Public health initiatives identified the great challenge as one of coaxing people to name the disease and to reveal their status, both of which were asserted as foundational to any fight against the stigma.

At the turn of the millennium, antiretroviral treatments remained difficult to access in Tanzania as in much of Africa. Confessional speech acts—public declarations of oneself as a person with AIDS—emerged as a technology for triaging the rapidly growing number of people diagnosed with HIV and for distributing the desperately inadequate supply of antiretroviral drugs.[35] This differential global access to ARTs reframed the HIV epidemic in Africa as a humanitarian emergency. The second phase of AIDS programming in Africa began in the early 2000s with the massive expansion of testing and pharmaceutical treatment through bilateral government and private philanthropic efforts.[36] In Tanzania, this shift led the government to establish the Tanzanian Commission on AIDS (TACAIDS), under the auspices of the Prime Minister's Office, to coordinate the efforts of the government and international organizations in planning, monitoring, and evaluating services. TACAIDS worked with the WHO to organize the local response to the WHO's 2003 "3 × 5" initiative, with the goal of treating three million people with HIV/AIDS by 2005. This initiative energized efforts in Tanzania, mobilizing resources toward the National Multi-Sectoral Framework on HIV/AIDS. Alongside the "3 × 5" initiative, other major funding initiatives were launched by the Global Fund to Fight AIDS, Tuberculosis and Malaria and by the US President's Emergency Plan for AIDS Relief (PEPFAR). These multilateral and bilateral efforts added to significant philanthropic efforts at the time by both the Gates and Clinton foundations. The focus on distributing ARVs fueled the collapsing of the "right to health" into the "right to pharmaceutical treatments," or what has come to be called the "pharmaceuticalization of health."[37] This concentration of programming on pharmaceutical distribution and the accompanying attenuation of support for HIV/AIDS groups and community-level responses marked the third phase of response.

Yet, "adherence time"—the temporal patterns required to conform to the demands of the medication, and of the clinic through which they obtain the medication—requires reorganizing individual and household activities. These drugs for life come to define the daily, weekly, and monthly rhythms of people's lives. Both people on ARTs and their health practitioners strive to coordinate "adherence time" with "project time," the tempos driven by funding cycles, the demands of evaluation, and the movement of experts.[38] There is a fundamental instability—if not incommensurability—between the limited time frame of projects and the demand of drugs for life. When the global financial crisis of 2008 abruptly slowed the flow of private and public funds into global health, AIDS programs were hit sharply.[39] Meredith Marten, in her article on "Living with HIV as Donor Aid Declines in Tanzania," argues that the sharp attenuation of investment in infrastructure set the stage for the third period of the programmatic response to HIV/AIDS in Africa, characterized by short-term, heavily biomedical interventions and a reduction of broader care services. As people listen to counselors who insist that they embrace their lifelong dependency on ARVs, the unreliability of these more recent time-bound projects distributing these ARVs further exacerbates the precarity they feel. The impact of this uncertainty has grown exponentially with the shifts in international health care funding from the broader investments in national infrastructures and primary health to the public/private enclaves of disease-specific, humanitarian efforts.[40] Susan Reynolds Whyte has referred to such temporal pressures as "global health time"—those times that emerge from shifts in political commitment, scientific interest, new methodological demands, and capitalist incentives.[41] Addressing HIV/AIDS does not generate a singular time, but rather, the management of HIV/AIDS—from individual adherence schedules and household reorganization to project cycles and funding mechanisms—generates different scales of experience, knowledge, and governance. The co-existence of these temporalities and the ways in which they rub awkwardly against one another become palpable as rhythmic tension in the lived experience of chronic illness.

In the throes of this tension, Tanzanians cannot take for granted that chronicity is a guaranteed product of pharmaceutical adherence. The labor of living with AIDS as a chronic condition is so intense (and relentless) that it makes chronicity itself feel precarious. In Kiswahili, people will say that they "drive life" (*kuendesha maisha*), but all know that they do not do so under conditions of their own choosing. This three-phase periodization draws attention to the practices that have shaped bodies, social relations, and physical environments in the wake of the epidemic. Outside the clinic, however, each of these temporalities is negotiated and coordinated in ways that shape lifetimes (biographical narratives) and reproductive times (the extension of kin through generations and across land).

### Livelier Times in the Near Future

As Wenzel Geissler and Ruth Prince argue in their beautiful ethnographic account, *The Land Is Dying*, the epidemic altered the shape of social life, the possibility of touch, and the nature of growth in East Africa.[42] In the pretreatment era, the prevalence of disease and death among those in their middle years ruptured the ways lineages organized the care of elders and children. As a result, the forms of communality through which people managed the fertility of bodies and lands were disrupted. Insofar as ARTs supported a return to productive and reproductive work, people with AIDS mobilized these drugs in an effort to stitch together ruptures in lineage and land caused by this sexually transmitted virus. From this perspective, efficacy lies not only in lower viral loads but also in the capacity to farm again, to feed one's elders, to send children to school. Medicine, with its intense focus on the life of the biomedical body, understands such physical capacities as desired impacts of treatment on lifetimes. And they are that, but not only.

Between 2015 and 2018, I talked with many members of TRMEGA, sometimes during gatherings at the headquarters in Maji ya Chai, an area on the eastern edge of the ever-growing town of Arusha and sometimes while walking through the gardens TRMEGA supported in schools, orphanages, homes for street children, and NGOs, as well as periurban neighborhoods. During these years, all members of TRMEGA who needed ART had basic access to the medication from a local clinic. Mama Nguya had an informal relationship with the private clinic of a Catholic AIDS project where many of the members received their ARVs. As will become evident below, Jane Satiel Mwalyego helps to inform people at the clinic about TRMEGA and even to facilitate a visit to Maji ya Chai for some. Mama Nguya also recounts instances in which the doctors at the clinic would refer patients directly to her. She positions TRMEGA as working with and across the care that the biomedical clinics offer.

John Ogondiek and Victor Wiketye, the scientists with whom I have enjoyed thinking and working since we met in 2008, also became particularly interested in the work of TRMEGA. For them, TRMEGA offered an example of an organization that promoted plant-based remedies entangled in clinical interventions, which existed neither fully inside nor outside of biomedicine. Both the plants and the organization stimulated questions that might be taken up at the Ngongongare Research Station of the National Institute of Medical Research, which they led nearby. They found conversations about the individual plant varieties in TRMEGA gardens particularly exciting. As we walked around Mama Nguya's garden, both would follow her invitations to taste or touch various plants. She was not opposed to these scientists imagining laboratory investigations. She also welcomed me asking them to assist with a series of formal interviews with members.

In May 2015, Mama Nguya called together the TRMEGA members. She asked them to bring a sample of a therapeutic plant that they knew and used. Jane cooked food from the TRMEGA garden that all could share, as John, Victor, and I cycled through extended interviews with each of the members. Some TRMEGA members spoke boldly and confidently about their diagnosis and their efforts to live with AIDS. Some spoke euphemistically, calling it *mgonjwa kawaida* (the common or usual illness), not indexing chronicity as much as referencing living in a world so marked by this virus that it was taken as a common or collective experience. Others were more oblique in their references, and a few turned away from any question that felt to them like it might implicate them directly as a person living with HIV/AIDS. Although at times, TRMEGA drew on the neglected social infrastructure of AIDS support groups, illness-based identities were carefully deemphasized in everyday interactions. Belonging was not mobilized through testimonial narratives. Instead, what provided continuity and connection were plants.

Jabari volunteered to share his story first, eager to talk about his transition.[43] He was two months shy of fifty years old and had been a member of TRMEGA for two years. He came to Maji ya Chai that morning from his home about forty-five minutes away on the other side of Arusha town, where he lived with his two children, next door to his parents. Jabari's wife had passed away. He originally came to TRMEGA through a support group that Jane organized and chaired for people with HIV, called Moringa Plus group. Jabari now served as this group's secretary. The group's name highlights the moringa plant, a tree whose small leaves have moved from South Asian curries, to components of international nutrition programs, to animal fodder in East African gardens, to "super food" status in the United States and Europe and an "immune booster" in herbal clinics in Tanzania.

Jabari, together with Jane and others, formed the group "in order to influence each other's opinions, share news of therapeutic treatments and nourishing eating, and organize different kinds of support, such as financial assistance." For Jabari, the withdrawal of AIDS programming after 2008 means that "it is harder to get help now" for the difficult work of altering the social-material conditions that define the structural inequalities driving the epidemic. Affiliating with TRMEGA offers new attachments for the members of Moringa Plus and extends their community. The connection was initially cemented through small gestures of material support. First, when Nguya received a donation of bicycles, she distributed some of them to the Moringa Plus group to help with informal entrepreneurial projects such as selling sheets and a small charcoal business. Later, when Jabari saw the benefits of plant-based treatments, he opened part of his home garden to the group. In turn, Nguya brought him cuttings, seeds, and seedlings from her garden to plant in his. His desire for these plants and his willingness to offer a portion of his land for the community garden evolved from his own journey with coming to recognize the vegetal relations critical to making a life with AIDS.

When Jabari reflected on his journey, it started with a troublesome ulcer. Each month, he went to the clinic to collect a thirty-day supply of antiretroviral medication. At that time, the clinic staff checked in about the effects and side effects of his prescription. Every six months, his visits also included blood work to assess his CD4 count. The previous February, during his appointment, he showed the doctor that a skin ulcer was growing slowly and becoming more debilitating. The doctor sent him to the hospital to check his blood sugar levels and assess whether he was diabetic. When the tests indicated that Jabari's glucose levels sat firmly in the "normal" range, the doctor prescribed an antibiotic. It helped a little, but after a while, another ulcer on Jabari's spinal cord had grown so large and painful that he reported being unable to bathe himself. He returned to the hospital and was prescribed a stronger and more expensive antibiotic. He found that this new medication would dry up individual eruptions, but the problem was persistent. He started being able to identify the sensation of an infection developing under his skin before it burst. When possible, he would start a course of the antibiotic early.

While the specific trigger for the ulcers remained elusive to him, Jabari noted that "if you have HIV, when sickness comes, the major cause will always be HIV." He reiterated the lectures of health care workers who had explained that his immune system is so suppressed that the body does not have the capacity to cope with bacterial or viral infections as it otherwise might. Both the virus and the ARTs taken to enable life with the virus profoundly reshape bodily capacities. Jabari and others know the virus is both form and context.[44]

In the midst of grappling with this stubborn condition, a friend recommended that Jabari visit Nguya. One day, when Jabari felt relatively strong, his friend helped him make his way to TRMEGA. As she had done with Jane, Mama Nguya gifted Jabari a canister of moringa. He returned home to drink the powder in warm water three times a day. Over the next few weeks, the individual ulcers not only shrank, but new ones did not follow in their wake: "I finished [the canister] and saw that the ulcers had continued to dry up. I did not have another problem [with more ulcers], but this mama told me that it is better to use this herbal medicine: 'It will continue to help inside the body.' So, she told me to use it in the soup, to use it in the tea. I took two [more] canisters of that medicine.... Then through TRMEGA, I have planted some at home." For Jabari, this account of skin ulcers is not only an account of managing a painful, acute condition and finding his way to more comfort. It is also the mundane, pus-filled struggle with everyday life through which he came to understand the reorientation that plant medicines invite.

Moringa, one of the more charismatic plants in formulating *dawa lishe* in Tanzania, potentiates different economic possibilities, different ecological relations, and different bodily engagements than pharmaceuticals do. It is, for instance, not just that moringa is sometimes sold and sometimes gifted, both enacting economies of exchange and debt between people, but that moringa moves into the garden through the sharing of cuttings. Plants draw attention to human–nonhuman relations: to hands in the dirt, to water pooling around shared cuttings, to the quality of the heat from the sun. As plants make space, they resist being reduced to raw materials for use by humans. They invite humans into a relationship with land and with other plants. Jabari explained:

> For example, maybe right now I am not using moringa, but in the past I was using it in my tea. You don't have to be sick to start using it. There is also this that they call lemongrass. I pick it [at home] because she [Mama Nguya] brought us seedlings and gave them to us in our [Moringa Plus] group. I went and planted it in my garden. So, I have been using *mchai-chai* [lemongrass tisane]. I cook the chai and drink it. I put this wide-leafed medicine in water until the water changes color. Then, I drink it regularly. I now drink this *mchaichai* every time I would have previously drank water. After discovering that herbal medicine [*dawa za asili*] can

be an alternative treatment [*tiba mbadala*] that helps my body by replen-
ishing its nutrients, I don't wait until I am hurting. As long as there are
no negative effects when using it, as with *mchaichai*, you don't need to
be sick to put it in your tea. Therefore, indeed, this is how I am using it.

Plant medicines do not work only or even primarily through reduction but
rather through addition.[45] They multiply desires, attachments, and rhythms,
generating a density of relations experienced as lushness.

John, the pharmacologist, asked Jabari, "So what did you learn about the
herbal medicine?" Jabari went beyond just answering John's question when he
responded, "I have learned [*nimejifunza*, literally I have taught myself] that
those alternative treatments [*tiba mbadala*] will not put poison in the body,
like those medicines from our colleagues in Europe. They [biomedical health
care practitioners] claim that their medicines are toxic to the body so it's better if we strive to use these plant remedies so that we can reduce the poison
of the medicine that they tell us we must use. [These plant remedies] help by
keeping the poison from building up in the body." Through a subtle redirection, Jabari indicates that the useful question is not what one has learned about
(*kuhusu*) herbal medicine itself but rather what one has learned through using
TRMEGA'S treatment. That is, through experiencing how plant(ing) remedies
affect the ways that bodies-on-ART move through the world; the ways that
they not only absorb but are put into motion by relations of life and lushness.
What Jabari learned was how to be transformed with and through plants, not
how to articulate the discrete agency of individual plants in relation to individual bodily ailments. What he learned was a mode of theorizing—a verbal
and embodied vocabulary—through which to shape the way flesh and world
are brought forth. By starting with his struggle with serial skin eruptions, Jabari
locates his story in the embodied practices through which TRMEGA invited
him to understand himself and his health differently. In so doing, he dislocates
the epistemological demands of the imperial pharmakon that forge pharmaceutical treatments and unsettles the double-bind it secures: that the toxicities
that enable life also simultaneously threaten life.

When he turned to try to address John's pharmacologically inflected question more directly, however, he found himself in a cramped space conceptually. He was left talking about what he did not know: "I am using it [herbal
medicine] often in order to avoid frequent visits to the hospital, and to avoid
the expense, because now I have been given the medicine to plant. Maybe I will
get another illness that these medicines do not treat. The ones that I have are
*mchaichai*, moringa and the scalloped-leaf medicine. I believe this medicine

enters the body and helps many things that I do not know. Indeed, I have not had fevers, or headaches, or painful joints." Jabari was left to talk only of what he believed.[46] The forms of evidence and demands of agency generated through the imperial pharmakon render the relations, practices, and reorganizations of planting, sharing, and taking moringa as nonknowledge.

> Maybe I should say it this way: I do not yet know much about the herbal medicine. I don't know which diseases are most appropriately treated with herbal medicine. That is, if you have a certain illness, you should use a certain medicine. While I do not understand this well, I know you don't need to be sick to use this plant medicine. I am using it often, I believe for me, that if it is put in the body it acts as protection (*kukinga*). It is like immunity (*kinga*). I don't know that if I use this medicine for a certain disease that it will cure that disease. Maybe I will be given that knowledge of a certain medicine treating a certain thing. But for now, I can't give the correct answer.

The logics of pharmacology leave Jabari knowing less. Plant(ing) remedies are less about mobilizing mastery over an object and more about responding to shifting relations between body and land. They work through the socioecological forces that constitute health. The shift away from temporalities of pharmacological knowledge to those of ecological relations is critical to countering the ongoing dispossessions of the imperial pharmakon.

Jabari takes his ART reliably. He organizes his life around adherence time, around the immediacy of the clinic. He coordinates the versions of his body salient in different spaces of his life—as grown son, as father, as person with AIDS, as friend, as neighbor, as community organizer, as farmer, and more—in ways that make space for the demands of ART. Yet he refused the presentism enforced by adherence time: "The shock will come here when they [doctors, nurses, and pharmacists] say, as I understand it, that their medicine puts poison in the body, Mmh! Now that I know that the medicine will poison me, but I have no other option, what should I do? Maybe when the grace of God comes, we will receive the knowledge about alternative treatments so we can stop using so many of these modern medicines." Jabari commits to ARTs as a condition for life. "I have no other option." But his "maybe" holds open a future that is otherwise. Is this best understood, in the words of the doctor quoted earlier in the chapter, as having "no conception of chronic disease"? Jabari does, I would suggest, refuse a version of chronicity that understands risks as knowable and treatment as interventions that hold risk stable. In so doing, however, he does not refuse the temporalities of the clinic. He multiplies them.

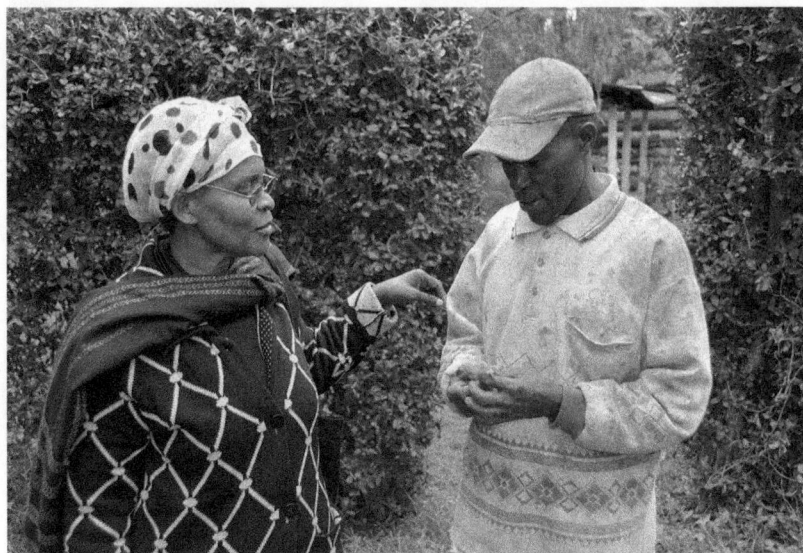

FIGURE 4.1 Helen Nguya advises Jabari on caring for some of the seedlings she brought from TRMEGA to be transplanted in his community garden, Arusha, Tanzania. Photo by author, 2015.

The practice of using plants therapeutically responds to broader trends about how health care drives the rhythms of life and living in the wake of AIDS in Tanzania. AIDS programming has become both short- and long-sighted. The promise of treatment as prevention has fed a vision of a future in which there is no AIDS. If all individual patients adhere rigorously to treatment, this logic contends, viral loads will be suppressed. HIV will no longer be transmitted. The children of today will be the last generation to live with HIV/AIDS. This discourse obscures the temporal possibilities between the functional immediacy of using ARTs and the utopic horizon of an end to the epidemic. It generates an "enforced presentism," on one hand, and a "fantasy futurism," on the other, and risks evacuating what Jane Guyer calls the "near future" in her influential 2007 essay "Prophecy and the Near Future."[47]

The near future is the horizon of infrastructural plans and research proposals as well as of building support groups, addressing stigma, cultivating soils and planting for crops, collaborating on small business ventures, pressuring national legislatures for better services, and reimagining property laws that support human rights. It is, as Guyer defined it, "the reach of thought and imagination, of planning and hoping, of tracing out mutual influences, of engaging in struggles for specific goals, in short, of the process of implicating oneself in the ongoing life of the social and material world."[48] She worried about the rise of a "public culture of temporality" in which the collaborative, creative, messy work of building collectives for social change in the near future is obscured in favor of long-term horizons. Guyer located this concealment in a "double move" driven by the macroeconomic policy changes of the last two decades of the twentieth century.[49] Neoliberal reforms of the 1980s and 1990s undermined government-controlled health care services, narrowing the "public" to which public health attended.[50] Donors started targeting specific diseases over primary health care and investing in time-bound projects rather than national infrastructure.[51]

The HIV epidemic energized a shift in temporal sensibilities in Africa as these macroeconomic changes redirected funding for health. The structure of emergency that shapes AIDS programming in Africa, and global health programming more generally, forces people to grapple with the evacuation of the near future in their everyday lives.[52] *How* Tanzanians grapple with global health's evacuation of the near future, however, is an ethnographic question.[53] The social and therapeutic projects that animate *dawa lishe* grapple by kindling relations with plants. Jabari describes how the substantive reality of the body is at stake in the ways that time is created in the space of this evacuation of the near future. Plant remedies collaborate in shaping his body as one with the capacity to resist ulcers. They reconstitute the body itself, rather than only address the infectious agent that might be causing the ulcer. TRMEGA's plant(ing) remedies address not only the wounds of poverty but also the foundational vulnerabilities constituted through dispossession's deformity of the social-ecological forces of health.

## Removing Chronicity

Jabari strove to lure the Tanzanian pharmacist interviewing him away from the strictures of toxicological logic. Plants, he suggests, reorient relations between toxicity and remedy. Kijani, another TRMEGA member, also worked through the relations of toxicity and remedy to open a space to consider the efficacy of plant medicines. When I met Kijani, he served as the gardener and guard at TRMEGA. He had been living on the plot in Maji ya Chai when Nguya bought it to build TRMEGA. He arranged with her to stay in exchange for maintaining the gardens. His knowledge of plant-based remedies, however, preceded his time with TRMEGA.

Before moving to Maji ya Chai, Kijani had worked farther east on the sisal plantations. During this time, he had used plant medicines. But not without concern. "With some other plant remedies, there is worry," he acknowledges. "How much should a person use?" He tells me that he knows "a lot of medicines." For instance, "if you dig out the roots of *mwarobaini* (*Azadirachta indica*, popularly known in English as neem) and *mjohoro* (*Senna siamea*, popularly known in English as the ironwood)[54] . . . now that is medicine!" He used these roots to support his health on the plantation. In particular, the conditions of the sisal plantation left him and other farm laborers continually exposed to malaria.[55] "There were a lot of mosquitoes there in the sisal water. You know they have a lot of foam. There were too many mosquitoes. The malaria was disturbing me. I started to drink neem. I used neem and I was better. My symptoms went away, but I cannot say I know (*kufahamu*) these herbs. . . . Many times when I listened to the experts on the radio, I heard that this medicine can cost you something; it can kill you." Kijani shares his ambiguous feelings about these plant medicines. He hesitates to say that he "knows" them given that his experience of their efficacy is unsettled by concerns about their toxicity and uncertainty regarding the quantities that should be ingested. He describes reports that he heard on the radio cautioning the listener about plant-based medicines. He concluded, "We do not know if it is very strong (*kali*, fierce) to drink. When you drink it, you are fearful. It is not standardized. *Sasa aina kiwango*, now there is no standard, you see. Even traditional medicine should be standardized."

But the medicines through TRMEGA, he argued, were different. Not because there had been collaborative efforts to examine the plants pharmacologically (although as mentioned above, Nguya was open to such an effort) but because a different sort of understanding, or knowing, of the plants was cultivated at TRMEGA. Whereas critiques of traditional medicine as dangerous because it is unstandardized enroll plant remedies in pharmacological arguments, at TRMEGA, Kijani came to supplement these critiques with activities driven by other logics. He no longer confined arguments about a plant's therapeutic value to its biochemical capacity for benefit or harm. A decade after working on the sisal plantations and struggling with chronic malaria, he broadened the lexicon through which he articulates plants as intervening into the ontological dynamism of the body. He came to focus more attention on ways of dwelling.

While managing the gardens at TRMEGA, Kijani watched the work of the organization. Jane slowly pulled him into the social life of the organization, and after some time, he joined officially as a member himself. While the organization forms loosely around issues concerning women and AIDS, it has never excluded either men or those not living with HIV. Gender and AIDS shape the worlds of all Tanzanians. Rather than organizing around a strictly disease-based identity politics, members built solidarity through an acknowledgment of the forces altering social roles, reconfiguring family dynamics, and impacting material livelihoods. The benefit of TRMEGA, Kijani argued, is that "here I learn about *madawa ya asili*," medicinal plants.[56] Through the everyday care of plants, building garden beds, composting, attending to the beehives, and coordinating the flow of water, he was drawn to the forms of dwelling—the ways of being on and through the land—promoted by TRMEGA. Working in Mama Nguya's garden in all the ways it resisted plantation logics, he came to consider how shifting relations with plants, water, and soil could intervene in the relentless, everyday violence that wears down bodies and diminishes vitality. The efficacy of the remedies promoted by TRMEGA, he argues, lies in their ability not only to remove pathogens (such as malaria parasites) but also to "remove chronicity."

While malaria was a constant threat in the sisal plantations, Kijani had suffered from chronic malarial fevers for years before working there. In our extended interview, he described how these cyclical fevers and chills debilitated him each year as the long rains started to wane. Malaria is endemic in much of Tanzania. Tanzanians differentiate between one-time cases and chronic malaria. Doctors concur that some people experience cycles of malaria. Subsequent fevers are not (necessarily) a reaction to reexposure. The person is asymptomatic much of the time. The parasites seem to lie dormant and may not be detectable in a local laboratory. The annual return of these fevers decreased, however, when he started living at TRMEGA. He confessed that this fact dawned on him slowly at first.

> I found myself reflecting on the fact that it has been a long time since I suffered with malaria. It reoccurred regularly in May and June each year. Starting in 1984, it came from nowhere and I had to go to the seminary office for treatment. Every May and June. Every year. Oh, I had so many receipts for medicine from the hospital. But now since this organization started (I think we started the planting at the end of 2004). . . . So since I started to use this medicine, I don't feel that malaria. I don't go there [to the hospital] anymore. I don't use those tablets anymore.

After some time, he realized the malarial attacks had not returned.

According to Kijani, the clinic understood the limits of treatments they had to offer in response to such long-term, recurrent malaria. Indeed, many have written of the ways that traditional medicine is recognized at the limits of biomedicine not only in Africa but beyond.[57] There is a hesitant openness at the edge of biomedicine's failure: "For sure, it has helped me with that malaria.

Even at the hospital, they were telling me to go and try medicinal plants. 'Go and try *madawa ya asili* because this malaria is chronic (*sugu*).' My medical file there is very long. Mmh, even until now, if they see me every one of those who work at the clinic will know me." In the face of the chronic, *madawa za asili* is worth a try. But beyond being just a desperate effort, what sort of response are plants to the chronic? Kijani argues that "the hospital cannot take the chronic out of malaria. They just cool it down, but later it comes back." Plants, however, open the possibility that chronicity can be removed: "In this organization [TRMEGA], I have been healed. For sure. Because I haven't gone to the hospital again. Even if I do come down with malaria, it is not with the regularity that I did before. Therefore, I must confess that I have benefited in that the chronic has been removed from the malaria (*marelia kuitowa sugu*)." Plants draw out an ecological mode of attention, calling on healing to address not only acute infection but also the injury of slower violences. Co-laboring with plants generates the needed times and tempos to articulate the temporality of structural violence. Plants and planting shift relations that are fundamental to life's continual becomings and, in so doing, offer the possibility of addressing seemingly intractable ailments. This approach, I am suggesting, is what defines *dawa lishe* therapies and their call for medicine to be nourishing, not only nutritious.

Kijani's experience illustrates how Tanzanians fold pharmacological logics into other ways of knowing, thereby multiplying the temporalities through which they manage illness and health. On one hand, knowledge of the therapeutic value of plants expands the options available in the face of illness. As he remembered:

> One time, I had pneumonia. I felt the burning in my side. But this moringa addressed it. And there are other medicines. For example, one day, I went to the clinic after feeling like the chronic malaria was coming back again. I went to be checked. I was diagnosed with typhoid and a urinary tract infection (UTI). I was told I needed medicine. I told them I did not have enough money with me and asked them to write down the prescription on a piece of paper for me so that I could go to find the money and then buy it later at a pharmacy. But it was not really because I did not have the money! Why should I pay because I know that we have the plant medicine to treat UTIs?

As he talks, however, it becomes apparent that plants are not only biochemical bundles that, if elucidated, might offer cheaper alternatives to pharmaceuticals. Their efficacy rather lies in the ways that they partner with humans, transforming the materiality of everyday life. Kijani points to the plant that he used to treat his UTI, saying, "It is right there beside the foundation." He shows me not only the shape and texture of the leaves but also their proximity. Efficacy resides in the dynamics of dwelling together. Healing requires not only sharing information but also transforming the ecology and economy of home.

Kijani describes how he planted lemongrass, moringa, and aloe vera at his mother's house as well. He diversified her garden. The lemongrass was ready for harvesting first. He showed her how to cut, clean, and dry the stiff leaves to prepare a tisane. That is, he is gathering new plant relations close to her, and he has started to demonstrate how to cultivate those relations. He confesses that he argues with his mother, trying to convince her that the lemongrass tisane does not need sugar. Lemongrass, he maintains, "helps to remove the poison (*kutowa sumu*) which enters through the food we eat, starting with sugar." Kijani encourages his mother to drink her tisane plain. Adding sugar, he warns, cancels the cleansing effects of the lemongrass. Later, when the moringa trees are bigger, Kijani plans to show her how to use moringa. When the aloe vera is large, he will teach her how to make juice.

Kijani's story shares how TRMEGA defines knowledge not as an awareness that is juxtaposed to attitudes and practices, but rather as a process of cultivating a density of relations and reorienting the attachments through which lush lives and lands are bodied forth. This approach is the reason, Kijani argues, TRMEGA grows slowly. "Do you know why it moves slowly? There are a lot of challenges. For example, to educate people, you cannot [just] give information, educating directly." Knowledge is not a thing that one has or gives, but a relation that alters you from what you had been. "You see, that is why we move slowly. Even this social worker who came said he would bring some groups here. Now he brought two or three groups. Don't you see? Knowledge (*elimu*) comes from here and goes to there." Knowing is in movement: in plants by the foundation, in cuttings moving to his mother's house. It is a trajectory of transformation. And it is slow. Knowing materializes not through any simple dissemination of information, but through the sharing of plants, the cultivating of a garden, the making of juice, the expansion of community. Knowledge is embodied in the social and material landscape.

The therapeutic potential of plants lies (at least in part) in their capacity to alter the way people dwell with and through the land, as well as with and through each other. They collaborate in the cultivation of spaces that hold, move, channel, and redirect *horu*, life force, in the Kichagga of waRombo who live on the eastern slopes of Kilimanjaro.[58] Descriptions that limit the role of gardens to the provision of nutrition are anemic, and those that account for herbal remedies as mere supplements in the face of shrinking public health programming are thin. TRMEGA membership draws people into new relations with plants and with each other as it invites them to locate healing in vegetal life projects that cultivate an ecological and social liveliness that biomedicine cannot envision. *Dawa lishe*, as a conceptual proposition, renders visible relations beyond the resource logics of pharmacology and the economies that they have enabled to flourish. Plants are more than biochemical bundles, proto-pharmaceuticals, and conspiring with them demands a rethinking of the relations between toxicity and remedy on which pharmacology (and its twin, toxicology) rest and a forging of the times needed for healing.

## After Adherence

Jane Satiel Mwalyego strives to be an example so that others, including her own children, can see that if they are diagnosed with HIV, it is not a death sentence. As described in the previous chapter, she has grown bold in her identity as a person living with HIV and clear in her commitment to taking her antiretroviral medications. After struggling against her diagnosis and against the losses that followed, she cultivated the capacity to adhere. Now, not only does her energy animate TRMEGA, but the providers at the clinic she attends also recognize her ability and they (informally) direct those who need more support than the clinic can provide to her. Therefore, when Jane says that she may be able to get off antiretroviral medications one day, she is not threatening noncompliance. Rather, she is reflecting on the indeterminate dynamism of bodies. She is aware that there is little space to express such reflections, however tentative and open-ended. She recognizes that with only an elementary school education, her reflections are more likely to be heard as ignorance or recalcitrance than as thoughtful theorizing.

One bright but chilly afternoon, Jane talked to John, Victor, and me on the porch of the office at the NIMR Ngongogare Research Station. She started with a warning: "I know this won't get into your head." In her experience, formally educated people find it difficult to conceive of that which she is trying to communicate. She suspected we would not have the capacity to really hear her. Yet, she plowed valiantly ahead.

> Learned people are the setback. When you are told the following, you say, "Ahh, no."[59] The thing which made me to be like this, with the doctor as my witness, is the use of traditional foods and nourishing medicines (*vyakula vya asili na tiba lishe*). ARVs are not at "issue" [uses the English for this word]. That thing is nothing, other than what it is doing. . . . It does not add any nutrients in your body. I tell you that it is not *dawa*

(medicine). It is designed to put the HIV to sleep and stop it from reproducing.

Traditional foods and nourishing medicines have "made me like this." We could see what "like this" meant: strong, vibrant, alive, able to care for children and grandchildren, able to build community and help others. This broadly lived version of health cannot rest on ARVS.

For Jane, the very definition of *dawa* (medicine) is at stake. The problem was ontological. As she spoke, Jane struggled anew to find the language to render visible the processes of unknowing obscured by pharmaceutical science. She mobilizes nourishment, *lishe* (to cause to feed), as a key conceptual frame through which to discern the nature of healing. ARVS, she tells us, are not nourishing. This does not mean they are ineffectual or unimportant; rather, their specific function is to repress, diminish, and interrupt the reproduction of the virus. The work of the active ingredients in these drugs is attenuation. Such work stands in contrast to *uganga*, which, as discussed in the Introduction, is that which increases strength, catalyzes maturation, stimulates reproduction, extends capacities, and invigorates bodies and worlds. ARVS, Jane argues, are critical but not sufficient, and to emphasize this point, she declares that they are "not *dawa*."

To be clear: Jane takes her ARVS, attends the clinic regularly, and has cultivated communities of support. She notes that thanks to her adherence to this pharmaceutical regime, "The virus is asleep within me." The point she wants to drive home, however, is that the limited focus of ARTS leaves the reality of her good health unexplained. As Jane notes, the nurses and doctors at the clinic "confess openly that this drug [the antiretroviral cocktail she takes] is designed to put the virus to sleep. So, when I take it, the virus does not reproduce, but it cannot give me health (*nisingepata afya*)."

Jane takes her health care providers seriously. She knows that ART does not "cure" HIV. Rather, it "puts the virus to sleep." One can live with a hibernating virus. As she discovered, when her own depression drove her to cast aside her ART, these drugs are a critical technology to help those with HIV survive (see chapter 2). They can stop the replication of the virus. Yet, replication is a very limited form of reproduction. Jane asserts that her doctors explicitly acknowledge that ART does not address the social, economic, and environmental vulnerabilities that fuel the replication of the virus within or among bodies. "Because it does not have anything extra, it does nothing more."

Healing requires being attentive to an expanded notion of reproduction, one that includes the diverse times and spaces through which the ongoingness of the physical body is sustained. Jane declares, "I am the way I am because of nourishing treatments (*tiba lishe*). I do not say traditional treatments (*tiba asili*). They are nourishing treatments. By this I mean, I eat until I am sated and through this I am also healed and protected. I get fed and at the same time I get cured. I don't have hypertension, I don't have high blood pressure either, in any situation." Jane mobilizes *lishe*, or literally "to cause to feed," in contrast to *asili*, which means traditional in terms of the "original." Governmental and scientific organizations refer to plants with therapeutic potential as *dawa za asili*. This reference to origin, *asili*, locates plants not only as part of traditional healing practices but as part of a traditional pharmacopeia, a collection of plants that may offer biomedical compounds for extraction, analysis, and production (for more details, see chapter 3). By assessing drugs in relation to their capacity to nourish, Jane expands the vocabulary through which medicine might be understood as efficacious.

Jane's church congregation knows that she is HIV positive. They are part of the community she has built to support her own adherence. When they discourage her from fasting on religious days, fearing it will weaken her, however, she waves away their concern: "If people are fasting for prayer, I fast. [They say,] 'Oh, you are using medicine, you will suffer from ulcers.' I told them here there are no ulcers. If it is three days, I fast too. If I fast for three days, you will only see me tired because the medicine I swallow each morning is exhausting. But I will not stagger because of these ARVs." Jane draws attention to her body's capacity to endure strain, deprivation, and stress. In so doing, she situates narrower evaluations of health as that which can be articulated through viral load. Health, for Jane, is the ability to participate, to extend relations, and to continue community.

With her focus on these broader forms of reproduction, Jane suggests that there is a time after adherence. Jane does not plan to stop her ARVs, but she asserts that "it is my hope that in the future my doctor may let me stop using it [ART]. I mean, if the virus is all gone from my body, they will say there is no reason to use it. I only have to be compliant until the doctor will tell me to stop." She does not offer this reflection as a challenge to biomedical authority, stating that she will continue to take her prescribed ART until the doctor tells her she no longer needs to take it. Rather, she shares the possibility (if not the conviction) that such a day could come.

By imagining a time after adherence, Jane pushes against the limits of a chronicity that locks up the capacity of bodies to change and that evaluates the efficacy of pharmaceuticals through their success in fixing the body in a stable relation to risk. By defining efficacy through *lishe*, she invites a reframing of the nature of the body, the meaning of *dawa* (medicine) and the possibility of healing. She challenged the pharmacist, veterinarian, and anthropologist to whom she was speaking not to turn too quickly to the denouncements of error, deviation, and failure that foreclose any alternatives to the continuous pharmaceutical management of risk demanded by biomedicine. Nourishment trains attention on strength through the dynamic qualities of the body to become other. Hearing her argument—allowing her words to "get into our heads"—requires considering alterability as a defining characteristic of the capacity to endure.[60]

## Making the Time We Need

The history and experience of HIV/AIDS in Tanzania are central to *dawa lishe*'s intervention into the chronic. Despite the consistencies and instabilities of AIDS programming, the existence of antiretroviral therapies has changed what it means to be diagnosed as HIV positive in Tanzania. These drugs have come to frame the possibility (even if interrupted and uneven) that this virus can be something that people are able to live with; a diagnosis is not necessarily a declaration of forthcoming debility and death. Over the past two decades, ART has forged new links between conceptions of the chronic and practices of pharmaceutical adherence that have implications well beyond HIV/AIDS. Rendering HIV/AIDS a chronic condition requires taking ARVs every day, for life. As one doctor told his patients, "This medicine is your food." He meant only to emphasize the rhythm and regularity with which they needed to eat their medicines, for life. Yet as we have seen throughout this book, *dawa lishe* plays in the fault lines of medicine and food. Practitioners ask, what would be the implications of demanding that medicine (*dawa*) be nourishing? Nourishment pushes back on the narrowness of a notion of efficacy limited to the attenuation of pathogens.

African therapeutics have long worked at the intersection of food and medicine. Thomas Cousins writes of the rise in popularity of commercialized "nutritive substances" with the intensity of the battle with the AIDS epidemic in South Africa.[61] These products moved north into sub-Saharan Africa and have stimulated new desires. During part of the time that I was conducting research in Tanzania for this project, Omega Wash was a particularly popular South African cure-all. In 2013 and 2014, many visiting the outpatient clinic of the hospital would mention to me having tried Omega Wash or being interested in trying it. In addition, people would walk into EdenMark asking for this remedy by name, and Romana or Jenipha would endeavor to steer them toward EdenMark "equivalents." Plant-based therapies commercialized in Tanzania are formulated in conversation with the "nutritive substances" arriving from South Africa and others from South Asia, the Middle East, and other parts of Africa. These products explicitly join the concept of nourishment with that of building immunity (Kiswahili, *kujenga kinga*, but often seen in English on labels). References to remedies that strengthen "immunity" politely hail Tanzanians with HIV/AIDS and with the broad set of conditions entangled with the ways this epidemic has changed gut health at a population level. Yet, *dawa lishe* makes its own intervention into the public debate over how to respond to the rising prevalence of chronic disease.

The majority of people seeking plant(ing) remedies from TRMEGA are driven by their desire to find alternative ways to address chronic conditions, often already diagnosed at the hospital. The alternative health clinics that I visited, and in which I observed, focused on hypertension, diabetes, rheumatism, kidney disease, cancer, and AIDS. The store shelves and garden beds included treatments for acute conditions. For instance, herbal soaps, pastes, and creams for skin diseases fill the glass counter that greets customers at EdenMark. Nearby, one of the most popular plants in the garden at TRMEGA offered a soft, scalloped leaf that, when chewed, relieves coughs and, when boiled, makes a tea that soothes sore throats. Many TRMEGA members announced proudly that they go to the clinic only for their ARVs and use plant-based medicine for all other health complaints. Still, most herbal treatments available at EdenMark and TRMEGA addressed one of the six biomedically defined chronic conditions previously mentioned or symptoms closely related to them. What distinguishes the social and therapeutic projects glossed as *dawa lishe* is that they are not offered as substitutes for pharmaceuticals; rather, they unsettle the relations that pharmaceuticals have cultivated between people and medicine. They draw attention to the chronic as the time of structural violence, the temporality of the burdens borne by bodies caught in the relentless depletions of economies built on extraction, alienation, and strategic abandonment.

*Dawa lishe* offers plant-based remedies for biological conditions in the service of kindling relations with land in all its forms. The charisma of plants lies in their invitation to labor in ways that generate the times needed to reimagine healing as a disrupting of relations of chronic depletion or slow death. *Dawa lishe* articulates alternative temporal arcs—otherwise obscured by economic logics—along which to reflect on, and intervene into, the material conditions of both plant and human life. Remedies loosen the grip of chronicity as it is linked with conceptions of risk and practices of adherence in biomedicine. When Kijani and Jane point to plants—both foods and medicines—that nourish, they reorient healing and offer a position from which to hold biomedicine accountable to other injuries. Nourishment is a response to the myriad traumas of structural violence, ecological destruction, social abandonment, and physical wasting—to the slow death that shapes life in the wake of the historical dispossessions that drive racial capitalism in the twenty-first century.

In this way, TRMEGA members reveal the risk of defining chronicity as a quality that can be achieved through compliance with pharmaceutical regimes. They draw attention to the ways that this definition forecloses any possibility that the body might become something other than what it is. That is, understanding efficacy as the stabilization of risk shortens the horizon of possibilities. Public health experts work for a future without AIDS, but their vision demands a kind of purity of viral extinction that only comes over the course of generations, never in the course of a lifetime. In contrast, Kijani's notion of "removing chronicity" and Jane's commitment to the possibility of life after adherence illustrate how *dawa lishe* apprehends healing as a process of the body becoming other than what it has been. That which is therapeutic is that which participates in shaping this transition in ways that are more (rather than less) desirable. Again, this is *not* an argument against pharmaceuticals. Jane is clear in her commitment to her treatment regime, and indeed, she helps many remain committed to their treatment regimes. Rather, it is an alternative mode of imagining how we go on when such toxicity is a condition of life.

While the therapeutic value of plants may, at times, be rendered visible through their correlation with biomedical measures like lower viral loads, TRMEGA members insist that plants' therapeutic value cannot be fully captured through such measures. Plants do not (only) battle pathogens and harmful agents inside human bodies but (also) collaborate in changing the field of play. Both timescapes and landscapes. Remedies stage encounters to remake both bodies and environments through acts of ingestion (eating, breathing, touching), digestion (decomposing, recomposing), and expulsion (discharging, excreting, exhaling). In these efforts to care, temporalities collide.

In its attention to specific chronic diseases and its simultaneous reorientation to the times of therapeutic work, *dawa lishe* highlights the risks of chronicity itself, that is, the risk that biomedical notions of the chronic focus in such a way that the cruddy, everyday slow violence—the continuation of colonial dispossession and the new configurations of postcolonial harms—is rendered invisible. Individuals working to survive in the harsh conditions of postcolonial neoliberalism are blamed for not thriving more. The temporalities of biomedical approaches to chronic disease separate present biological functions from the long histories that give rise to vulnerability, uncertainty, and pain of bodies struggling to go on. Chronicity, then, comes to articulate a particular relationship to the past, present, and future. But this raises questions: How else is the relationship between the past, present, and future articulated? What ways of being with the past in the present come to be considered therapeutic?

Those you meet in these pages, I argue, re-theorize relations among the past, present, and future—ancestors, living bodies, and forces of fertility—through action and story. Their social and therapeutic projects put pressure on the concept of the chronic by both revitalizing the near future and developing an experiential understanding of endurance as the force through which relations continue to unfold. Their plant(ing) remedies expand the scope of healing by dislocating, dismantling, and reinventing the (post)colonial logics embedded in medicine's land relations. Ultimately, as they work on the body and on time, they also work on knowledge. *Dawa lishe* is offered to capture this work and build solidarity among projects seeking to expand the scope in which biomedicine locates reproduction and imagines reproductive justice.

I find in TRMEGA's projects—and others like them at EdenMark, WODSTA, and Dorkia Enterprises—a generative conversation with the concept of "collective continuance" as it emerges in the work of indigenous scholars and environmental justice advocates.[62] They locate collective self-determination in a capacity to reflect on, care for, and labor with the forces of life and liveliness—human and nonhuman—that nourish possible indigenous futures. Collective continuance is a refusal to surrender to the apocalypse-that-is-now: the constant negations, the persistent depletions, and chronic injury of modern nation-state governance, as well as the configurations of science, economy, and law that support it. It considers everyday practices of renewal and responsibility, the never-ending hard work of healing and building solidarity. Such concepts also refract through Donna Haraway's concerns about ongoingness, even if not always explicitly in her citations.[63] Collective continuance recognizes an open-ended horizon of social, cultural, biological, and territorial ongoingness. It does not require a return to the past or mark a fundamental or ontological rupture with it. Rather, it acknowledges the dynamic relations that animate processes of composition and decomposition over time. Such ongoingness allows us to hear Jane's effort to contrast the constant replication of the virus that might be slower or faster but remains stuck in a mimetic loop, with a story of reproductive time that allows for radical ontological dynamism. Replication, Jane seems to want to get into our heads, is a useful time for the working of a pharmaceutical, or a moment of diagnosis, but if held too tightly, if allowed to obscure all forms of reproductive time, it forecloses versions of the future that do not simply replicate the present. Nourishing the forces of continuance and ongoingness reimagines the scope of reproductive justice and draws on healing to recognize alternative forms of bodily and territorial sovereignty in the service of futures otherwise. *Dawa lishe* is an invitation to join this collective struggle for therapeutic sovereignty: for the ability to engage in actions that will bring alternative objects of therapeutic practice (e.g., appetites), alternative bodies (e.g., distributed body-land relations), and alternative times (e.g., plant times) into being.

# 5

# Properties of
# Healing

*Dawa lishe* challenges the forms of knowledge and modes of engagement that have emerged at the intersection of science, law, and capital in postcolonial Tanzania. This does not mean that projects highlighted in this book eschew all connections with scientists. Nor that they seek to work in some ideal space untainted by "the market." Rather, TRMEGA, EdenMark, WODSTA, Dorkia Enterprises, and others promote forms of care, modes of exchange, and habits of reciprocity that tie together liveliness and livelihoods in lands drained and bodies left vulnerable through the extractive logics of economic development. Their work cultivates a space of creative infidelity toward the scientific and legal technologies through which plants become available as resources for generating wealth. The techniques through which their projects organize, produce, and distribute their remedies trouble the appropriating arrangements through which the scientific elucidation of plants serves to reduce them to exploitable properties and subordinate them to regimes of ownership.[1] In this way, *dawa lishe* names efforts that take the reinvention of economy and its relationship with ecology to be the site for therapeutic work.

Understanding *dawa lishe* as creative infidelity—as a strategic engagement with science, the state, and the market that simultaneously subverts logics of extraction—requires understanding how these extractive logics have come to define modern plant–human relations in the first place. Chapter 3, "Registers of Knowledge," described the historical and institutional moves through which traditional medicine has come to be a site where, as the Basque philosopher Michael Marder would say, "the economic effaces the ecological." Marder wants to resist the sharp constriction of the ground of politics that this effacement facilitates. For instance, when national and international debates about plant life close around the question of whether plants (their substance, molecular inspirations, or knowledge about their uses) are private property or public domain, they undermine the ability to collectively imagine the forms of agency available through more animated ecologies and the public ethical debates they call into being. He argues for a conception of "ecology" that is not tethered to the economic (even as its other), an ecology that does not rely on a romanticization of "eco" when articulating nonappropriatable relations between people and plants, relations such as the "love for plants" cultivated by Mozambican gardeners about which Julie Soleil Archambault writes in Ihambane, or the ancestral presence in the *vihamba* of Kilimanjaro. Rather, he works to elucidate the constitutive incommensurabilities that sustain the zones of ignorance in European social theory.[2] Highlighting, rather than blackboxing, these incommensurabilities accounts for the obtuseness of those steeped in formal scientific ways of knowing that Jane articulated to John, Victor, and me in the previous chapter ("I know this won't get into your head"). Marder's arguments draw on art, literature, and philosophy to consider forms of attachment, affection, aesthetics, and play that are obscured by the economic. He reveals how power works by concealing the complex emergent qualities of people–plant relations. The focus on "the West" in Marder's work—and in much of the emerging field of critical plant studies—is challenged, however, by scholarship rooted in Black and indigenous geographies. This work has not only illustrated the racial politics of economy's effacement of ecology but also highlighted the potential for healing and reinvention possible in a recognition of modes of dwelling in which not all relations with plants are exchange relations.[3] This book draws the provocation of *dawa lishe* into these conversations by attending to how both entrepreneurial initiatives, such as EdenMark's and Dorkia Enterprises', and nongovernmental nonprofits, such as TRMEGA and WODSTA, take up plant remedies as more than resources for ethnobotanical inquiry, neoliberal reclamations of tradition, or distinctly African solutions to inadequate access to food and medicine.

Extending the methodological engagements of critical plant studies through ethnography embraces efforts striving to kindle alternative relations with plants as sites for grounded theorizing. Such theorizing might offer speculative reimaginings of economy and ecology. In the gardens drawn together by *dawa lishe*, plants are neither fully digestible into regimes of value (nutraceutical, novel molecule, agricultural commodity) nor fully indigestible (spirituality, magic, ecology). Producers engage with the institutions through which the qualities and capacities of plants are made available to be secured as property. They instrumentally use them to establish their legitimacy and develop their remedies. But not only. They simultaneously expose how these institutions structure the "unknowing" of alternative modes of dwelling with and through plants. Producers draw attention to the co-laboring of people and plants and experiment with the forms of property—intellectual and territorial—that might be generated from within this multispecies collaboration. In this way, *dawa lishe* challenges the "rampant economism" that overdetermines dominant ways of articulating impediments to human flourishing both locally and globally.[4]

This chapter explores how a particularly charismatic banana reveals the tensions between the economic and the ecological as it is caught up in efforts to heal land and bodies in postcolonial Tanzania. Bananas are a fertile site to experiment with the ways that the co-laboring of people and plants might disrupt and rework contemporary notions of property. It is not only that bananas are a staple crop in many areas in East Africa, supplying calories and supporting livelihoods. Bananas also offer uniquely evocative ways to think about dwelling— the entanglement of economies and ecologies. Bananas embody reproductive power for those who call Kilimanjaro home, as well as many from Mount Meru, and those farther north along Lake Victoria, where Mama Nguya was raised. They encircle homesteads, nourishing the forces that support fertility. They soften when steamed or boiled over the fire, creating the smells associated with sharing plates with kin and fortifying connections. They are a key ingredient in local beer through which they generate communality, mark marriages, and celebrate initiations. Banana beer poured back into the soil appeases ancestors, solicits spirits, and animates efforts to sustain health and, still today in some places, brings rain.[5] As placentas are buried at the foot of *migomba*, banana trees, they embody the movement of bodies and lands through each other. They hold lineages and carry them into the future. In short, bananas locate health in continuance.

A growing number of social and therapeutic projects in Tanzania are finding bananas creative interlocutors in experiments to seed new modes of dwelling that work in the folds of the economic and the ecological. They offer companionship in efforts to articulate the harms of promoting livelihoods at the expense of liveliness. And they do more than this. Bananas embody a form of life that both exceeds the ways they have been altered by capitalism and remains open to continued alteration as they grow through their entanglements with kin and community, as well as ecological, (post)colonial, racial, gendered, and infrastructural histories. Through this entangled multiplicity of relations and the frictions that inhere in them, bananas resist the disciplining that dissolves their futures into those of the economic (for instance, as a nutraceutical, food supplement, or herbal medicine).

Bananas repeatedly emerge at the core of *dawa lishe* initiatives, exemplifying alternative ways of generating therapeutic, ecological, and economic value (and mediating their entanglements). Their complex social and material lives in Tanzania make them apt collaborators in efforts to move back and forth between regulatory tracks for food and drugs. By unsettling the ontological distinctions that ground differences between agriculture and medicine, both producer and plant are able to elude pressures to conform to toxicological techniques of managing harm and political techniques of managing the traditional. *Dawa lishe*'s political potency as a potential category of knowledge, practice, and collective organizing is a product of practices that both use and misuse, expose and elude, translate and mistranslate, the ontics of the state.

Yet, the work is precarious. The weight of the links between scientific and economic logics, as well as their history and institutionalization through traditional medicine, continually threatens to absorb *dawa lishe* into existing regimes of value. As a proposition, then, rather than an institutionalized category of practice, *dawa lishe* is a way of thinking and acting that is always on the verge. The verge of becoming. Of becoming other. Of disappearing. It is a commitment to working toward new objects of knowledge in the service of land relations that heal—that is, land relations that body forth people and lineages, plants and soils, that are fuller, more mature, more diverse, stronger, or otherwise better able to go on (*uganga*). In so doing, producers engage institutions that reify plant knowledge as property (e.g., laboratory work to confirm the composition of a formula or the presence of particular nutrients) and render plants and their packaging capital (e.g., trademarked formulas or copyrighted company names or blurbs on labels). But they simultaneously undermine the fixity of these configurations (e.g., sharing a cutting of the plant from which the formula is composed, cultivating a network of growers and gardens). As producers navigate regulatory regimes, they seek multiple relations. Even as they subject themselves and their remedies to the disciplining relations of property and the logics of economy, they simultaneously rekindle plant relations that invite the becoming of alternatives. *Dawa lishe*, as a proposition that understands healing and health in the cultivation of such alterlives,[6] exists only insofar as producers and users move their projects forward by explicitly loosening the links between practices that articulate the properties of plants and practices that structure the possibility of their appropriation.

This chapter explores the work of a small business called Dorkia Enterprises, whose two founders are advancing a specific banana varietal as "therapeutic." Their efforts are animated by connections to international food sovereignty projects striving to innovate new forms of property and new avenues to claim rights to food and its means of production. The economizing logics of commercializing this banana varietal, *kitarasa* (Kiswahili), as a remedy, however, threaten to overwhelm *dawa lishe*'s mode of theorizing. Dorkia Enterprises strives to engage, but remain unfaithful to, the economic—that is, to contribute to livelihoods but not to overwhelm alternative forms of liveliness in the process. Uncertainty over how to navigate this tension raises both ethical and practical questions for the founders. Struggles over control of the business emerged as struggles over how to harness the labor of *kitarasa* for livelihoods versus how to work with its unruly liveliness. The divergence of the founders' paths (at least for a time) embodies the tenuousness of the "alterlives" that *dawa lishe* seeks to glimpse. They were up against dominant forms of success, knowledge, and status that all pull toward relations rooted in the alienation of land and labor and their appropriability of resources. There was no position from which to act that was pure, no move that was uncontested. *Dawa lishe* is offered in an effort to build common cause among people and projects committed to generating more liveliness by exploiting the fissures and fault lines of capitalism. Under such conditions, solidarity between people working together, no less among projects, even for a time, is hard won.

To capture this precarity as central to the potency of *dawa lishe*, this chapter moves between two analytic frames: the discussion of Dorkia Enterprises and their work to elaborate *kitrarasa* through partial engagements with the state and the market, as well as an effort to think with *kitarasa* and the liveliness in its insistent excesses of these logics. *Kitarasa* invites us to think along temporal trajectories that extend well beyond the lifespan of a business or its founders. Bananas have transformed landscapes, the matter of soils, and the substance of bodies. Biomedicine's articulation of the therapeutic as biological effects firmly circumscribes *kitarasa*'s liveliness within the "body proper," or what Povinelli calls the "skinned existent."[7] This move to the nutrient and its cellular (or epigenetic) effects is critical to creating economic value. The radical narrowing of analysis to the scale of the human is also critical to sustaining the illusion of bodily sovereignty. *Dawa lishe* recognizes bananas not only as an agent outside the body that might have an effect on it but also as the movement of forces that give rise to embodiments (human and nonhuman) of life. In other words, *kitarasa*'s agency is not only seen as working at a particular scale (the body) but also by generating scale (person, lineage, home, etc.).

"Scale," as Timothy Clark teaches us, "does not constitute some sort of background to experience: it inheres in and effects its basic structure, categories, and openness to phenomena."[8] *Dawa lishe* mobilizes *kitarasa* to generate alternative structures, categories, and sensibilities through which wellness and illness, strength and weakness, might be experienced. I am interested in how therapies fold the scales of bodies and lands into one another in ways that make ecological relations sense-able and dwelling otherwise a place of experimentation. This chapter explores these processes through contests over therapeutic claims, the possibilities of storytelling, and the labor of human and plant remembering. I work with *kitarasa* to make available—even if for just a time—propositions that leverage the liveliness of the plant beyond the ways that it is disciplined by the state as food or medicine and of plant–people relations beyond their legibility in late liberal logics. Together, *kitarasa*, Dorkia Enterprises, and I experiment with generating the spaces (and times) we need to rework modern notions of toxicity and its relationship with remedy, as well as to heal from the chronic depletions and persistent injury of living in a toxic world.

### Charismatic Bananas

Alex Uroki first mentioned Dorkia Enterprises' work to me, although at the time, he did not know the name of the company or its founders. Uroki was in the middle of experimenting with drying green banana peels. We were leaning over his large, wood-fired, industrial dryer filled with pieces of banana peel that he would later grind into a powder and package. He commented that someone else was also making banana flour. I made a mental note to look for this flour in Moshi, where he thought it was being produced. Fieldwork for this book involved following such hints: tracking down collaborators in a variety of vegetal life projects, following the clues on the packaging of therapeutic foods and herbal medicines, and tracing the information offered by *duka* owners about the products they sold. Three months later, I found myself in the offices of the small, women-run business in the Small Industry Development Organization (SIDO) complex in Moshi.

Dorkia Enterprises rented the unit on the end of a short row of brick offices close to the entrance of the industrial park. Pumpkin, red and white amaranth, sweet potato, papaya, and passion fruit welcomed visitors as they passed through the dense, verdant demonstration garden, looking for the door. The founders, Rukia Seme and Dorcas Kibona, shared a commitment to gardening without the use of commercial pesticides and herbicides and sourcing their materials from farmers who only used organic fertilizers. To the side of the small courtyard stood a well-constructed solar dryer. On the first morning I arrived, its shelves were lined neatly with banana slices, *mlonge*, and *matembele* (sweet potato leaves). A glass case in their office held samples of colorful packaged products: *mlonge* leaves and seeds, ground papaya seed powder, avocado seed powder, and sweet potato flour, as well as the banana flour that had inspired Uroki. I would come to hear these two women consistently describe all of these products as *dawa* (that is, products that might be engaged in the service of nourishing, fortifying, building strength, and thickening relations).

I met Rukia Seme first. She offered me lemongrass tea and shared with me her motivations for the business and hopes for the work. In 2015, Rukia was still employed full-time as a government procurement officer in the regional block. She initiated this small business as part of investing in connections that could carry her into and through her retirement. The focus of the business, however, responded to a larger problem: toxicity as a condition of modern life. She began,

> The reason [we work to develop these products] is definitely that we see how much people are affected (Kiswahili, *athirika*[9]) by hospital medicine. Someone uses pharmaceuticals prescribed in the clinic and afterwards [they develop pains]. When they come for a test, they are told that they have stones in the kidney. "Now," they ask themselves, "when did I swallow a stone?" But these stones are a side effect of the drugs they have taken. The hospital medicine enters and hardens. It forms solid things inside the kidney. They accumulate somewhere inside and solidify forming something like stones. But if someone uses our remedies, these are natural. They are grown with a lot of manure. We don't use industrial fertilizers with chemicals. We apply [cow] manure, organic fertilizer. Therefore, there are no "side effects" [used the English for this phrase]. Therefore, there is no harm [Kiswahili, *madhara*] to the body.

The therapies formulated by Dorkia Enterprises are envisioned as an intervention into the consequences of modern medicine. Like Jane, Rukia is not proposing that people turn away from pharmaceuticals entirely; rather, the challenge is to develop the new relations and entanglements that make it possible to survive the toxicities inherent in healing and eating in Tanzania today. She suggests mediating the slow violence of pharmaceutical toxicity by making the move to food. Yet this move is not straightforward.

> If they drink this, they will not experience any side effects *because it is food* [italics added]. Eee, these bananas are eaten as food. But they are a remedy [Kiswahili, *tiba*]. For instance, this lemongrass that you are drinking, it has a good flavor and it is also a remedy. You feel good in your body [after you drink it]. The fatigue ends. The toxins in the body are diminished. The toxins are the result of the maize. Most grow maize with urea fertilizers. The toxins are from the poison that we put on the plants, pesticides. These chemicals enter the food slowly. Then it [the poison] is filtered by your body.

Through the "natural," Rukia dissolves the epistemic boundary between medicine and agriculture. She begins to cultivate the ground to hold both to account. Medicine should be nourishing; food should be healing. The friction she creates draws attention to—demands attending to—land relations.

Because she worked full-time for the municipality, Rukia knew she needed a partner in this small business. Dorcas Kibona was young, energetic, and smart. She did not have another job. They joined efforts through Dorkia Enterprises. Rukia offered vision and financial support for the founding of the business. Dorcas managed day-to-day tasks and committed herself to its development and growth. Neither had the land to cultivate the large quantities of plants needed to make their business sustainable. From the beginning, they worked through church groups and community connections, organizing small-scale farmers. They developed particularly thick relations in Kibosho, a banana growing area up the slopes of Mount Kilimanjaro from Moshi town and Dorcas's *nyumbani kabisa* (natal home). Because Dorcas attended to the everyday running of the business, over the years that followed, I came to know her most intimately. Dorcas invited me into her work and life as she struggled to create more linkages with farmers, as she strategized to secure land to grow on, as she navigated bureaucratic hurdles to certify their products, and as she sought creative connections to introduce these products as therapeutic. We visited farms and government offices together. I witnessed as Dorcas's energy, competence, and humor captured the attention of many, including district officials, café owners, public health specialists, and the Slow Food movement. In fact, the Slow Food Foundation for Biodiversity accepted their demonstration garden as a member of the 10,000 Gardens in Africa project in 2015, only eighteen months after Rukia and Dorcas officially incorporated Dorkia Enterprises.

Their banana flour emerged as the most charismatic of Dorkia's product line. It drew national attention in agricultural fairs and gained international exposure through food sovereignty projects. Dorkia Enterprises' flour, unlike Uroki's, is made out of a single varietal: *kitarasa*. When this green cooking banana is cut, it oozes a distinctive sticky orange sap that stains the pale flesh of the fruit. *Kitarasa* is said to be indigenous to the slopes of Mount Kilimanjaro, one of dozens of varieties of bananas and plantains that grow on the mountain. Women walk down the slopes of Kilimanjaro each morning, entering the town of Moshi carrying meter-long, shallow, rectangular baskets on their heads, laden with 20 kg or more of various banana varieties. I never saw *kitarasa* being carried to market, however. *Kitarasa* is rather mushy when boiled. It does not maintain the texture and integrity of the cooking bananas prized by many in the region. Some people from farther north are downright scornful of its consistency. In the dense banana groves that surround homes on the slopes of Kilimanjaro, however, *kitarasa* are still found. Their suckers spread out from the corm, the rhizomatic tuber at the base of the plant, birthing new plants. In the past, the texture of the cooked *kitarasa*, as well as its potency, lent itself to being prepared as soup for infants and the elderly. In particular, during rainy seasons when it was cold, and bodies grew achy and joints rebelled, this soup satisfied and soothed. I also have heard that breastfeeding mothers used a soup of *kitarasa* to increase milk production. Over generations, many have continued to make affordances for *kitarasa* plants, even if they do not cultivate them in large numbers, find that their yields are relatively low, and see little commercial benefit.

Dorkia Enterprises' *kitarasa* flour was conceived through narratives from both Rukia's and Dorcas's childhoods. This Dorkia product emerged from their engagement with the worlds and ways of eating that grew them—that is, from the ways mothers and grandmothers fed them, cultivated their growth, saw them and their siblings and cousins mature, and from the extension of these techniques of cultivating strength through gifts to their neighbors.

> We were given medicines. Every three months or so, mother would boil herbal medicines. These leaves are just out there. She grew them there in the farm. She would pick them and give to us. There was no suffering with the flu. There was no feeling cold. There in the village there was no feeling cold. The body was "strong" [English] because of traditional food and because of those plants that mother was giving to us. At the time, there, some people were wondering why my mother's children did not fall ill, even when they went to the fields. We were cultivating where

there were many mosquitos, but when we went there, even when we were bitten by mosquitos, malaria did not catch us (*malaria hayatushiki*). Completely, completely. Therefore, people were saying ok, this woman is good. She has things that she is giving her children. But mama was not selfish (*mchoyo*) she gave them medicine (*dawa*). She told them to gather the leaves and to boil them. Because she grew them, you did not need to go search for the plant. She grew them in the farm. You just pick them up in the farm. For this reason, I am grateful. Her actions gave me the confidence to do what I am doing now.

Dorcas's childhood recollections involve being cared for before sickness, treated before a complaint arose. *Kitarasa* was part of this way of living with, and through, plants to cultivate bodies that are "strong."

Dorcas has vivid memories of her mother cooking *kitarasa* when she was a child on days when she was able to buy fresh fish. Others remember that *kitarasa* was used in times of hunger, not just out of a more desperate need to fill stomachs but because it offered fortified bodies in trying times. For people living on Kilimanjaro and later Mount Meru, where *kitarasa* spread, eating *kitarasa* was a response to depleting contexts such as hunger, diarrheal conditions, childbirth, and old age. Now, Dorcas understands everyday life as depleting. The toxicity of foods grown with pesticides, modern chickens raised with hormones, pharmaceuticals consisting of poison, and more compels her to build on the memory of her mother by cultivating remedies that support the notion of nourishing and healing as an approach to living.

Dorcas learned from her mother how to prepare banana flour. In Kibosho, her home area, however, this flour was made with a different varietal, *ndizi ng'ombe*. A German missionary introduced Dorcas to the process and desirability of making flour from the banana with distinctive orange sap: "I have been improving [what my mother did]. By this I mean, my mother just instructed people to gather [the leaves of] a tree, to pick from that tree over there and. . . . But now I have entered a process that is not like my mother's. I have entered into a process of improving [the plants]. I am putting them in a state that will enable them to reach [abroad]." The potencies of *kitarasa* draw on its capacity to gather strength across these different temporalities and spatialities.

Dorcas points to *kitarasa*'s orange sap as evidence of its power to heal. She argues that eating porridge made from the *kitarasa* flour she produces will help people manage diabetes, hypertension, rheumatism, obesity, and a range of other conditions. "*Kitarasa* has medicine in the sap. This banana has sap that is different than that in all the other bananas in the world," she exclaims. "The sap

FIGURE 5.1 *Kitarasa* banana broken in half to display orange sap, Moshi, Tanzania. Photo by Sabina Mtweve, 2021.

is the medicine. The sap is insulin. There is a large amount. That is why when someone with diabetes eats it, even if their sugar level starts at 30, it goes down to 6. This person will not have used a shot. If a doctor asks him, 'What have you used?' He will tell him, 'I ate food. I ate *kitarasa*.'"

## Claims and Labels

Between 2014 and 2016, as *kitarasa* attracted more attention and Dorkia Enterprises' ambitions expanded, the company faced increasingly complex issues of regulation. The process was frustrating. The Tanzania Food and Drug Administration (TFDA) municipal officer seemed critical and dismissive, if not openly seeking a bribe. While the regional officer who visited the facilities was more encouraging, he recommended structural changes that took time and resources. The Tanzania Bureau of Standards (TBS) fees were manageable, but the TFDA registration fees felt burdensome. Navigating each new bureaucratic hurdle trained Rukia and Dorcas to articulate the value of *kitarasa* in the lexicon of the state. The TFDA's requirements in particular drew Dorcas's ire, because they obscured what she and Rukia saw as *kitarasa*'s therapeutic value: "The challenge now for us is that TFDA as TFDA wants us to do it as food. They don't want you to do it as a therapeutic food (*chakula dawa*). They don't want that. There is even our brochure which says it is nutritious, curative, and preventive. They are saying, 'We don't want that. If it is a nutritious food, we want that it remains just a nutritious food.'"[10] For regulators, this distinction between the nutritious and the therapeutic is ontological—and therefore consequential. That which is nutritious must by definition be governed as a food. That which is therapeutic must by definition be governed as a drug. Claims determine which regulatory track producers must follow, which staff reviews their applications, which forms of evidence are required, and which other institutions must be involved.

Claims become contentious because they insist that producers coordinate their engagements with legal, scientific, and economic institutions through their commitments to locating efficacy "*in* that banana" (emphasis added). As the staff in the Complementary Medicines Program at the TFDA explained, "When someone claims that this *kitarasa* maybe it has this, this, this, it could be true. But it is necessary that the person has to go further. They must say what component it has, which [component] has the ability to help with those issues, because for you to be able . . . for something to be able to help, say for instance with diabetes or whatever, it means that there is necessarily a certain component in that banana which does that. Now for us in Tanzania it is very expensive for someone to do such a study." Categorizing *kitarasa* as a food *or* a medicine, and circumscribing its claims accordingly as nutritious *or* therapeutic, shapes the way that their product relates to laboratory science. The following exchange in the TFDA headquarters in Mabibo, Dar es Salaam, between the same staff member and John Ogondiek, my colleague from the National Institute of Medical Research, illustrates this dynamic:

> TFDA: But again . . . in light of our regulation and guidelines, we can't just allow someone to just write or say that if you eat *kitarasa* banana, "you will be cured . . . it will help with diabetes." This claim raises many questions.
>
> JOHN: So the issue is labeling; that is the major problem.
>
> TFDA: It's a major problem.
>
> JOHN: But how do you advertise it? Again, if you want to promote it, those are the things which attract the customers.

TFDA: But also, to be allowed to promote something, you must have data and evidence. That fact, when I say this—it has this effect [must mean] I have already done this and this and this in relation to what I claim. Maybe I have introduced [the treatment to] people with this problem. They did it for a certain period. I monitored their health and I concluded that. As you know, to come to a conclusion there are a lot of things to do. Research entails a process of exclusion of everything [except the treatment] so that you can conclude directly that this banana really brings about this effect. It is a big study and it is difficult. When you give someone that [banana] but at the same time also advise on dietary changes, it is a lot of things. Stop meat. I don't know don't do this or that. Do not eat fat, but eat this banana. Now you cannot tell what helped him directly to be where he is. The banana is not the only factor. There are also these things he has left or whatever. So, it his hard. That is why we normally do not accept claims without having proper data, scientific data, which shows someone did this and this.

As our TFDA colleague explains, "Research entails a process of exclusion." Claims, she insists, must speak to a direct response between a specific chemical element in the banana and an observable, repeated, and therefore measurably predictable effect in individual biological bodies upon exposure. The TFDA adjudicates the legitimacy of claims on labels based on their ability to harness scientific studies isolating reactions in laboratory settings. Those that refer to scientific studies authorizing the effects of active ingredients are endorsed as therapeutic; those who do not, or cannot, must restrict their claims to nutritional content. This capacity to harness published scientific work establishes a boundary between the therapeutic and the nutritional, between drugs and food.

In contrast, *dawa lishe* operates through a process of multiplication rather than exclusion. As Rukia and Dorcas suggest, the orange of the sap is one sign of potency. It might be or contain "insulin." But for them, *kitarasa*'s efficacy extends beyond isolated biochemical compounds. The ways that *kitarasa* and its orange sap move through human and nonhuman communities in the region locate efficacy in its ability to transform relations beyond the body proper. Their efforts to commercialize *kitarasa* flour build on these strategies for motivating change. Eating *kitarasa* may, for instance, mean not eating other foods. Buying the banana flour may put one in touch with people like Dorcas, who follow up. It may draw one into a community of support and care. It may encourage a person to shop in a store where they find other foods that are beneficial to their health. It may make them think about their health more broadly, or it may connect them to their grandparents, extended kin, and the land in ways that influence their attachments, the people they depend on, and those they care for. It might draw them into farming or reinvigorate and diversify the farming they already do. These relational effects—what the TFDA does not recognize as legitimate responses to the banana itself—engender a fundamentally different approach to understanding efficacy. Multiplication as method for creating knowledge has the potential to reveal the processes of "unknowing" embedded in the exclusions inherent in scientific research. It establishes an approach that consistently decenters the forms of remembering and forgetting that sustain the state's institutionalized narratives.

Labels, as John skillfully highlighted in the conversation above, are one site where this struggle plays out. The criteria through which the TFDA distinguishes between "promotion" and "fraud" rest on a singular ontological understanding of plants as bundles of biochemistry. Extensive discussion on such epistemic distinctions between truth and falsity, knowledge and charlatanry, defines the field of science and technology studies. Since Barnes and Bloor's early arguments about the need for symmetry in sociological studies of scientific knowledge through the many challenges, corrections, and modifications of this theory over the past forty years, examinations of how claims to truth are evaluated expose the limits and politics of science.[11] This focus on symmetry shapes one version of a story about the struggle over labels. *Dawa lishe* orients us a little differently, however. Practitioners are not committed to critique as an end goal or to historicizing epistemic boundaries as the heart of their work. They want to unsettle the exclusions embedded in the way that the state organizes relations between science, economy, and law. Rukia and Dorcas argue that *kitarasa* is therapeutic, not by disciplining the banana but by expanding the notion of healing and multiplying the possible trajectories along which bodies and ecologies might alter, as well as the possible futures toward which people and plants might flourish together.

Importantly, Dorkia Enterprises does not offer their *kitarasa* flour as a pharmaceutical equivalent (sap with insulin) to facilitate eating increasingly homogenized diets high in salt, sugar, and fat on increasingly smaller parcels of land. That is, Rukia and Dorcas do not offer *kitarasa* as a salve to the wounds of economies and ecologies built from the alienation of labor and land. They refuse to position their *kitarasa* flour as (only) a technical fix for the health problems created by industrial food systems and land dispossession. Instead, they offer their flour to animate the diversity of *vihamba*, energize the collective work of women in rural areas, expand organic cultivation, and re-member relations that nourish the fertility of people and soils simultaneously. This represents a fundamental challenge to the state's regulatory thinking. Whereas our interlocutor at the TFDA pointed us toward research as the mechanisms through which a plant or aspects of it are taken up as the protagonist in a story that serves to define the ontics of food and drugs, Rukia and Dorcas pointed us to stories of childhood, of eating, and of illness that reveal healing as inherently relational. In so doing, they destabilize the ontics of the state and, I would suggest, the relations between economy and ecology they embody.

The next section considers the forms of research that authorize different types of protagonists in plant stories (nutrients vs. active ingredients) and that articulate different forms of biological engagement (the nutritious vs. the therapeutic). I explore how this research exposes the limits of the stories that can be heard and those that might not be able, in Jane's words, to "get into your head" (or at least the heads of the TFDA staff). Dorcas, Rukia, and Jane strive to cultivate relations with plants that can neither be dissolved into the economic (effacement) nor be cast only as refusals of the economic (resistance). The work to secure claims to *kitarasa* as therapeutic is an effort to destabilize the ontological commitments that sustain the intellectual claustrophobia created by tethering all thought and politics to the economic.

### Testing, Testimony, and Other Ways of Storying

Testing and testimony have emerged as two dominant modes of storying relations with plants within the circumscribed work of the economic. These approaches represent different ways of understanding how plants work and what counts as evidence of their effects within commercial and regulatory contexts. On one hand, testing involves scientific trials of proof that repeatedly identify and isolate internal elements of the plants. For foods, nutrients are identified and measured. For drugs, more intensive trials are required to trace and evaluate their effects as caused by particular, discrete, identifiable phytochemical configurations (as "active ingredients" listed on labels). The type of testing differentiates the horizons of claims: nutritional or therapeutic. As evidenced in the struggles over the labels of Dorkia Enterprises' banana flour, state institutions that test plant products have the power to validate and invalidate claims about a product's benefits. On the other hand, testimonies are experiential narratives of people who have used the plant and sensed changes in their lives. Testimonial statements may be offered as complementary evidence, personalizing claims formulated through scientific testing. Testimonials may also arise outside of any relationship with, or reference to, testing; they may rather be posed as an alternative way of knowing and a different form of evidence. Testing and testimony offer a productive site to examine how *dawa lishe* providers try to unsettle dominant relations between the economic and ecological, inventing a path that would alter ways of being.

Nutritional studies articulate the chemical composition, as well as physico-chemical and functional properties, of banana pulp and peel flour in an effort to develop new applications.[12] A study comparing the antioxidant compounds in Cavendish banana peel and pulp extracts—the variety of banana found in almost every supermarket in North America—found that the content of anti-oxidant compounds was higher in the peel than in the pulp.[13] These findings have spurred speculation about possible ways to utilize banana peel as an innovative ingredient in various food and nutraceutical products.[14] Whether articulating difference across banana varietals or parts of the banana, these studies specify a material's properties (and relevance) through an identification of its internal characteristics. The replication of results necessary to scientific knowledge supports the circulation of products necessary to bodily, organizational, and monetary economies (including those of academia). For Dorcas, testing promises to establish *kitarasa* as a stable entity, consistent over time and space, and therefore durable enough to support commodification and accompanying property claims.

To strategically navigate the power of both testing and testimony, Dorcas has cultivated a powerful endorsement from Dr. Shao, a former vice provost of Tumaini University and director-general of Kilimanjaro Christian Medical Centre (KCMC), one of the top three teaching research hospitals in the country. This institution dominates the landscape of Moshi town, and Shao's reputation for driving KCMC toward research and expanding international collaborations remains powerful even after his retirement. When his tenure at KCMC came to a close, he continued clinical work at the Moshi Occupational Health Centre and started the Better Human Health Foundation to ensure that the medical research collaborations he had cultivated stayed in Moshi. Dorcas touts Shao's use of Dorkia's products among patients in his hospital and his support of their work. Yet the properties of his testing (qualities internal to plants and bodies) and those of his testimony (relations that body forth individuals, institutions, and ecologies) are very different.

Dorcas has also invested in gathering experiential accounts of *kitarasa*'s effects elsewhere. For a couple of years, she invested time teaching the cooks of a canteen in the municipal offices how to prepare a porridge from her *kitarasa* flour, and she spent hundreds of hours in the café talking to middle-class government officials who dined there. Later, she joined up with a retired municipal health officer who had started an NGO focused on home care for the elderly.

Dorcas traveled with her for outreach. She talked with groups of elders, making ties between their childhood memories of eating and her new products. She worked to cultivate a taste for eating in ways that have therapeutic value, as well as a taste for the *kitarasa* itself. In the process, she has collected numerous testimonies: narrations of the intersecting vulnerabilities that *kitarasa* (and other products) have helped people move through. Dorcas does not oppose these testimonies to the possibilities of testing, nor does she fold them into the logics of testing as qualitative data. She does insist on keeping them alive and in conversation with the pressures that drive articulations of therapeutic value considered legitimate by regulatory bodies like TFDA.

When Dorcas gathers together the expressions through which people share their experience about eating *kitarasa*, it is not with the intention of establishing the ways in which they replicate each other (or at least not only). Rather, these testimonies are powerful as they layer on top of each other—and indeed, once she starts recounting testimonies of clients, they come in a flood! Their potency is in the ways that they sound and resound across one another. Each story comes to exceed any individual account of eating *kitarasa*. The formulaic qualities of testimony can raise different questions than the details of individual trajectories.

Similar to how Luise White teaches us about rumor, I would like to suggest that testimony is more than a recollection of experience; it is a genre of storytelling. Testimonies of eating and being healed by *kitarasa* might animate "stories to map the landscape outside of lived experience."[15] The power of layered testimonials lies not only in how they render visible the forces shaping one as a subject but also in how they invite engagement with forces that might enable one to be otherwise. That is, testimonials for *kitarasa* and other remedies can be read for evidence of that which is bearing down on the speaker(s), the forces of control or dependence that organize being and acting through both submission and resistance (e.g., the pharmaceuticalization of the right to health). Testimonials of healing through plant(ing) remedies can also be read as evidence of—and further invitation to cultivate—forms of affection, ways of knowing, modes of reflection, and kinds of labor that nourish alternative forces of lushness. As Dorcas gathers and shares testimonials of *kitarasa*, she seeks not only to expand who can speak for, with, or about plants but also to draw others into relations, connections, multiplicities, incorporations, and excorporations that might reshape the speaker(s) and the listener(s).

Testimony, however, also has a history—a history entangled with missionization, postcolonial development, and the binaries of modernity. Not only does the narrative form of testimonies to the healing effects of plant-based remedies resonate with the form of narratives that parishioners offer in churches throughout East Africa as they bear witness to God and to their subjectification, but also testimonies to the healing effects of plant-based remedies are themselves offered in the church. Uroki's pastor invited him to speak about the benefits of EdenMark products from the pulpit. His pharmacy shelves hold products he has purchased from a Kenyan pastor, whose remedies inspire his work. Mama Nguya and Jane informally support the work of a Catholic HIV/AIDS clinic as TRMEGA welcomes their patients who need additional support to benefit from the ARVs the clinic distributes. As argued earlier, such moves are not coincidental but part of efforts to cast *dawa lishe* projects as ethical. They evoke connections between the religious and therapeutic in both the form and content of testimonies to the efficacy of plant-based remedies.

Early on, HIV/AIDS programs in Tanzania explicitly collaborated with churches to expand testing and organize support groups. In "Antiretroviral Globalism, Biopolitics, and Therapeutic Citizenship," Vinh-Kim Nguyen describes the care and support provided by faith-based organizations in HIV/AIDS programming in Africa. His account provides a fuller picture in which to apprehend the resonance that emerged between Christian narrative forms of testifying to being a subject of God and public health narrative forms of testifying to having HIV/AIDS.[16] Testing programs and their companion support programs sought to "empower" people through support groups that called for declarations of one's status (not "hiding," in the name of fighting stigma) and ongoing claims to positive, healthy living.[17] Group members learned to fashion themselves as particularly worthy recipients of very scarce medications and to advocate for resources from both governmental and nongovernmental programs. Participants' experiences in these programs illustrate the ways that testing and testimony have become intimately entangled in public health discourse. Testing establishes the truth of the body in laboratory results (objective knowledge), and testimony establishes the truth of the individual through experience (subjective knowledge). The exemplary person with HIV/AIDS is an ideal forged through these complex knowledge–power relations that divide subject from object, the modern from the traditional, science from culture, while also structuring their entanglements. The resulting programming creates the modern ethical subject through identity and adherence.

Jane expertly mobilizes the genre of testimony shaped by the politics of HIV/AIDS in Tanzania. Yet in the clinic, in her visit to parliament, and in her leadership of member groups for TRMEGA, she also begins to loosen the knot of testing and testimony tied so tightly in HIV/AIDS programming. She—and, as we have seen, many others throughout this book—moves with plants. Seeds travel in pockets, cuttings are given away, canisters of herbal formulas are sold and gifted, and people are pulled through space, noticing, touching, and at times tasting the vegetation. By locating work and healing in this movement with plants, Jane embodies *dawa lishe* as an experiment in an alternative style of narration. Her stories do not escape but rather unsettle the dominant role plants play in the history of biomedicine. Similarly, as Dorcas gathers stories of *kitarasa*, she layers them together in an effort to unsettle the forms of control that link health and governance through the TFDA and other state institutions. Collecting and recollecting summons alternative webs of attachment with and through *kitarasa*. She repeats the stories to me, to state officials, but perhaps most importantly to those she advises in her office, in the municipal café, in the elder groups, in the shops where her products are sold on consignment, and to *wagonjwa* (people who are ill) elsewhere. This continual telling and retelling gives the stories of *kitarasa* space to bounce off each other. They play across ways of knowing and healing that invite us to listen differently—that is, to listen for something other than evidence of an empowered (or unempowered) subject or an effective (or ineffective) object.

As stories of *kitarasa* extend out into the world, turning and returning around each other, they render visible the forces of injury and depletion that caused this varietal to recede previously and to resurface now. These tellings become evidence of the slow violence of colonialism, dispossession, industrial agriculture, and environmental degradation that disfigured relations through which the forces of life move. In these layered stories, there is also a composing and (re)composing of attachments, as well as an experimental conjuring of new and old ecologies of being and belonging. *Kitarasa* offers glimmers of living that do not invest in the stabilization of person as agent, body as object, and plant as resource, distinctions that structure the bodily and territorial sovereignties of liberal rights. Stories of its therapeutic value, its "indigeneity," its culinary history, its use in treating sick or wounded cows, its connection to the health and recovery of women after birth, are choreographed and rechoreographed by NGOs and small-scale entrepreneurs for international audiences. Below, I am particularly interested in how this memory work resonates with the food sovereignty movement and how Tanzanians are leveraging these connections to create spaces to cultivate alternative forms of sovereignty in Africa today.

Before describing this memory work, however, and its global alliances, I want to pause to discuss the ways that the Tanzanian state strives to control these stories of healing with, through, and alongside plants. Whole, raw plants are unruly. They slip through and around the rampant economism that shapes the logics of the state. In so doing, they offer a vocabulary through which to share glimpses of that which cannot be dissolved into the economic. The TFDA and the Ministry of Health have devised a unique way of managing this unruliness of relations—those that can neither be disciplined by testing nor captured by testimony. By operationalizing "rawness" as a legal term and tasking the Baraza, as it was called into being by the 2002 Traditional and Alternative Medicines Act, with its management, they intervene in healing through radical localization (spatially and temporally) of plant medicines (for more on the Baraza and the history of the institutionalization of traditional medicine, see chapter 3).

## The Work of Rawness

Dorkia Enterprises is not alone in grappling with the obstacles that testing poses to the therapeutic claims it makes for its *kitarasa* flour. Rukia's and Dorcas' frustrations highlight the power in the separation of food, drugs, and whole, raw plant-based remedies. The last category strives to capture the many plant-based remedies that do not fit clearly within the regulatory responsibilities of the TFDA as food or drugs. It was called into being through the Traditional and Alternative Medicines Act.

Chapter 3 described the government's research units that focus on traditional medicine in the Tanzanian National Institute of Medical Research and in the Muhimbili Center for Health Science, Tanzania's primary teaching and research hospital. Both units mobilize the laboratory as a technology to attune matter to the logics of the economy. Plants become bundles of biochemistry to be managed and capitalized on. Processes of extraction, synthetization, and formulation emerge as grounds for claims to intellectual property. Yet, the use of whole plants, no less combinations of whole plants, throws a wrench in the linkages that facilitate these translations. While I have written elsewhere about the implication of legal technologies that enable the owning of plant life, here it is enough to say that even with these legal innovations, whole plants present a challenge to property regimes.[18]

Tanzania strives to manage the potential of whole-plant remedies to disrupt the processes through which the economic assimilates the ecological bureaucratically. The state can channel some whole-plant products through regulatory agencies as food or food supplements. In a country where herbal medicine is used widely, however, this strategy is only a partial solution, perhaps most immediately because herbal medicines could overwhelm the system. But this is not just an issue of state capacity. African healing has long posed a challenge to sovereign powers. Modern witchcraft ordinances, passed down from colonial legislation, recognize traditional healing practices as a potential mode of resisting state authority. The Tanzanian government wants to control both the toxicity of plants to protect citizens and plants' potential to organize against state authority itself. Furthermore, designating whole plants with therapeutic potential as food limits the ability of the state to protect indigenous plants and local knowledge. As Tanzania and many "Least Developed Countries" negotiate with the World Trade Organization over compliance with international property law regimes, they strive to protect such knowledge by reimagining both private property and the public domain. Whole plants raise a dense thicket of issues. "Rawness" has become a key innovation for the Tanzanian state as it navigates this complicated legal and political terrain and strives to manage the therapeutic value of plants under the sign of traditional medicine.[19]

The 2002 Traditional and Alternative Medicines Act tasks Baraza with registering all local raw plant-based remedies for which program officers at the TFDA cannot gather the sufficient scientific evidence. The requirements to register a traditional medicine with the TFDA entail kinds of scientific evidence that are not readily available for local remedies that are widespread in practice. Therefore, to avoid the criminalization of local healing practices, the act authorizes the Baraza to register locally produced "raw" remedies. In contrast, international remedies for which there is not sufficient scientific evidence to be registered through the TFDA are technically not legal to import. Yet, despite the clarity of the law, because resources for enforcement are severely limited, there are many small shops throughout the country filled with herbal remedies from India, Yemen, the United Arab Emirates, China, Kenya, South Africa, and elsewhere.

By providing a legal mechanism through which to legitimate local healing practices, "rawness" offers the state the ability to continue developing institutional capacity and legal structures around plant medicines as "drugs," without holding all traditional healing practices hostage to scientific evidence that is not forthcoming. The TFDA works with government research institutions in relation to a few plants and products. For these scientists, the causal links that inhere in local, "raw" remedies are theoretically knowable. Yet, the number of biochemical possibilities is beyond their capacity to elucidate with the technologies and funding at hand. For this reason, the Tanzanian state, in the name of local knowledge, facilitates the possibility of producing plant therapies despite not yet knowing which biological mechanisms or processes are responsible for generating the therapy's effects, if they are efficacious in relation to specific diseases, and if any individual formulation falls within the technical thresholds of safety, *as long as they are not "scaled up."* That is, the law sharply circumscribes the circulation of "raw" therapies. They are to move only from the healer to the patient. They must remain specific and specifiable, until the property or properties of the plant material can be identified. Rawness becomes a conceptual category through which plants are held out as appropriable in the future. The scale of patient–healer interaction becomes both irreducible and unscalable, until scientific studies can articulate molecular matter that can be separated from the healer through property regimes and made to circulate more widely. The local is sacrificed in lieu of a future global. In the process, all nonappropriable matters are also strictly localized.

*Dawa lishe* insists on cutting differently across these pairings; producers strive to put whole plants and combinations of whole plants into circulation well beyond the localized exchange of healers and patients. As producers source their plant material in volume, organize production, and aim to distribute products widely, they consider the kinds of land relations at stake in the therapeutic. Rawness offers a less dynamic conceptual category in these endeavors than the prospect of storying plants. I am interested in the ways that producers have found common cause with food sovereignty movements and the memory projects that animate activism within them as they seek to slip out from under the disciplining pressure of rawness.

Remembering becomes a method within projects to experiment with innovating just relationships between economies and ecologies through commercialization. In fact, this chapter is an experiment itself, an attempt to remember *with* (not just about) *kitarasa*. *Kitarasa*, a banana said to be indigenous to Kilimanjaro, has arisen as a particularly charismatic actor in the rise of therapeutic foods and as a site for the projects of *dawa lishe*. The charisma of plants like *kitarasa* is a product of their ability to move (and facilitate movement) between scales of knowing and acting—local and global, indigenous and cosmopolitan, corporeal and ecological, chemical and historical, mythical and analytical, sensory and scientific.[20] Thinking with *kitarasa* and its interscalar capacities, I draw together an account of the techniques for re-membering that are mobilized by activists working to create alternative stories. They seek narratives that open arguments—rather than close them—over the properties of therapeutic potency in an effort to establish conditions for innovative forms of property claims. An account of *kitarasa* and its collaboration with humans might share the banana's rhizomatic expansion—continually reaching out, reproducing, regenerating in ways that subvert economic logics.

Dorkia Enterprises refuses to be restricted by the scalar limitations of the state's category of rawness. They want to circulate their flour broadly. As they expand their efforts and negotiate their claims to existing forms of property (both material and intellectual), they also simultaneously organize the relations of their flour's production and circulation in ways that draw users into its nonappropriable relations. Dorcas works to teach people to love *kitarasa*. She works to story it through recipes new and old that cultivate a taste for its flavor, and she invites people to marvel at its orange sap. As Dorcas reaches out to women's groups to source *kitarasa*, those who still cultivate *kitarasa* in their *kihamba* remember its connection to the diversity and multiplicity of childhood home gardens. When others desired to contract with Dorkia Enterprises to sell *kitarasa*, this history was extended to them from women in their church groups, together with small sprouts for their gardens. The soils of the other *vihamba* (plural of *kihamba*) came to remember *kitarasa*. Accounting for this way of linking livelihoods to liveliness requires an ethnographic analysis that allows notions of the therapeutic to be moved by different temporal imaginations (and by the futures they might engender).[21]

Only by training attention on the co-laboring of people and plants have I apprehended *kitarasa*'s participation in debates over what it means to heal in a toxic world and to account for how it remakes the times and spaces of the therapeutic. In caring for the affordances that constitute strength for dwelling in a toxic world, *kitarasa* expands the political beyond debates over access (i.e., human access to plants or medicine or markets), representation (i.e., which people can speak for which plants), and thresholds (i.e., how much harm is worth enduring for how much benefit). By emphasizing *kitarasa*'s uniqueness to the region of Kilimanjaro, its "indigeneity," Dorcas gestures to a broadly popular understanding in Tanzania of the inalienability of the lineage–land relations in this region. Even for those outside Kilimanjaro, buying the flour and eating *kitarasa* gently attuned their bodies to the flavors of a lushness that had enabled response to periods of hunger in the region. While rawness produces a localness that continues to reinforce histories of dispossession and obscures their embodiment through chronic vulnerabilities and persistent illnesses, *kitarasa* remembers the forces through which wellness, strength, fertility, and the capacity to heal are nourished. This "food's" therapeutic qualities are elucidated through the ways that eating *kitarasa* orients us differently by re-membering; it gathers in its substance the movements that made people and landscapes what they are today.

## Memory as Method

With persistent work and accommodation, Dorkia Enterprises succeeded in moving their *kitarasa* flour through the Tanzania Bureau of Standards (TBS) and the TFDA and registering it as a food. Dorcas responded to their structural demands and incorporated the sorts of testing required to establish standards. In 2016, she traveled to Dar es Salaam to receive a national award on behalf of Dorkia Enterprises from the TBS for their *kitarasa* flour, which was named one of the fifty best Tanzanian brands. The same year, their *kitarasa* flour was accepted into Slow Food's Arc of Taste, a project establishing a plant-based catalog of "culturally significant foods in danger of extinction." Finding common cause with food sovereignty rather than access-to-medicine movements invites subjection to a different set of social-material arrangements. This reorientation opens up another set of connections along which to organize. In the process, it gestures to an alternative relationship between the properties of therapeutic efficacy and those of political and economic value, and a different conception of health and healing.

Toward the end of the Terra Madre gathering for Slow Food in September 2018, I attended the gathering of the African delegation. The Arc of Taste project was raised as one of these lines of organizing. The goal was ultimately not in the impressive display made every two years in Turin, Italy, or the polished catalogues of items from around the world, but rather in the process of talking with people to gather their stories. This memory work led to the identification of *kitarasa* and the recognition of Dorkia Enterprises. These practices of remembering initiated Dorkia's relations with Slow Food, highlighted the company's work, and motivated the move through national regulatory agencies. Their success and Slow Food's focus has generated broader interest in *kitarasa*. The Slow Food convivium organizer for Kilimanjaro is now working with two other women's groups in Rombo, farther up the slopes of the mountain, to expand *kitarasa* production and organize the processing in ways consistent with agroecology principles and an economics that strives to sustain communities.

This project pushes against traditional articulations of food sovereignty, which is about the right to determine what one grows and eats, the conditions under which it is cultivated, and how it is distributed. In the Arc of Taste, however, the question of sovereignty expands to include re-membering (in the sense of reassembling and reconfiguring) relations between plants and people. For some involved in the movement, this process serves as a tactic to establish connections between people and (specific) plants that might then be used to advocate for land rights and property claims.[22] When Tanzanian producers of *dawa lishe* experiment with such alliances, the stories they tell help to re-collect and to elaborate relations that might articulate therapeutic value otherwise.

These efforts resist the ways that modern agriculture mobilizes notions of toxicity and its relation with remedy to focus debates on thresholds of harm that

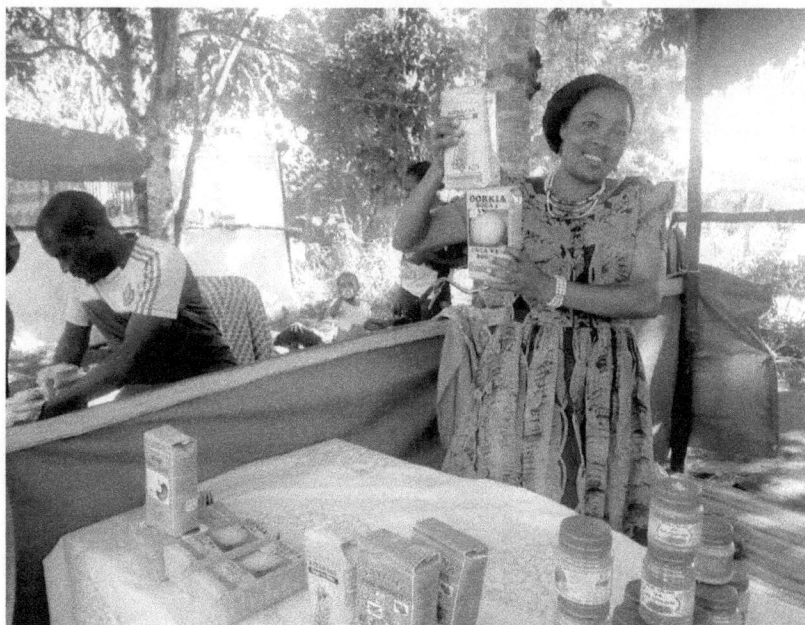

FIGURE 5.2 Dorcas Kibona displaying Dorkia Enterprises *kitarasa* (*top*) and pumpkin (*bottom*) flours at Nane Nane, the national agricultural fair, Dodoma, Tanzania, 2016. Photographer unknown.

might be born in the service of productivity. When considering the therapeutic value of plants, this resistance raises fundamental questions about health. For Dorkia Enterprises, attributions of healing are assessments of how bodies and plants are put into motion through each other, through the multiplicities of the soil, and through certain social and economic institutions. For this reason, testimony emerges as more than a data point or evidence but as a movement: a process of hailing embodied subjects through relations that counter depletion, deterioration, and weakening.

Human stories of plants, however, may also be absorbed by—even extended through—the market. After all, for some, food sovereignty is about expanding which humans have control over food. In these initiatives, the central question is one of human rights, not of ecological assertion. In their emphasis of excess, resistance, recalcitrance, and nonhuman agencies, however, memory projects may—at times and for some—also provide redirection. The question, then, may become, what might plants re-member? What stories might emerge if we listened not only to human memories of *kitarasa*, but also to those memories carried in *kitarasa*'s relationships with other plants and planting techniques as well as with soil, water, and other species? How might we account for the worldings produced as bananas and humans are put into motion through each other? How might attending to plant memories transform our understanding of sovereignty itself?

## Plants' Re-Membering

Neither political, economic, nor affective arguments are enough to engage these questions of what *kitarasa* remembers and what it has forgotten.[23] Bananas have changed landscapes and bodies, linked continents and hearths. Bananas embody—in their substance, in their capacities, in their vulnerabilities, and in their distribution—the pressures and frictions of the past. They remember in their fertility and in the narrowing of their diversity that people and bananas have emerged together over millennia. In Kilimanjaro, bananas and humans have supported each other's flourishing, even as what it means to flourish has changed with shifting environmental, social, political, and economic conditions.

These histories of East Africans and bananas working through each other, creating spaces to dwell, are thought to be an extension of a longer history. Some archaeologists trace the first instances of human cultivation to bananas.[24] Ten thousand years ago, hard dark seeds made the fruits of the wild bananas in Southeast Asia and the Asia Pacific inedible. The most widely held hypothesis is that an unusual mutation among plants with fruits that had smaller or no seeds caught the attention of those foraging for corms of the large herb in the forest. The particularly fleshy fruits compelled people to dig up and transplant some of the suckers—taking them out of the forest to plant in more open, more accessible spaces. Thinking with these longer temporalities requires methods that articulate a different scale of re-membering than activists' oral histories can. These archeological hypotheses root analysis in the movements that altered what humans and bananas had been. When foragers trained their attention on plants that had become other than themselves and drew these around them, they also tapped into and reproduced the potential to make themselves other than they had been.

Bananas traveled well. Wild banana plants moved, reached out, produced new suckers. This movement met with human movements, and together they entered new spaces. They survived long sea voyages. They accompanied humans down rivers, away from valleys, up mountainsides, and deeper into forests. One preceding the other: bananas reaching out, humans taking them up and traveling with them only to transplant them in a new locale, so they might reach out again, extending, collaborating in the transformation of lands, as they reached out again rhizomatically, sending suckers from their corm entering a space, transforming an ecosystem. As they filled landscapes, they also filled stomachs, energizing legs and arms, signaling the sort of satiation that supports reproduction. Their decomposing bodies and fruits changed the composition of soils as well as altered the quality of light and shade, shaping patterns of growth beneath them.

Archaeologists disagree about whether bananas arrived in West Africa and spread east or vice versa. The traces of tiny phytoliths sifted out of tons of African soil led some to place their arrival 3,000 years ago and others significantly earlier.[25] The micro-remains of ancient bananas suggest that they joined with yams, sorghum, and millet on the continent. These newcomers collaborated with people in ways that changed the material conditions of existence for both. When humans came together with bananas—moving, sorting, traveling, transplanting, eating, composting—they altered the forms of endurance that inhere in the banana and its capacity for change. As the banana also did for humans.

The slopes of Mount Kilimanjaro, where *kitarasa* are said to be "indigenous," have been inhabited by humans for at least 2,000 years. The ancestors of those on the mountain now likely came to Kilimanjaro from the northeast about 300 years ago.[26] Bananas and their humans settled together, shaping each other's preference for living at altitudes of 3,500 to 6,000 feet. The dense banana grove created spaces that were qualitatively different from the farmland in the lowlands and the higher plateaus, where people cleared fields to cultivate annual crops, such as finger millet, sorghum, beans, and later maize, and sometimes grazed livestock. Bananas were central to dwelling; they co-created places for living.[27]

The *vihamba*, the home gardens that prevail on the slopes of Kilimanjaro, evolved over several centuries. Shifts in climate, droughts, excessive rains, changing compositions of soils, impacts of grazing, disease, and pests encouraged people and plants to experiment with new strategies for survival. Some evidence of water management, manuring, and terracing can be tracked. It is still possible, for instance, to find farmers irrigating their fields today with furrows that date back to the seventeenth century.[28] In the nineteenth century, when the Germans wanted to punish and remove those they reified as "the Chagga," who had expanded westward to Mount Meru, they burnt down banana groves and repatriated their wives to Kilimanjaro.[29]

Today, these home gardens are relatively small (0.5–1.7 ha) and densely green. On the slopes of Kilimanjaro, banana plants seem to cocoon around homes, holding them, sheltering them in their shadows. Narrow, twisting paths wind their way among the home gardens, one leading into the other so that a newcomer likely cannot distinguish the boundaries. When walking through these areas, it was not infrequent to turn a corner and find myself in the middle of a courtyard. Homes feel tucked into the landscape, enveloped by bananas, fed by vegetal growth. These homesteads are highly diverse spaces, dwellings that feel lush for the density of human, animal, and plant species. They generate economies of many kinds, but not only.[30] In these gardens, affordances are made for the continuance of many things, and these affordances are not (only) driven by an immediate instrumentality. These home gardens co-create worlds that sustain in time frames that engulf individual lives. Elders are buried in their home gardens. Graves are not just tended to but lived with. These banana groves carry lineages across space and time, exerting both possibility and pressure on each new generation. *Kitarasa*, a banana that is not brought to market, not even considered desirable by many, continues in this context, and this is what it remembers, what is carried in its distinctive orange sap.

When Dorcas works with municipal officers at the canteen, groups of elderly Tanzanians through a home care NGO, or patients in the Moshi hospital to cultivate their taste for *kitarasa* flour, she invites this process of remembering to move through people. When Dorcas organizes women through church groups in Kibosho and Rombo, talking with them about farming practices, compost, and organic fertilizers, and contracting with them ahead of time to sell her the *kitarasa* they harvested from their *vihamba*, she invites people to dwell in *kitarasa*'s relations.

### Therapeutic Trajectories Through Ecology and Economy

Michael Marder teaches that "economy authorizes us to speak about dwelling while forgetting what it means to dwell."[31] Turning plants into resources dissolves their labor-to-dwell in the economic; they come to dwell only *for* us rather than *with* us. *Kitarasa* suggests that imagining what it means to dwell requires remembering differently. Plants are hosts and guests, kin and community. Reducing plants to the material through which people make livelihoods rests on the tactics of unknowing explored in the previous four chapters, elaborated here through the ways in which the epistemics of testing and the ontics of rawness obscure healing as land relations. Remembering differently means apprehending the ways in which plants constitute social worlds and generate not only what we mean by "ecology" but also what we mean by "society," "politics," "knowledge," and "economics." *Dawa lishe* providers collaborate with *kitarasa* to question whether economic development is an adequate lexicon for navigating tensions between accumulation and reproduction. *Kitarasa* attunes them to the instability of the commodity, and together they work (in) the space where the grasp of the economic is slippery. Rather than stabilize *kitarasa* as a commodity through the pharmakon, *dawa lishe* highlights its instability as an aspect of its therapeutic potency. Together producers and plant co-create projects that experiment with the times and spaces needed to reimagine healing in ways that might support an ongoingness that exceeds the bounds of modernist progress and might provide ways of reckoning with its self-devouring growth.

The specificity of plants matters, both when healing with them and when thinking with them.[32] Bananas and plantains are the world's fourth most important food crop (after wheat, rice, and maize). They have made and toppled governments, justified wars, and forged brutal political alliances. They are a major player in global industrial agribusiness. When Slow Food's Arc of Taste project reaches out to Dorkia Enterprises to work together to develop and highlight *kitarasa*, it is undoubtedly thinking about the potency of *kitarasa*'s dwelling in relation to that of the single varietal cultivated in commercial plantations, the Cavendish. The banana industry has exploited the rhizome's capacity for asexual reproduction. Each banana plant is a clone of the parent. The plantations coerce timelines for planting, harvesting, and ripening—and, thereby, create enormous economies of scale. Their capacity for uniform production, however, depends on the vulnerability of extreme similitude. The monocultures of industrial production have nurtured the virulence of Panama disease, a root fungus.

While the Cavendish banana, so common in our supermarkets, is now entering Africa through commercial plantations, Africa's smallholder farmers foster a great diversity of bananas and plantains. Central African farmers cultivate the most varieties in the world. Uganda is the second-largest producer of bananas in the world after India, and Tanzania is the eighth-largest producer. Yet, neither East African country exports a significant amount of its crop. Bananas are a staple food crop in this region, and the fruits are consumed locally, providing some seventy million Africans with more than a quarter of their calorie requirements.[33] Yet, the pathogens that plantations gather do not stay within their borders. The thriving fungus that threatens the world's commercial crops could spread to and devastate the cultivars that supply African local markets. *Kitarasa*'s world-making efficacies—and its liveliness as part of exceptionally biodiverse spaces—hold out hope that trends of disease and depletion might be disrupted.

If we join in reconsidering the modern notion of toxicity through Bantu conceptions of *uchawi* and *uganga*—that is, through East African histories of reckoning with relations that harm and those that heal, respectively—then our conception of toxicity cannot turn only on poisons and pollution. Toxicity is about social-material configurations that foster depletion, infertility, and harm. When *dawa lishe* producers shift between regulatory tracks for foods and drugs, they expand on the frictions between testing and testimony and experiment with alternative ways of apprehending plants. Their stories illustrate that pharmacological and toxicological logics do not circumscribe the layered notions of the therapeutic that still press upon people today. In her study of toxicology in Senegal, Noémi Tousignant recognizes a strong association between toxicity and waste in Africa: "both as literal waste that is dumped, as an external, material assault on Africans, and as symbolic of the continent's superfluity in the global political-economic order, that is, of Africa *as* waste."[34] Resisting this slippage between efforts to track uneven exposure to harm and images of Africa as laid to waste, a wasteland—as well as Africans as wasting away—is not only a rhetorical exercise, but Tanzanians suggest it is also an ontological one. By refusing to limit their reflections on and labor over the toxic dumping of capitalism's by-products, pollution, and waste, Tanzanians highlight that liveliness, fertility, and growth are both depleted by *and* potentiated by toxicity.

Producers and users of *kitarasa* exemplify *dawa lishe*'s appreciation of experiments parsing, (re)mediating, and inhabiting the toxicities that simultaneously enable contemporary life and deplete the very conditions of its ongoingness. Dorcas's assertion that *kitarasa* is therapeutic is also an assertion that healing in a toxic world demands an openness to the possibility that people as well as plants will become other than what we might have been—altered to ourselves. It works against the logics of the plantation. The coming together with food sovereignty movements, the evocation of healing as a condition of sovereignty, focuses attention on the ability to shape the relations involved in the unfolding of plants and human bodies together. Accounting for the therapeutic potencies of these remedies requires considering how accounts of the historicity of toxicity open our analyses to the ways that concepts of toxicity articulate differently across time and space. In Tanzania, such differences create frictions that not only lead to misunderstandings and failed treatment programs on the continent but also structure forms of violence that obscure critiques of the contemporary moment and their visions for ways to move into the future.

## Precarity in the Pull of the Economic

Bananas in northern Tanzania are community and ecology; they are gendered and infrastructural. They are caught in the generation of global wealth and threatened by the intracontinental migration of viruses. Tanzanians collaborate with bananas to speak of a liveliness that exceeds livelihood. *Kitarasa*'s remembering opens a space to question whether economic development is an adequate lexicon for navigating tensions between accumulation and reproduction that shape well-being today. And Dorkia Enterprises' efforts illustrate the possibility and impossibility of these efforts to reframe relations between economy and ecology within global capital.

During fieldwork, I strove to attune to the moments in which *kitarasa* and other *dawa lishe* remedies managed to interrupt pressures to conform to pharmacological techniques of managing toxicity and political techniques of managing the traditional, by troubling regulatory boundaries between foods and drugs. This unruliness, I argue, animates Tanzanian experiments with alternative ways of generating therapeutic, ecological, and economic value (and mediating their entanglements). The pull to the economic, however, is strong in a world where its logics have remade both medicine and the environment. The space in which to apprehend the ecological in modes that are neither assimilated within the economic sphere nor tethered to it as that which is defined as a break or an interruption, is very cramped.[35] Dorcas and Rukia navigated this precarity together for a period. After Dorcas married, she moved with her husband to Mbeya in southwestern Tanzania and founded Dorkin Organic Farming Co Ltd. While the two women moved apart from each other, both continue with *kitarasa* as well as other plant remedies.

Navigating the precarity of *dawa lishe* as a radical political project requires remaining both aware of the forces of economism and experimental in the forms through which projects organize people–plant relations. When is the shape of the legal entity (small business or nonprofit), the structure of human relations (organizational charts whether very hierarchical or not, kin, friendship, etc.), the modes of nonhuman relations (*vihamba*, community gardens, markets, and the ways that the intimate experiences of plants are storied), the formulation of products and the space of their circulation (picked and prepared at home or dried, packaged, and sold nationally and internationally) overwhelmed by the economic and its extractive logics? How do we hold each other against this inexorable pull and through which collectives might we question our responsibilities as well as create new possibilities? There are no easy or one-time answers to these questions. They are part of the collective work.

I believe that anthropology, in its dexterity with multiple forms of evidence at the intersection of the humanities and the sciences, might be made to contribute to this work—in part by capturing emergent objects of knowledge and exploring the value of the collectives they make possible, if even for a moment. That is, if anthropology is put in the service of speculative projects that not only historicize "against the horizon of the current techno-economic permutation of metaphysics" but also nourish alternative ways of being, knowing, doing, and relating for which we have no horizon.

## Conclusion  *Therapeutic Sovereignty*

*Dawa lishe* emerged as a phrase to describe a loosely linked set of social and therapeutic projects in northern Tanzania. Mama Nguya and I sought a way to identify that which animated the partial connections among TRMEGA, WODSTA, EdenMark, and Dorkia Enterprises and to find vocabulary to describe the lines along which they were building solidarity. Not everyone found the phrase as compelling. Alex Uroki scowled at my use of it. He preferred to downplay the shapeshifting between traditional healer, entrepreneur, yoga practitioner, and myriad other positions through which he built his practice. Drawing attention to the way his efforts pushed back against the categories of the state could bring sanctions. In contrast, Dorcas was happy to playfully take up the phrase as she tried to cut a path for her life and work. Over the past decade, I have found that most Tanzanians understand the sorts of products that such a phrase is meant to encompass, although they might also initially include a range of therapeutic foods and herbals whose production, circulation, and use do not seek to cultivate alternative land relations. The herbals industry, after all, is on the rise and draws explicitly on the histories of science and nation that rendered nature digestible for profit and Africa as a site of extraction. When I used this phrase in conversations with regulatory officials, however, they consistently recognized it as that which jams up their postcolonial bureaucratic technologies. *Dawa lishe*, then, is neither a state category nor what science and technology studies scholars like to refer to as an "actors' category." Rather, Mama Nguya offered it as a proposition. I took it up as an invitation to think and work collectively in an effort to reconceptualize healing within the frictions between medicine and agriculture. *Medicines That Feed Us* extends the commitments embodied in the ways a specific set of projects disrupts and reinvents the categories through which health is known and linked with governance, as well as seeds new solidarities. Through the writing of this book, *dawa lishe* has been a theoretical provocation, and it has transformed into a broad analytic.

Drawing together the term *dawa* (often translated as "drug," but more faithfully a "pharmakonic substance") and the term *lishe* (used in nutrition for "fortified food," but literally "to cause to feed") creates a sense of irony. It forces a recognition that the existing fields of medical and agricultural science and governing bodies regulating drugs and food systematically divide bodies from land. Even when the consequences of this separation are harmful, there is no way to work outside of these institutions. Medicine and agriculture are powerful forces organizing contemporary modes of dwelling. They transform humans and nonhumans, shape economies and ecologies, affect intimate engagements in and with the world, and forge political, legal, and humanitarian regimes. *Dawa lishe* is an effort to delineate a conceptual space to stage encounters between these fields of knowledge and practice. By moving between regulatory regimes for drugs and food, *dawa lishe* mobilizes the ways of knowing and objects of knowledge central to each field to open up the other. It strives to kindle plant relations that intervene in the chronic injury and relentless depletions rendered invisible by maintaining medicine and agriculture as separate fields of knowledge, and bodies and land as separate objects of practice.

We might think, then, of the fundamental provocation of *dawa lishe* as lying in the space between the words of the phrase. This space between embodies an invitation to generate alternative ways of problematizing our moment and experimenting with solutions. It holds the possibility that the social-ecological-therapeutic projects highlighted in this book and others in Tanzania and beyond will find solidarity in the work of decentering the specific configuration of science, law, and economy through which modern states manage the constitutive ambivalence of the *pharmakon*. The power of *dawa lishe* comes from its challenge to reconceptualize toxicity and its relations with remedy and memory through vocabularies forged in African histories of healing. In East Africa, the philosophical tensions between *uchawi* (the mobilization of forces to harm, often translated as witchcraft) and *uganga* (the mobilization of forces to heal) remain palpable in the word *dawa*. In this book, I have used these tensions to describe the rich, affective vocabularies available in Kiswahili (and related Bantu languages) to articulate the relentless depletions and slow death of worlds where livelihoods depend on a system of social disposability and ecological abandonment.

*Dawa lishe* creatively taps these philosophical tensions and histories of healing to rearticulate the modern concept of toxicity and its relationship with remedy and memory. The work of TRMEGA, WODSTA, EdenMark, and Dorkia Enterprises encourages apprehending toxicity as a drastic thinning of relations. Healing, then, is a thickening of relations among humans and between humans and nonhumans that is experienced as lushness: a liveliness that supports ongoingness. What I offer the collective work that is already happening in Tanzania to dislocate the imperial pharmakon, to reinvent body-land relations, to recognize other forms of sovereignty, is a story that shapes how we see and connect these projects. For our challenge ultimately is collective. *Medicines That Feed Us* began with the following question: What does it mean to heal in a toxic world? The book concludes with an invitation to find common cause with social-ecological experiments seeking to answer this question: *How* do we heal in a toxic world? How do we heal in a world premised on extraction, depletion, and the disposability of humans and nonhumans? How do we counter the techniques of unknowing that obscure the fact that the thinning of social and ecological relations continues without regard for repair, reckoning, renewal, or reparation? How do we thicken relations—with the living and the dead and those to come—to cultivate the forms of lushness that support ongoingness?

These questions share concerns with others who write about toxicity, multispecies ethnography, futures after capitalism, and ontological and epistemic decolonization. This book, however, does not start from these literatures, even if it has found companionship in them. This book starts from a deep commitment to, and therefore a rooting in, place. As I wrote, I was called to be faithful to the form and liveliness of the *kihamba*—an agroforestry system among the Chagga on Mount Kilimanjaro that spread to the Meru on Mount Meru, which harnesses the fertility and liveliness of land and lineage through a specific mode of dwelling—more than any academic theory. The *kihamba* is where ancestors are buried, and offerings are made in iterative reckonings with the past. It is where newborns' placentas were buried before the practice of delivering babies in hospitals became common, and disrupted such land relations. The *kihamba* embodies the social-ecological relations critical to ongoingness, and its thriving highlights techniques of gathering, channeling, and continually nourishing these relations. It is a fundamentally inalienable property. While the pressures to sell the valuable land on Kilimanjaro have been high, and the average size of *vihamba* [plural of *kihamba*] has shrunk steadily over the past hundred years, the heart of a *kihamba*, marked by the *isale* plant (*Dracaena fragrans*) and the site of graves, cannot be sold. As the force of a lineage's past, present, and future, any desperate measure to treat it as a fungible asset threatens to catalyze dire consequences for generations.[1] Yet, the *kihamba* embodies the concept of body-land without a suggestion that this forecloses the possibility of thriving markets. (As mentioned in the Introduction, markets around Mount Kilimanjaro and Mount Meru preceded colonialism.) It invites consideration of the intimate relations formed through the co-laboring of plants and people without reiterating the closed loop and economizing logics of permaculture. It shows how notions of harming and healing—toxicity and remedy—in this area, are also tied, as Derrida insisted, to memory, but without understanding memory as the purview of mind or psyche.

The potency of the *kihamba*'s embodiment of an inseparable body-land was evidenced during COVID-19. In the face of a global pandemic and a weak government response, Tanzanians thickened relations with kin, neighbors, and plants.[2] While I was "locked down" in Ithaca, New York, far from Tanzania, I learned of its unfolding in the region only through WhatsApp messages and newspapers. When I returned in early 2023, I found that people spoke of their experience during that time most powerfully through the presences and absences in their *vihamba*. As the social-ecological basis for health and healing—the space out of which ongoingness is possible—*vihamba* continued to embody the relations that had supported survival and care in the previous two years. *Vihamba* offered a space from which to respond to fevers, difficulty breathing, coughs, and congestion. People had planted, gathered, and shared leaves, fruits, barks, and roots that were thought to be helpful. They recounted to me how no one wanted to deny a request of friends, family, or neighbors in the urgency of the pandemic, as they stood next to lemon trees, still struggling to recover after having been stripped bare of both their fruits and leaves for therapeutic teas and steam treatments. Herbs were also shared across networks, with cuttings given away and transplanted in *vihamba*, rhizomatically extending a pharmacy of herbal options for those who might fall ill. In one *kihamba*, as an elderly man regaled me with stories of his family and the configurations and reconfigurations of his *kihamba* through the generations, he brought me to see not only his parents' graves but also that of his younger brother. The latter's grave was marked with a cross that read "2022." It was covered with Tulsi plants, popular in that year for addressing respiratory symptoms. COVID-19 was a time when the social-ecological dynamics of health were palpable. Lushness enabled response, and cultivating lushness around and through others was healing. Interestingly, as I talked with youth in 2023, I found that more were interested in plant(ing) remedies now than had been when I started this research in Kilimanjaro in 2013.

As I walked through *vihamba*, reconnecting with friends and place—finding my own grounding in that postpandemic moment—I saw many of the plants that scaffold *dawa lishe* plant(ing) remedies, available and actively used in these home gardens. The resonance between the plants of the post-COVID *vihamba* and those of TRMEGA, EdenMark, Dorkia Enterprises, and WODSTA illustrates a broad shift in how people are apprehending that which constitutes the therapeutic as they grapple with intertwined health and environmental crises. The projects in the preceding chapters are both catalyzing change and harnessing it. Through their efforts to rekindle plant relations by spreading practices of growing, sharing, healing, and eating, the body-land grows more palpable in everyday life.

In the wake of twenty-first-century economism, the space for such work outside of home gardens is cramped, and victories are hard won. *Dawa lishe* strives to capture collective efforts to manifest understandings of healing as land relations not only in individual remedies but also in the sorts of collective initiatives that can push for a reconfiguration of science, economy, and law. Each of the projects at the heart of this book is limited by the lexicons available in institutionally separated fields of medical and agricultural knowledge, as well as by the economic, legal, and political systems that offer discrete locations from which to orient. *Dawa lishe* may be rooted in histories of body-land relations such as those of the *kihamba*, but it unfolds from the intersection of the nongovernmental organization and the small business, the nonprofit leader and the entrepreneur. Both positions are limited: The forms of argument and action they facilitate are circumscribed by the same terms that make them legible. To gain some intellectual leverage, they draw on African therapeutics to innovate modes of observations, forms of engagement, approaches to knowing, arrangements of agency, and ways of being in the body and in the land to reconceptualize health and its relationship to governance.

The concerns that drive these projects are not unique to Tanzania. The organization of land and labor as resource rather than relation—as fuel for capital accumulation rather than as the entanglements through which collective continuance is sustained—has energized extractive economies worldwide and catalyzed modes of living that have altered the very substance of the earth.[3] Insofar as development (at the level of the community, nation, or region) depends on the economization of life, labor, and land, well-being is tightly tied to the forces that undermine it. This is the double-bind that defines the twenty-first century and resonates with debates internationally over the Anthropocene and its alternatives (e.g., the Plantationocene, Capitalocene, and Chthulucene).[4] Arguments over when our current epoch started point to the centrality of the violence driving the alienation of land and labor central to the organization of human life and its impact on the matter of the earth. Both also ask us to consider the modes of dwelling, a living with and becoming with, as potentially generative of wounded, toxic relations, as well as flourishing, fertile relations. The former focus attention on the long tentacles and myriad effects of the relations between science, war, and economy; the latter highlight the fundamental racialization and extractivism that remade landscapes, reconfigured human-plant relations, and established the condition through which the chemical industry came to be seen as "needed" in (plantation) agriculture and central to strategies to ensure food security. In this book, I have taken this debate as evidence of a moment of shared reckoning with the ways that modern forms of human dwelling and the kinds of knowledge and modes of politics that shape it have fundamentally undermined our ongoingness. People, soils, plants, waters, mountains, glaciers, and pollinators are wounded. What draws TRMEGA, WODSTA, Dorkia Enterprises, and EdenMark together—what members seem to see in each other—is a commitment to developing remedies that attend to the healing of such body-land wounds. *Dawa lishe* indexes projects through which Tanzanians are reinventing the concepts needed to act response-ably in relation to these threats to ongoingness. Remedies rekindle ways of knowing, being, and laboring with plants; interventions attend to barrenness; and solidarity is built across the fraught politics of lushness.

While toxicity is a condition of modern life, bodies and lands do not bear these harms of the world wrought by racialized capitalism evenly. *Dawa lishe* teaches us, among other things, the ways in which nineteenth-century notions of toxicity and its relation to remedy help to hold in place dispossession and dispossessing narratives. Therefore, rethinking toxicity through African healing is a political act as well as a therapeutic one. As dominant articulations of body and land, economy and ecology, health and environment start to unravel, what we understand as knowledge, medicine, and justice lose their ground. Power, as Foucault insisted, is in the "order of things."[5] TRMEGA, EdenMark, WODSTA, and Dorkia Enterprises are all engaged in pedagogical projects that approach the challenge of toxicity by cultivating spaces for a collective (re)articulation of how bodies and environments are folded into one another. They are working to reorder things. In so doing, they seem to suggest that what is needed in this moment is not just a new modality of healing, an alternative or additional form of care, but what we might call, drawing on philosopher and physicist Karen Barad's work, an alternative ethico-onto-epistemology.[6] While the phrase is not particularly elegant, its cumbersome quality expresses the difficulty in staking out terms of debates that refuse the subject–object, mind–body, and social–material divisions structuring modernist thought. Barad argues that ethics (how we value), epistemology (how we know), and ontology (what we know about) emerge together in practice. *Dawa lishe*, as I have extended it in this book, is a speculative project. It aims to unsettle the entanglements of ethics, epistemology, and ontology that have rendered dispossession invisible, capitalism inevitable, and the forms of bodily and territorial sovereignty it enables natural. And it experiments with alternatives.

*Medicines That Feed Us*, then, is not an account of "medical pluralism."[7] The effort is born of a particular place and time—periurban northern Tanzania in the first quarter of the twenty-first century—when the recognition of the ways that economy has fundamentally refigured ecology and living undermines the possibility of future life and in specific interactions—the growing of therapeutic plants, the extension of gardens, and the animation of networks of older peasant organizations and HIV/AIDs support groups. But *dawa lishe* does not name a culturally specific alterity that can be compared symmetrically with other medical systems. It is not a local or indigenous knowledge whose strategies for making truth claims can be juxtaposed to those of a global or cosmopolitan biomedicine. Rather, *dawa lishe* is a grounded provocation to challenges that define our ecological and medical moment. It reworks the very terms through which we might understand different forms of knowledge and local or global practice. *Dawa lishe* is a call to recognize that working toward a future in which we disentangle societal health and economic development from the very practices that threaten the possibilities of bodily and ecological ongoingness requires new ways of knowing and alternative foci of healing interventions.

Central to the aspirations of *dawa lishe* is the decentering of the body proper with its individual boundaries and the refocusing of attention on body-land becomings. The social and therapeutic projects described here experiment with practices that articulate the object of healing as a distributed body. Therapies address this new distributed configuration of body through its incorporations and excorporations, compositions and decompositions. Located in eating and animated by appetite, the boundaries of this body range over the spaces of all that it takes in and the spaces that then absorb its waste. The body-land that *dawa lishe* articulates includes the fields that produce food crops and therapeutic plants; the water that flows through the large foreign-owned coffee estates before reaching these fields; the ecosystems transformed by the widespread spraying of DDT; the soils in which pharmaceuticals, plastics, and pesticides have been improperly disposed; the seeds of companies that have bred out the possibility of local exchange; and more. This is an ontology that can begin to orient healing toward body-land becomings as both objects of intervention and the space of health.

This distributed body motivates weaving back and forth across the institutionalized tracks for food and drugs and appealing to both traditional medicine and food sovereignty initiatives. It helps Tanzanians evade the rapid economism that has been spun out from empire and racial capitalism. Through the cultivation of appetites that nourish body-land becomings, the recognition of chronicity as the temporality of slow violence, the insistence that knowledge is embodied in collective action, and the extension of gardens, producers identify (and expand) cracks in the dominant logics of living. They establish the more-than-human relations through which other ways of being, forms of reasoning, modes of engagement, and approaches to organizing collective life, can thrive.

Interventions are invested in rekindling relations with plants that rebuild people's capacity to co-labor with them. In tracking everyday engagements with plants, I have come to attune to the ways that this collective work exceeds utilitarian engagements in which humans harness plants' labor as a resource for environmental or curative projects. Throughout the writing of this book, a small moment—really just an image—has returned to me over and over again, breaking through my thoughts and arguments. It has become an origin point of sorts in my own questioning of what kinds of plant relations are at the heart of *dawa lishe*. One hot afternoon, when walking between a strip of shops and stalls of vegetable sellers, my colleague Romana bent down and plucked the leaves of a scrubby, short *Euphorbia hirta* plant pushing up through the dry, packed earth along the curb in front of EdenMark. The singular weed growing in this heavily trafficked area created a brief pause, drawing her and me into recognizing it and its striving. As she rubbed the sticky white fluid between her fingers for a moment, she remembered out loud its reputation as helpful with digestive and respiratory complaints, reminding herself and teaching me.

At times, the co-laboring of people and plants can be in direct service to the healing of bodies, communities, waterways, and soils. Romana's momentary turning to life's striving in this cramped and compacted ground suggests we might also think about the ways in which co-laboring can be in service to thought. If knowledge is located in action, and if healing is a process of co-laboring with plants to shape body-land becomings, then work is both physical and intellectual. What if we viewed Romana's gesture as a citational practice? What questions might be asked if we were to cite plants, not just studies about them? What arguments might be crafted through the joint authorship of people and plants? Perhaps in our efforts to think the impossible with others—to render visible the revolutionary dimension of modes of attunement and ways of being outside of extractive logics of capitalism—we also need to ask, whose citational practices are legible? This would suggest that not only do citational practices continue to marginalize communities of people who have been dispossessed, but they also fundamentally emerge from histories that mark forms for knowledge that constitute the relations of dispossession, including histories that articulate knowledge as the sole purview of humans. *Dawa lishe* poses modes of attention and engagement that support alternative citational practices in the service of reproductive justice for body-lands.

I find the provocation of *dawa lishe* generative and TRMEGA's work particularly inspirational. Perhaps we might think of gardening itself as a possible mode of working together with plants, a practice to reinvent ways of generating propositions and constructing arguments about living in ways that support mutual flourishing. TRMEGA's rhizomatic extension of gardens through the periurban is offered as a response to explore paths forward in the face of the inadequacy of nineteenth- and twentieth-century articulations of toxicity and the claustrophobia of the political positions they afford. And yet gardens-as-form, like lushness-as-aesthetic, are not innocent. The epistemological and ontological grounding of modern botanic gardens emerged through colonialism and shaped the plant relations that animated the development of modern medicine and agriculture. It motivated the rise of the forms of lushness that inhere in the plantation and the greenhouse, the cut flower industry and the ornamentals of the church garden. Therefore, the garden as revolutionary form may be provocative, but it is not self-evident. The work of TRMEGA lies in the specificity of the relations it stages, and their elevation to a pedagogical project that strives to cultivate collectives that will reshape the ground of politics.

Locating health and healing in land relations, as well as directing remedies toward body-land becomings, challenges the forms of corporeal and territorial sovereignty that determine who can make an argument, what counts as justice, and the lines along which equity is measured. *Dawa lishe* points to the idea that alternative articulations of sovereignty already exist in Tanzania, and they are being actively worked out, elaborated, and extended in the service of both bodily and ecological healing. Therapeutic sovereignty is not limited to the ability to choose what sort of healing or medicine is desirable; it is, rather, the right to experiment with practices that compose and decompose an ethico-onto-epistemology of healing. *Medicines That Feed Us* offers *dawa lishe* as an analytic for thinking through such radical possibilities, for thinking in ways that will change everything: the social, physical, political, and economic structures through which we live. It is not a solution so much as a tool to begin formulating what Suman Seth calls a "revolutionary remedy."[8] *Dawa lishe* is an invitation to search for how to live into a collective responsibility to attend to the chronic injury, relentless depletions, patchy barrenness, and chemically saturated liveliness of our moment.

In the Introduction, I suggest that *dawa lishe*—as a modality of healing, and therefore also as a grounded theory and an embodied provocation—may be ephemeral. The value of studies of emergent forms, however, does not lie in prescient predictions of what will be lasting. Rather, their value lies in the way they work to hold open, even for a moment, the possibility of being and dwelling otherwise. Drawing out and learning from such emergent speculative forms is one way of joining with diverse (human and nonhuman) others to render thinking the end of capitalism slightly less impossible and of working toward alternative futures for lushness.

For this reason, as I finish this book, I have been collaborating with colleagues to conceive and implement an innovative interdisciplinary, land-based project to bring the animating force of *dawa lishe* into a major teaching research hospital in Tanzania. This is not (only) the abstract work of "disseminating knowledge" (through conferences and talks) or the practical work of moving things (plants, medicines, techniques) to create a teaching, research, healing garden, but it is also the transformative work of centering relations. TRMEGA—Mama Helen, Jane, and Rose—have been central to the Uzima Project, as we call it, at Kilimanjaro Christian Medical Center. In Kiswahili, *uzima* means wellness, maturity, wholeness, and fertility; it connotes an alternative lexicon to organize reproductive justice for body-lands. Through the Uzima Project, TRMEGA works collaboratively with faculty, students, doctors, nutritionists, and environmental health specialists, as well as with other local experts in permaculture and rainwater harvesting engineers. Reconfiguring science, law, and economy means reconfiguring the relations through which each is practiced and through which they come to be entangled. The method is one of thickening relations and experimenting together.

This coming together is not always easy. After all, we are not "filling a gap" in knowledge; rather, we are "trying"—in the sense of so many parents, grandparents, aunts, siblings, pastors, imams, neighbors, NGOs, entrepreneurs, doctors, and nurses who tended to the sick, weakened, and depleted during COVID with foods, herbals, baths, steam, drugs, and love. Uzima is a speculative project, an experiment with new relations, new comings together, new ways of dwelling. Our land-based teaching, research, and healing continually asks, how can we manage the *pharmakon*—toxicity, remedy, and the past-with-us-in-the-present—through relations that are nourishing? How can a medical center with a tertiary hospital, a medical school, and an international research center nourish body-lands? What are the implications for teaching medical students, doing research, and treating patients? The work to answer these questions is the work of therapeutic sovereignty. It requires forging common cause and building solidarity. *Karibu*, you are welcome to join us.

# Acknowledgments

Books are exercises in gratitude. They are a process of making space with others. This one is the result of decades of relationships that allowed for working, living, growing, and thinking together. The arguments—and the dreams that shape them—emerged while planting, harvesting, composting, eating, cooking, listening to stories, sharing laughter, whiling away hot afternoons, and sitting with those who were sick and sometimes healing; they took shape through myriad conversations, clinic visits, lectures, conferences, roundtables, meetings, co-readings, long drives, and shared walks. In fields, homes, hospitals, gardens, school courtyards, orphanages, and on roadsides. The production of this book and the reproduction of my life, my family, and my communities on both sides of the ocean are inextricably entwined. My gratitude for all who have touched me and touched all I love cannot be fully expressed here.

I remain profoundly honored by those who welcomed me into some of the most difficult struggles of their life as they faced illness, debility, loss, and pain. Thank you also to the healers, herb sellers, shop owners, and farmers who brought me into their confidence, and to the colleagues and friends who carefully pulled me into the intimacies of the work that came to ground my thinking. All are with me in these pages and will remain with me, as this book is not an ending but an offering to go on together. Thank you for sharing your stories, fears, courage, and care.

My deepest gratitude for the time, companionship, camaraderie, and insights of Helen Nguya, Jane Satiel Mwalyego, Rose Machange, Dorcas Emily Kibona, Rukia Seme, Alex Uroki, Jenipha Muro, and Romana Damasi Masawe. They are at the heart of this work and these stories. Both my mentor Binti Dadi and my friend John Wilfredy Ogondiek passed away during the writing of this book. The loss of both will always be deeply felt, even as the gifts of their lives' work continue to generate blessings and their memories bring joy.

Many others helped me understand the landscape in which therapies were being assessed, developed, controlled, and valued. My special thanks to the Tanzanian Ministry of Health, especially those in the Traditional and Alternative Health Practice Council; the Institute of Traditional Medicine at the Muhimbili University of Health and Allied Sciences (MUHAS); the Tanzanian National Institute of Medical Research (NIMR), particularly the scientists in the Mabibu and Ngongongare laboratories; the Business Registrations and Licensing Agency (BRELA); and the Tanzanian Food and Drug Administration (TFDA). Hans-Martin Hirt and Peter Feleshi revealed a different trajectory of international engagement through anamed. My thanks to all those to whom Helen Nguya, Dorcas Kibona, and Alex Uroki introduced me in Slow Food International and global food sovereignty movements. My special appreciation for those who invited me into their offices and labs for an extended time: John Ogondiek, Victor Wiketye, Rogasian Mahunnah, Naomi Mpemba, Kenneth Hosea, Catherine Gwandu, Daniel Kisangau, Amina Msonga, and Gloria Mbogo.

During this research, I was affiliated with the Kilimanjaro Christian Medical Center. My colleagues in the Regional Dermatology Training Center hosted me, welcoming me into clinics, on outreach visits, and for weekly meetings, reviewing the status of patients. They leaned in even when my anthropological questions took a different shape from their clinical ones, and they invited me to give lectures and to co-advise students' research. A special thank you to Henning Grossman, John Masenga, Daudi Mavura, and Rachael Manongi.

My colleagues at Cornell provided me with a warm and generative atmosphere for both research and writing. Our collective work in the university shaped how I think about the efficacy of knowledge and held space for me to research and write. Thank you to those who thought with me through presentations in the Department of Anthropology, the Institute of Comparative Modernities, the Society of the Humanities, the Institute for African Development, and the Botanic Gardens. My particular gratitude to those colleagues who read all or parts of this work: Lori Khatchadourian, Alex Nading, Lucinda Ramberg, Suman Seth, and Matt Velasco. Thank you also to Adam Smith, Durba Ghosh, and Nerissa Russell, who went above and beyond to protect space for me to write. My students at Cornell are a gift. Our readings and discussions in my seminars on Toxicity, Anthropology of Body, and Postcolonial Science have energized me in this work. A special thank you to those graduate students whose work has entwined with mine over the years: Hannah Ali, Charis Boke, Rebecca Ciribassi, Hayden Kantor, Aparajita Majumdar, and Ashley Smith.

While this book was researched at Cornell, my thinking and my commitment to Africanist scholarship will forever be shaped by my time at the University of Florida. Thank you especially to Leo Villalon, Renata Serra, Alioune Sou, Abdoulaye Kane, Hansjoerg Dilger, and Luise White. Parts of this scholarship were presented at American Anthropology Association and African Studies Association meetings over the years, as well as at the "Legislating Gender and Sexuality in Africa" conference at the University of North Carolina, the Consortium for Humanities Centers and Institutes (CHCI) Medical Humanities Summer Institute at Dartmouth College, the series of workshops at Northwestern University expertly facilitated by Helen Tilley that led to the special issue of *Osiris* on "Therapeutic Properties," and the series of gatherings organized by Keiichi Omura, Atsuro Morita, Shiho Satsuka, and Grant Jun Otsuki for *The World Multiple* volume. Thank you also to colleagues who engaged with me during talks at the University of Toronto, University of Pennsylvania, Weill Cornell Medical School, Kent University Law School, African Studies Workshop at Harvard University, the UNC–Chapel Hill Working Group on the Moral Economies of Medicine, and Centre de Recherche, Médecine, Sciences, Santé, Santé Mentale, Société (Cermes3) dans Ecole des Hautes Etudes en Sciences Sociales.

This work was supported by a Mellon Foundation New Directions Fellowship, a National Science Foundation Scholar Award, a Wenner-Gren Foundation Research Award, and an American Council of Learned Societies Fellowship, as well as small grants from the Cornell Institute for the Social Sciences, Cornell Department of Anthropology, the Mario Einaudi International Studies Center, and the Society for Humanities at Cornell University, and a Cornell REFRESH grant. A blissful year was spent at the Society of the Humanities gathered together with my "Skin" colleagues Gemma Angel, Andrea S. Bachner, Erik Born, Naminata Diabate, Pamela Gilbert, Alicia Imperiale, Gloria Chan-Sook Kim, Karmen MacKendrick, Timothy Murray, Kevin Ohi, Emily Katherine Rials, Elyse Semerdjian, Samantha Noelle Sheppard, Daniel Smyth, Alana Staiti, Ricardo Wilson II, Nancy Worman, and Seçil Yilmaz. The short May weekend writing retreats the Society hosts and the Making Time to Write Groups of Cornell colleagues have brought the support of being in a community of writers.

I have benefited greatly from being part of several collectives thinking and working together: the Central New York Humanities Corridor Health Humanities working group on Medicine, Disease, Disability, and Culture supported by the Mellon Foundation that I co-organized with Lois Agnew and Andrew London of Syracuse University; the "Making Workshops" I co-led with Margot Lystra at Cornell University in which we drew, danced, walked, and explored the ways our bodies know and share in coming to know together; the Law, Organization, Science and Technology (LOST) in Africa group led by Richard Rottenberg; the Qualities of Life working group supported by the Cornell Einaudi Center; and the Ecological Learning Collaboratory I co-organized with Neema Kudva and Rachel Bezner Kerr. Between 2014 and 2021, Judith Farquhar, Carla Nappi, and Volker Scheid organized a series of delightfully creative and rejuvenating Translating Vitalities workshops that drew together artists, social theorists, anthropologists, and Chinese Traditional Medicine experts to co-create, including Sue Cochrane, Vincent Duclos, Jen Foell, Dianna Frid, Jim Hevia, Larisa Jasarevic, David Luesink, William Mazzarella, Barry Saunders, Cinzia Scorzon, Clare Twomey, Alexa Wright, and Peng Yu. In more recent years, my colleagues in the Uzima Collective worked together to conceive of a decolonial healing garden that nourished an extension of these ideas. Thank you to Mary Mosha, Sabina Mtweve, Rachel Manongi, Florida Muro, Rehama Mavura, Gloria Damian, Jeanne Moseley, Rhoda Maurer, and Sebastian and Mama Kiondo.

Outside of these groups, many others have also contributed to both the joy and possibilities of my scholarship. Thank you now and always for those conversations that made me change direction and for the companionship to keep going: Gregory S. Alexander, Sean Brotherton, Sandra Calkins, Tim Choy, Tatiana Chudakova, Emilie Cloatre, Joe Dumit, Judith Farquhar, Steve Feierman, Jeremy Foster, Hanna Garth, P. Wenzel Geissler, Sherine Hamdy, Linda Helgesson, Nancy Rose Hunt, Uta Hussong-Christian, Rachel Bezner Kerr, Gwynn Kessler, Neema Kudva, Lili Lai, Oskar Liivak, Julie Livingston, Tony Liwa, Ramah McKay, Sarah McGaughey, Laura Meek, Lauren Monroe, Tara Mtuy, Simon Mtuy, Projit Bihari Mukharji, Muna Ndulo, Abigail Neeley, Vihn-Kim Nguyen, Anne Ouma, Rob Peck, Rachel Prentice, Ruth Prince, Lisa Richey, Richard Rottenburg, Sophia Sarafova, Shiho Satsuka, Barry Saunders, Suman Seth, China Scherz, Emma Shaw, Rebecca Stoltzfus, Leah Sweet, Harris Solomon, Helen Tilley, Anna Tsing, Claire Wendland, and Mei Zhan.

This book is significantly better for the generosity and insight of Julie Livingston, Michelle Murphy, Emilia Sanabria, Lucinda Ramberg, and Rebekah Ciribassi, who committed a weekend to an intensive book workshop. Each not only carefully engaged with the version of the manuscript available then and brought their brilliance to it, but each also embodied the compassion, generosity, and commitment to collective thinking that makes being read a true gift. Also, a special thank you to Margot Lystra for hand drawing the maps in this book and the ways she has helped me to think differently about moving, mapping, and making space.

The work and attentive care of research assistants in Tanzania and the United States contributed to this work. Sia Mrema has been a cherished interlocutor and friend since our first days in Moshi. She offered both intellectual insight and practical skills to the project over many years, in addition to coordinating a team of transcribers and translators, including Julius Raymond, Raymond Mrutu, Tuma Mweta, Sahmim Said, Fadhili Terri, Eunice Chibanda, Susan Mpemba, and Samira Sadiki. At Cornell, Hanna Haile and Sarah Brewer helped with archiving and coding.

Thank you to Megan Harris and Alex Djedovic who made the editing and revising of this book fun and taught me more about writing in the process. I am also extremely grateful to two fabulously generative anonymous reviewers for Duke University Press and to Courtney Berger for her consistent support and guidance of this project. Permission for this research was granted by the Tanzanian Commission for Science and Technology (COSTECH), National Institute for Medical Research (NIMR), Kilimanjaro Christian Medical Centre (KCMC), and Cornell University Institutional Review Board.

A much earlier version of chapter 1 was published in "A Politics of Habitability: Plants, Healing and Sovereignty in a Toxic World," *Cultural Anthropology* 33, no. 3 (2018): 415–43; parts of chapter 3 were published in "Properties of (Dis)Possession: Therapeutic Plants, Intellectual Property, and Questions of Justice in Tanzania," *Osiris* 36 (2021): 284–305; and a photo essay appeared in "Cultivating Vitality: A Photo Essay," *Anthropology News*, January 24, 2018. I am also indebted to the parts of my thinking that developed through the writing of "Healing in the Anthropocene," in *The World Multiple: Politics of Knowing and Generating Entangled Worlds*, ed. Keiichi Omura, Atsuro Morita, Shiho Satsuka, and Grant Jun Otsuki (Routledge, 2018), 155–72.

I am deeply grateful for the family that keeps me connected with the past and the future: John and Marianne Langwick, Leslie Weaver and family, Kirsten Langwick-Temples and family, Laura Bianco and family, Judy and Mark Bercuvitz and all my Bercuvitz kin, and our daughter's two "other mothers," Binti Dadi and Monica Daniel. For nineteen years, Steven Valloney and my yoga community have grounded me, and more recently, my rock-climbing pals have brought sanity and joy to mid-days. Diana Levy and Bruce Fabens, and Jessica Allison and her girls, brought Ithaca to Moshi. Many at Tikkun v'Or have held our family in sickness and in health. A special thank you for those who have treated us like kin in Tanzania: the Mtuys, the Buckleys, Julie Atkins and Solo, John Ogondiek and Victor Wiketye and their families, Mariamu Mpende and Yanini, Sia Mrema, Sadiq and their family, and Maryam Gonga.

This book grew with our daughter, Tsadia. She was conceived as the initial work in Dar es Salaam began and received her learner's permit to drive as the manuscript was completed. Tsadia, you nourished so many of the relationships in this book. Thank you for all you did to make Moshi your home, for your openness and joy everywhere you went, for your patience and compassion as you visited clinics and gardens with me, for your reflections and insights as this research shaped the movements of your childhood. And for your faith that I would finish. *Asante sana.*

All of this, the making houses homes year after year, the gardens of lush plants, and the friends and families that gathered in them, would not have happened without Jeff. Thank you, *mpenzi.* Years of conversations, holding both joy and grief together, reflecting on the things that really matter, shape this book. Your own commitment to regeneration runs through its pages. Your keen eye for what is really at the core of things drew you to insist that the phrase *dawa lishe* should be taken up as a central frame for this exploration. In addition, your hours of reading, feedback, and editing helped bring the book to life. You were the one who told me to hold on ... and the one who did so with me. *Nashukuru.*

# Notes

## INTRODUCTION: HEALING (IN) A TOXIC WORLD

Ideas that continued to evolve in the introduction and other chapters of this volume developed through the writing of "Healing in the Anthropocene," in *The World Multiple: Politics of Knowing and Generating Entangled Worlds*, ed. Keiichi Omura, Atsuro Morita, Shiho Satsuka, and Grant Jun Otsuki (Routledge, 2018), 155–72.

1. Million, "There Is a River in Me," 33. In fact, assertions that some languages are "naturally" abstract are a way of privileging some locations for thinking and the voices that are most often found there. See also Diagne, *The Ink of the Scholars*.

2. Murphy, "The Experimental Otherwise," 105–9.

3. Research and Markets, "Herbal Medicines—Global Market Trajectory and Analytics," and Bareetseng, "The Worldwide Herbal Market: Trends and Opportunities," 575–84. Bareetseng works in the South African Council for Scientific and Industrial Research, in the Advanced Agriculture and Food Cluster.

4. Grand View Research, "Nutraceuticals Market Size, Share and Trends Analysis Report."

5. Street, "Food as Pharma," 361–72.

6. For precolonial Africa, see Schoenbrun, *A Green Place, a Good Place*; Kodesh, *Beyond the Royal Gaze*; Tantala, "The Early History of Kitara in Western Uganda"; Janzen, *Ngoma*. The review essay by Feierman, "Struggles for Control," and his book, *Peasant Intellectuals*, exemplify work on the colonial period. Also see Livingston's work on cancer, *Improvising Medicine*, and development, *Self-Devouring Growth*.

7. See Arnold, *Toxic Histories*, which examines poison and poisoning in India, another anglophone postcolony.

8. Taylor, "Age of Disability."

9. Nguyen first developed the concept of "therapeutic sovereignty" in the context of HIV/AIDS programs in West Africa. His deeply committed ethnographic and clinical work enabled him to account for the complex ways in which individuals and collectives sought treatment beyond the state when national health care systems were insufficient. He captured the practical pressure of triaging resources when the need for antiretroviral therapies far outstripped local capacity and described the resulting mobilization of "confessional technolo-

gies" to shape moral economies of care. His concept has since aided me and many others in gaining leverage on how dominant versions of sovereignty are lived—phenomenologically, biopolitically, and ethically—amid the complex configurations of national and international power in the late twentieth and early twenty-first centuries. And yet, as indigenous scholars, among others, have taught us, alternative sovereignties have been and are being thought and lived in the fissures of colonial, racial capitalism. Provincializing the ways in which modern medicine has come to consolidate specific concepts of bodily and territorial sovereignty that were originally forged in the juridico-political regimes of early liberalism opens a space to render visible the alternative versions of sovereignty embodied in different articulations of that which is therapeutic. I offer an approach to therapeutic sovereignty that conceives it as a space of debate, struggle, reimagining, and reinvention.

10. Feierman, "Struggles for Control."

11. Through a philosophy of mathematics in Africa, others have also explored how scale and length are thought and enacted in various parts of Africa. See Eglash, *African Fractals*; Gerdes, *Geometry from Africa*; and Verran, *Science and an African Logic*. In particular, I find myself wondering over Eglash's statement: "There is no way to connect fractals to the idea of dimension without using infinity, and for many mathematicians that is their crucial role" (18). Does it follow, then, that the ways that fractals are lived in Africa require actions informed by a particular notion of indefinite ongoingness? Might this be a way to inspire more practical work animated by commitments to continuance neither salvific nor apocalyptic?

12. Povinelli, "Toxic Late Liberalism," and Ahmann, *Futures After Progress*. On uneven, gendered invisibility of toxicity in sub-Saharan Africa, see Stein and Luna, "Toxic Sensorium."

13. Hecht, "The African Anthropocene."

14. Wetsman, "Air-Pollution Trackers Seek to Fill Africa's Data Gap."

15. Some of the results of which are contributed to WHO's publication by Mudu, *Ambient Air Pollution and Health in Accra, Ghana*.

16. Tousignant, *Edges of Exposure*. In relation to history of agriculture specifically, see also Tousignant, "Toxic Residues of Senegal's Peanut Economy," 5–8.

17. Stein and Luna, "Toxic Sensorium."

18. Nixon opens his book *Slow Violence and the Environmentalism of the Poor* with Lawrence Summers's now infamous confidential memo to staff during his tenure as chief economist of the World Bank (1991–93), in which he asserts that "countries in Africa are vastly under polluted." Summers then goes on to blithely explain the "impeccable" economic logic "behind dumping a load of toxic waste in the lowest wage country" (18–19).

19. Much of this literature is inspired by feminist, queer, critical race, and postcolonial theory. See Krupar, *Hot Spotter's Report*, in which she reminds us that the impossibility of purity is not an issue of the so-called developing world. Yet other ways of articulating boundaries (and managing the flow of substances as they constitute insides and outsides) remain important. For instance, see Roberts, "What Gets Inside," 592–619. The impossibility of purity does not mean that all entanglements are welcome.

20. Shotwell, *Against Purity*. For Latour's argument, see Latour, *We Have Never Been Modern*.

21. One might argue that at the heart of the interest in alternative modernities lies an attraction to the creative maneuvers of people outside of the United States and Europe to

reject forms of purification that would marginalize them from discourse while simultaneously translating the institutions of modernity.

22. Fortun, *Advocacy After Bhopal*; Petryna, *Life Exposed*.

23. For examples of such efforts at argument and of actions taken in such deeply fraught spaces, see Tilley, "Ecologies of Complexity"; White, "Poisoned Food, Poisoned Uniforms, and Anthrax," 220–33; Hecht, *Being Nuclear*; and Stein and Luna, "Toxic Sensorium."

24. Agard-Jones, "Chemical Kin/Esthesia."

25. Cone, "Should DDT Be Used to Combat Malaria?"

26. See Schmidt, "Unfair Trade E-Waste in Africa"; Redfern, "EU, US Dumping Toxic Waste in Africa"; and Minter, "The Burning Truth Behind an E-Waste Dump in Africa."

27. Clapp, "The Toxic Waste Trade with Less-Industrialised Countries," 505–18; Ntapanta, "Polarized Cityscapes," 227–43; Ntapanta, "'Lifescaping' Toxicants," 7–10.

28. Campbell, Dixon, and Hecky, "A Review of Mercury in Lake Victoria, East Africa," 325–56; Belem, "Mining, Poverty Reduction, the Protection of the Environment, and the Role of the World Bank Group in Mali," 119–49.

29. Tilman et al., "Forecasting Agriculturally Driven Global Environmental Change," 281–84; Mwegoha and Kihampa, "Heavy Metal Contamination in Agricultural Soils and Water," 763–69.

30. While pesticide and herbicide use is prevalent on small holdings, many report feeling ill after application. See Ngowi et al., "Smallholder Vegetable Farmers in Northern Tanzania," 1617–24.

31. Mwita, Hosea, and Muruke, "Assessment of Genetic Modification." The Tanzanian government's 2006 National Biosafety Framework requires the monitoring of genetically modified organisms. This study to assess GMO contamination in both imported maize stocks and processed soybeans in the country was conducted to support the development of policies consistent with this framework.

32. Fisher et al., "Pollution and Toxicity," 1–4.

33. Malhi, "The Concept of the Anthropocene," 77–104.

34. Crutzen and Stoermer, "The Anthropocene," 17–18; Crutzen, "Geology of Mankind," 23.

35. Baskin, "Paradigm Dressed as Epoch," 9–29.

36. If the impact of a flat, rather homogeneous "humanity's" growing impact on "nature" leads us only to strategize about the nature that "we" see ourselves as creating, then the Anthropocene supports a regrouping around the status quo. The environmental crisis is seen as a great equalizer (Latour and Porter, *Facing Gaia*). Yet, while the Anthropocene continually inscribes itself in all our bodies—we *all* have endocrine disruptors, microplastics, and other toxic things chugging through our metabolisms—it manifests differently in different bodies. Those differences, along with the histories that generated them, matter a great deal—both in relation to inequities of the burden and to humanity's relationship with the planet (Yusoff, *A Billion Black Anthropocenes or None*).

37. For Plantationocene, see Mitman, "Reflections on the Plantationocene." For Capitalocene and Chthulucene, see Haraway, "Tentacular Thinking." For Ravenocene, see Thornton and Thornton, "The Mutable, the Mythical, and the Managerial."

38. The United Nations has called for "connecting global priorities" by recognizing the links between biodiversity and health.

39. The effects of climate on health are myriad. Of relevance to the issues at the core of this book is recent research on the effect of increased carbon dioxide on the nutrient value of food crops. This early-stage research is beginning to raise alarms that environmental changes may lead to the "great nutrient collapse." Evich, "The Great Nutrient Collapse."

40. The One Health concept that emerged as the World Health Organization, the Food and Agriculture Organization, and the World Organization for Animal Health (OIE) collaborated after the highly pathogenic H5N1 avian influenza in the early 2000s directs sustained attention to the interaction of humans, animals, and ecosystems. Schools of public health are linking with veterinary colleges, but in places like my home institution, Cornell University, they are also being created from the center of the veterinary college.

41. Horton and Lo, "Planetary Health," 1921–22. Calls for a program on planetary health, addressing the ways that human impacts on Earth's system have created health challenges for (and between) human and natural systems, have been taken up by a range of universities, nongovernmental organizations, and governmental research groups.

42. Haraway, *The Companion Species Manifesto*.

43. Ghosh, *The Great Derangement*.

44. Livingston, *Self-Devouring Growth*.

45. Livingston, *Self-Devouring Growth*, 9. See also Chao and Enari, "Decolonising Climate Change," 32–54.

46. Vimalassery, Pegues, and Goldstein, "On Colonial Unknowing."

47. Nading, "Living in a Toxic World," 209–24.

48. Hoffman, "Toxicity"; Todd, "Fish, Kin and Hope," 102–7; Jain, *Malignant*; Livingston, *Improvising Medicine*; and Langston, *Toxic Bodies*.

49. For examples in Africa, see Hecht, *Being Nuclear*; Thomas, "Beauty"; and Hoffman, "Toxicity."

50. Murphy, "Alterlife and Decolonial Chemical Relations," 494–503.

51. Povinelli, *Geontologies*.

52. See Alaimo, *Exposed*; Chen, *Animacies*; and Shotwell, *Against Purity*.

53. Foucault, *The History of Sexuality*.

54. Ndlovu-Gatsheni, *Epistemic Freedom in Africa*.

55. Comaroff and Comaroff, *Theory from the South*.

56. Diagne, *The Ink of Scholars*.

57. For more on the in-between as a place from which to think postcoloniality, see Bhabha, *The Location of Culture*.

58. Langwick, "Witchcraft, Oracles, and Native Medicine," 39–57.

59. For more on thinking through "belief," see Pigg, "The Credible and the Credulous," 160–201. See also Good, *Medicine, Rationality, and Experience*.

60. Haraway, *Staying with the Trouble*.

61. For more on "the enmeshment of flesh with place," see Alaimo, *Exposed*.

62. Shapiro, "Attuning to the Chemosphere," 372. In the interpretive social sciences, the demands of grappling with toxicity have incited the theorization of exposure. Mitman, Murphy, and Sellers, eds., "Landscapes of Exposure: Knowledge and Illness in Modern Environments"; Fortun, *Advocacy After Bhopal*; Petryna, *Life Exposed*; Fortun

and Fortun, "Scientific Imaginaries and Ethical Plateaus in Contemporary U.S. Toxicology," 43–54; Murphy, *Sick Building*; Choy, *Ecologies of Comparison*; Alaimo, *Exposed*; and Haraway, *Staying with the Trouble*. In addition, see Hecht and Gupta, "Toxicity, Waste, Detritus."

63. For an extended argument on relations as method, see Liboiron, *Pollution Is Colonialism*. Also within this text, see particularly insightful articulations from a graduate student, Edward Allen, through his doctoral work, especially pages 21–22, 126–27.

64. "Arusha Stingless Bee Honey."

65. Soini, "Land Use Change Patterns."

66. For reference to the FAO's program recognizing "globally important heritage agricultural systems" and for the inclusion of the *kihamba* in particular, see https://www .fao.org/giahs/giahsaroundtheworld/designated-sites/africa/shimbwe-juu-kihamba-agro -forestry-heritage-site/en/. Last viewed April 2, 2024.

67. Maghimbi identifies early reports indicating small coffee exports from German East Africa as early as 1899. Maghimbi, "Recent Changes in Crop Patterns in the Kilimanjaro," 73–83.

68. Myhre, *Returning Life*.

69. Rogers, "The Kilimanjaro Native Planters Association," 94–114; Mhando, "Conflict as Motivation for Change," 137–54.

70. Hunt, *A Colonial Lexicon*.

71. For more on schooling in Kilimanjaro, see Stambach, *Lessons from Mount Kilimanjaro*.

72. For a link between houses and the commodification of beef, as well as the corresponding shift in the ontology of cows, see Livingston, "In the Time of Beef," 35–60.

73. Maro, "Agricultural Land Management Under Population Pressure," 273–82; Soini, "Land Use Change Patterns and Livelihood Dynamics on the Slopes of Mt. Kilimanjaro," 306–23.

74. Tagseth, "Oral History and the Development of Indigenous Irrigation," 9–22.

75. Minde, "Law Reform and Land Rights for Women in Tanzania," 64–66.

76. O'kting'ati et al., "Plant Species in the Kilimanjaro Agroforestry System," 177–86; Hemp, "The Chagga Home Gardens," 203–9.

77. Hemp, "The Banana Forests of Kilimanjaro," 1193–217.

78. Pignarre and Stengers, *Capitalist Sorcery*.

79. Scholarship inspired by Sylvia Wynter's distinction between the plantation and provision grounds is provocative to think with in relation to the *kihamba* in that it poses critical questions about the co-constitution of complex human-plant configurations. But the plantation and the provision ground are also formations unique to the economics of land caught up and transformed through the Atlantic slave trade. The *kihamba* formation and transformation bear witness to different pressures. See both DeLoughrey, "Yam, Roots, and Rot: Allegories of the Provision Grounds," 58–75; and Castellano, "Provision Grounds Against the Plantation," 15–27.

80. Kimambo, "Environmental Control and Hunger," 71–95; Hakansson, "Politics, Cattle and Ivory," 141–54.

81. Giblin, "Land Tenure, Traditions of Thought About Land, and Their Environmental Implications in Tanzania," 1–56.

82. Manji, "The Case for Women's Rights to Land in Tanzania," 11–38.

83. Oguamanam, "Plant Breeders' Rights, Farmers' Rights and Food Security," 240–68.

## 1. FUTURES OF LUSHNESS

A much earlier version of this chapter was published in "A Politics of Habitability: Plants, Healing and Sovereignty in a Toxic World," *Cultural Anthropology* 33, no. 3 (2018): 415–43.

1. I thank Steve Feierman for this point. Villages in this region comprise a cluster of household-based farms dominated by banana (the staple food crop) and coffee (the primary cash crop). For more, see Weiss, *The Making and Unmaking of the Haya Lived World*.

2. Mama Nguya's grandmother would have been unlikely to have access to farmland defined by the cultivation of bananas by men. Mama Nguya's establishment of TRMEGA echoes long-standing strategies by women in Kagera to generate possibilities within the frictions of patriarchal systems of land ownership. See Weiss, *The Making and Unmaking of the Haya Lived World*.

3. For more on multispecies ethnography and archeology of ethnic formation in Africa, see Schoenbrun and Johnson, "Ethnic Formation with Other-Than-Human Beings." In relation to the *kihamba* and Chagga identity specifically, see Fisher, "Chagga Elites and the Politics of Ethnicity in Kilimanjaro, Tanzania." For more on the consolidation of Chagga political identity on Mount Kilimanjaro in the 1940s and 1950s, see Bender, "Being 'Chagga.'"

4. Van Der Plas et al., "Climate-Human-Landscape Interaction in the Eastern Foothills of Mt. Kilimanjaro."

5. Spear, *Mountain Farmers*.

6. Based on her observation of members' success, Mama Nguya has hypothesized that *imarisha* works against the development of drug resistance, allowing people with HIV/AIDS to continue with one drug regimen for longer than otherwise might be possible.

7. Anwar et al., "*Moringa Oleifera*."

8. Rockwood, Anderson, and Casamatta, "Potential Uses of *Moringa Oleifera*"; Gopalakrishnan, Doriya, and Kumar, "*Moringa Oleifera*."

9. Strathern, *Partial Connections*.

10. For ways that the toxicity of pharmaceuticals is problematized elsewhere in Africa, see Hamdy, *Our Bodies Belong to God*, and Livingston, *Improvising Medicine*.

11. I am inspired in *The Use of Pleasure* by Foucault's ontological claim for ethical substance as the matter of the self on which ethical and moral discourse reflects. Ethical substance, he argues, is that which demands reflection and work. Attending to this substance is what makes ethical subjects. In ancient Greece, reflection on and labor over the substance of bodily pleasure forged ethical subjects. Today, in Tanzania, the potency of *dawa lishe* resides in the way that it captures an active material-social space to work on an alternative ethical substance: toxicity.

12. Crane, "Lush Aftermath."

13. The power of Crane's articulation of the forms of lushness that are fostered by colonialism, violence, and dispossession rests in part in the ways that it pulls together the discovery of the plantation, the greenhouse, and the lawn. See, for example, Zarate, "Maintenance," and Majumdar, "Recalcitrant Life Worlds." The former focuses on those

laboring in the yards and gardens of Orange County, California, while the latter extends our analysis of the violence of extracting labor to that of plants' labor in her story of the *Ficus elastica*, which failed the British as a plantation crop in India but was made lucrative as a houseplant through the chemically saturated greenhouses in Florida.

14. Saas, "How a Garden Changed My Life."

15. For a discussion on how affective encounters between people and plants in Mozambican gardens inspire reflections and reorientations that generate the possibility of new futures, see Archambault, "Taking Love Seriously in Human-Plant Relations in Mozambique."

16. Olarinoye et al., "Exploring the Future Impacts of Urbanization and Climate Change on Groundwater in Arusha."

17. Weiss and McMichael, "Social and Environmental Risk Factors in the Emergence of Infectious Diseases."

18. Tsing, Mathews, and Bubandt, "Patchy Anthropocene."

19. Alpers, "Gujarat and the Trade of East Africa, c. 1500–1800." To give a sense of the scope of trade and therefore the extent of contact, Alpers suggests that by 1595, Gujarati traders were censured by the Portuguese because they had created such a lucrative role for themselves in maritime trade that they threatened Portuguese profits. See also Subrahmanyam, "Between Eastern Africa and Western India, 1500–1650."

20. Alam and Subrahmanyam, "A View from Mecca."

21. Majumdar, "Recalcitrant Life Worlds."

22. For an important argument on the political and ecological significance of native plants, see Mastnak, Elyachar, and Boellstorff, "Botanical Decolonization."

23. Hemp, "The Banana Forests of Kilimanjaro."

24. Noe, "Reducing Land Degradation on the Highlands of Kilimanjaro Region."

25. Agamben, *Nudities*.

26. Roitman, *Anti-Crisis*.

27. For ways of thinking with and about compost that push against a metaphysics of purity, see also Jones, "(Com)Post-Capitalism."

28. Reetsch et al., "Traditional and Adapted Composting Practices Applied in Smallholder Banana-Coffee-Based Farming Systems."

29. Here, I join: Hall, "Toward a Queer Crip Feminist Politics of Food"; Puig De La Bellacasa, "Making Time for Soil"; Haraway, *Staying with the Trouble*; and Lyons, "Decomposition as Life Politics."

30. For Roitman, *Anti-Crisis*, this forgetting is understood as that which articulates crisis as the point that requires radical epistemological revision, the rupture that justifies intervention and something new.

31. Abrahamsson and Bertoni, "Compost Politics," 125.

32. Haraway, *Staying with the Trouble*.

33. For more on "collective continuance," see Whyte, "Food Sovereignty, Justice, and Indigenous Peoples," 347.

34. Turner and Somerville, "Composting with Cullunghutti."

35. Abrahamsson and Bertoni, "Compost Politics."

36. Afonso and Imbassahy, "Feeding the Earth: Composting and Compost in an Indigenous Garden in Rio de Janeiro," 10.

37. Holbraad, Pedersen, and de Castro, "The Politics of Ontology." This is also the logic of ancestors, of a dying in which one might return, but not as one's self, not as human, not as embodied. Ancestors lay claim to the living. They demand feeding, coaxing, appeasing, and being in relation with. They compel actions that remember that embodied life was born from them. Compost proposes a theory of nonhuman ancestors to be fed, appeased, and coaxed into new relations that will protect and nourish.

38. Cadena, *Earth Beings*.

39. About anticommodities, see Glover and Stone, "Heirloom Rice in Ifugao." For more on agrarian resistance to capitalist forms, see Scott, *The Art of Not Being Governed*.

40. In "From the Anthropocene to the Planthroposcene," Myers writes, "Plants entice entire ecologies of other creatures to participate in their care and their propagation: they have the know-how to entrain others in the service of their rhythms, their wiles, and desires" (297). See also Hustak and Myers, "Involutionary Momentum."

41. This account joins work in the emerging field of critical plant studies, which Stark, "Deleuze and Critical Plant Studies," describes as challenging "the privileged place of the human in relation to plant life" (180). See also Marder, *Plant-Thinking*.

42. Such gardens grow and extend, but they are not scalable (as with matsutake in Tsing, *The Mushroom at the End of the World*). Rather, they invite us to connect the histories of the making of economies of scale with the making of ecologies of scale. For gardens as storied space, see Hall, "My Mother's Garden."

43. Povinelli, *Geontologies*.

## 2. EFFICACY OF APPETITES

1. Tanzania National Bureau of Statistics, *National Sample Census of Agriculture 2019/20*.

2. My use of "propositions" here follows on Latour's reading (and elaboration) of Whitehead, *Process and Reality*. As Latour writes, "Propositions are not statements, or things, or any sort of intermediary between the two. They are, first of all, actants. . . . They are not positions, things, substances, or essences pertaining to a nature made up of mute objects facing a talkative human mind, but *occasions* given to different entities to enter into contact" (italics in original). Latour, *Pandora's Hope*, 141.

3. See, for example, Farquhar, "Eating Chinese Medicine."

4. For more on "nutritive substances" in South Africa, see Cousins, "A Mediating Capacity."

5. Good, "How Medicine Constructs Its Objects," in *Medicine, Rationality and Experience*, 65–87.

6. Clare, *Brilliant Imperfection*. Such grappling is also apparent in efforts to decolonize the clinic. For example, see Smith-Morris et al., "Decolonizing Care at Diagnosis."

7. In fact, Alex Uroki is the only vegetarian Mchagga I have ever met. As many contemporary diets in northern Tanzania, those in Kilimanjaro tend to be meat-centric.

8. Uroki's insistence on not being restricted by single identities, by living through addition and incorporation, resonates with Homi Bhabha's articulation of the power of refusing to purify one's claim to identities and insisting on acting from the interstices. In highlighting hybridity, liminality, or mimicry, Bhabha argues for the productivity of

ambivalence—between the divisions that define modern race, sexuality, class, and, in this case, we might say *kabila* or "tribe." Uroki, however, is concerned not only with cultural production but also with the material realities of bodies and worlds. Bhabha, *The Location of Culture*.

9. Tanzanians refer to mild and pleasant tastes, those that are easily palatable, as *baridi* (cool) or *tamu* (sweet), and those that are more complex or harsher as *kali* (sharp or fierce) or *uchungu* (bitter).

10. Bitterness is a quality whose therapeutic capacity is at stake in a range of small experiments in Tanzania. See Meek, "Fugitive Science."

11. Farquhar, "Eating Chinese Medicine," 483.

12. I owe much to the discussion of throughness to conversations that happened among the Translating Vitalities Collective's members. See also Scheid, "Promoting Free Flow in the Networks."

13. The production facilities are just slightly south of the highway, tucked away in a quieter area. Another small shop is in Arusha on the grounds of the national museum, and Eden-Mark medicines sit on the shelves of shops in Moshi and farther way in Dar es Salaam and Nairobi. Another factory primarily for the production of baobab oil remains in Dodoma.

14. Digby, Ernst, and Muhkarji, eds., *Crossing Colonial Historiographies*.

15. Wendland, *A Heart for the Work*.

16. The forms of sovereignty that Uroki and other *dawa lishe* producers imagine break from the "modern constitution," as Latour has accounted for the divisions that structure the governance of scientific knowledge. Latour, *We Have Never Been Modern*. Striving for new possibilities, new kinds of sovereignty by explicitly moving across the ruptures and divisions so critical to modernity and the state, they implicitly challenge biomedicine to reimagine itself through alternative ways of accounting for being, knowing, and relating. For the import of this work in re-reading philosophical texts, see Mol, *Eating in Theory*.

17. Landecker, "Being and Eating."

18. Landecker, "Postindustrial Metabolism."

19. Landecker, "Food as Exposure."

20. See also, Bitar, *Diet and the Disease of Civilization*.

21. Landecker, "Being and Eating," 257. For additional anthropologists' writings about metabolism, see Hardin, *Faith and the Pursuit of Health*; Hatch, *Blood Sugar*; Mendenhall, *Rethinking Diabetes*; Weaver, *Sugar and Tension*; and Moran-Thomas, *Traveling with Sugar*.

22. Political-economic realities support the creation of private wealth and the extension of international institutions through initiatives into African welfare. See Redfield, "Bioexpectations: Life Technologies as Humanitarian Goods." See also Scott-Smith, "Control and Biopower in Contemporary Humanitarian Aid." For more on the political economy of postcolonial nutrition, see Nott, "'How Little Progress'?"

23. Livingston, *Self-Devouring Growth*.

24. Roberts, "Food Is Love"; Solomon, *Metabolic Living*; and Yates-Doerr, *The Weight of Obesity*.

25. For a recent example of the use of this language in relation to global health, see Gomez-Temesio, "Outliving Death."

26. Bayart, *The State in Africa*.

27. For more on *amandla* (Zulu for strength, power, efficacy), see Cousins, "A Mediating Capacity."

28. Janzen, *Ngoma*.

29. Weiss, *The Making and Unmaking of the Haya Lived World*, and Myhre, *Returning Life*.

30. For a description of similar practices as they have been observed farther up the eastern side of Mount Kilimanjaro, see Myhre, "*Horu*," 98–145.

31. Langwick, "Healers and Their Intimate Becomings," in *Bodies, Politics, and African Healing*, 87–120. The stories of the appetites of ancestors, spirits, or a range of devilish others that are told in Tanzania resonate across sub-Saharan Africa. Casey, "Eco-Intimacy and Spirit Exorcism in the Nigerian Sahel"; Johnson, ed., *Spirited Things*; Masquelier, *Prayer Has Spoiled Everything*; and Stoller, *Embodying Colonial Memories*.

32. Mol, *The Body Multiple*. Healers are those not only with the "knowledge" of how to heal but also with bodies that heal. Becoming a healer means that the relations that materialize as body are ones that heal. In the south, where I previously conducted fieldwork, a healer's hands and spit were said to be agentive. Therefore, in these healing traditions, the body is not an epistemic object to be known but a complex gathering of relations to be hosted, engaged, mediated, and coordinated.

33. Pigg, "The Credible and the Credulous." For a broad discussion of the problem of belief in the subdiscipline, see Good, "Medical Anthropology and the Problem of Belief."

34. Cohen, "A Body Worth Having?"

35. Mol, *Eating in Theory*.

36. Cousins, "A Mediating Capacity."

37. Tironi and Rodríguez-Giralt, "Healing, Knowing, Enduring," 100.

38. Here I use the word "enact" in line with the practice-oriented ontology described by Mol in *The Body Multiple*.

39. Cabnal first introduced the concept of body-land in an effort to describe the interconnections between colonial control of indigenous bodies and lands as territories ("Acercamiento a la construcción del pensamiento epistémico"). Others have expanded on this concept; see, for example, Altamirano-Jiménez, "Indigenous Women Refusing the Violence of Resource Extraction." I lean into these indigenous scholars' work and draw it toward *dawa lishe* projects in Tanzania in an effort to reinforce their insistence that existence is always collective, individual, interelemental, and interspecies and to invite solidarity among initiatives that redefine healing as practices that nourish these relationships and health as being in good relation.

40. The QMRA is part of a fleet of technologies currently circulating in East Africa that Tanzanians debate quite fiercely. Diagnostic and curative claims about these or "alternative" technologies generate curiosity and skepticism, hope and derision. Uroki, for instance, rejected a popular foot tub said to extract toxins from the body. He experimented with pranic healing and acupressure. He worked with local craftspeople to create massage rollers. He purchased electric leg compression machines and vibrating massagers to increase circulation in the extremities. When he returned from a trip to India with me and other colleagues, in which we had visited an ayurvedic hospital, he designed and hired carpenters to build a massage table and steam bed. In 2020, during COVID, he built an aromatherapy steam room.

41. Dilger, "Healing the Wounds of Modernity."

42. Bicego, Rutstein, and Johnson, "Dimensions of the Emerging Orphan Crisis in Sub-Saharan Africa."

43. See, for example, Urassa et al., "Orphanhood, Child Fostering and the AIDS Epidemic in Rural Tanzania"; Nyambedha, Wandibba, and Aagaard-Hansen, "Changing Patterns of Orphan Care Due to the HIV Epidemic in Western Kenya"; and Evans, "'We Are Managing Our Own Lives. . . .'"

44. Appleton, "'At My Age I Should Be Sitting Under That Tree.'"

45. Cohen, "The Paradoxical Politics of Viral Containment," 29.

46. Nguyen, "Government-by-Exception."

47. Nguyen, *The Republic of Therapy*.

48. While gesturing toward structural issues, Uroki has organized his interventions as commodified herbal medicines. Even as he is motivated to see others heal, he is driven by the hope of making a living—to amass enough wealth to send his children to school; support his home and life, including travel and experimentation; and grow EdenMark so that he might pass it on to his middle son. Insofar as this effort extends capitalist forms, EdenMark is bound by the same inability to address structural issues for which others have critiqued humanitarian interventions into severe malnutrition cases. See, for example, Redfield, "Bioexpectations," and Scott-Smith, "Control and Biopower in Contemporary Humanitiarian Aid."

49. Ralph, "What Wounds Enable." See also Ralph, *Renegade Dreams*.

50. Simpson, *As We Have Always Done*.

51. Mol, *Eating in Theory*.


### 3. REGISTERS OF KNOWLEDGE

Parts of this chapter were published in "Properties of (Dis)Possession: Therapeutic Plants, Intellectual Property, and Questions of Justice in Tanzania," *Osiris* 36 (2021): 284–305.

1. For further discussion about colonial misinterpretations of *uganga* and *uchawi*, see Chanock, *Law, Custom, and Social Order*.

2. Arnold, *Toxic Histories*.

3. Africanist scholarship also has a rich engagement with questions of the unknown, unknowing, and strategic ignorance. For foundational work, see Last, "The Importance of Knowing About Not Knowing." For key aspects of this conversation, see Littlewood, ed., *On Knowing and Not Knowing in the Anthropology of Medicine*. The politics of this blindness are being articulated with increasing precision in such work as Geissler and Prince, "Active Compounds and Atoms of Society," and Geissler, "Public Secrets in Public Health."

4. Cadena, *Earth Beings*.

5. Langwick, "Partial Publics"; Livingston, *Self-Devouring Growth*.

6. The highest rates in the world of the biomedical condition known as albinism are in sub-Saharan Africa. See Hong, Zeeb, and Repacholi, "Albinism in Africa as a Public Health Issue." The World Health Organization has estimated that 1 in 1,429 infants are born with albinism in Tanzania.

7. Activists and academics have been drawn to cataloging the beliefs around albinism as the first step in initiatives to counter them. See Hong et al., "Albinism in Africa as a Public Health Issue."

8. Maron, "Witchcraft Trade, Skin Cancer Pose Serious Threats to Albinos in Tanzania."

9. *Wachawi* (plural, Kiswahili, witches and sorcerers) have long been known to gather in cemeteries. They may talk, plan, play, and eat on the graves of those buried there. Or they may dig up the bodies, taking parts of them to make their medicines. These tendencies structure the fear that witches will move from the horrors of disrupting dead bodies to the terror of dismembering live ones. The circulation of body parts and their incorporation into medicines that facilitate the deep inequalities of power and wealth have indexed the metaphoric cannibalisms of slavery, colonization, or new forms of capitalism.

10. "Tanzania Sentences Albino Killers to Death," *Daily Nation*. The Tanzanian government has expressed anger in addition to disgust and embarrassment, including in its commentary on a recent documentary film. See Wulfhorst, "Tanzania Criticizes Film Documenting Attacks on Albinos for Witchcraft."

11. Under the Same Sun, a Canadian-Tanzanian nongovernmental organization dedicated to advocacy for people with albinism, provides regular updates of the "Reported Attacks on Persons with Albinism." As of June 17, 2021, the organization reported that, in Tanzania, there have been seventy-seven killings (most recent May 2021) and twenty-five grave violations. In addition, of the ninety-four Tanzanians with albinism who have survived attacks, "all are deeply traumatized and some severely mutilated" (most recent November 2019); one person is missing, and three people have been granted political asylum (in Canada and the United States). Under the Same Sun, "Reported Attacks of Persons with Albinism."

12. Langwick, Wiketye, and Ogondiek, "Working with Traditional Healers."

13. There may be several reasons for the greater success with the registration of healers in areas that have seen more incidents of violence against people with albinism. Some district-level Traditional Medicine Coordinators—those in the Ministry of Health who are most immediately tasked with reaching out to traditional healers and facilitating a successful registration process—articulated registration as a technique for social control. It identified healers and collected information that made them accessible to state officials. Therefore, Traditional Medicine Coordinators may see this work as particularly urgent in the Lake Zone, which saw greater violence against people with albinism. Additionally, healers may have found that registration provides some legitimacy in the eyes of the state, making it worth the effort and expense.

14. Malebo and Mbwambo, *Technical Report on Miracle Cure Prescribed by Rev. Ambilikile Mwasupile*.

15. Mattes, "The Blood of Jesus and CD4 Counts."

16. "News Archives—February 2011," Ngorongoro Conservation Area Authority, https://www.ngorongorocrater.org/archive2_11.html.

17. "Two Cabinet Ministers Meet Loliondo Ex-Pastor," IPPmedia.Com.

18. Feierman, *The Shambaa Kingdom*; Kodesh, "History from the Healer's Shrine"; Janzen, *Lemba, 1650–1930*; Janzen, *Ngoma*; Schoenbrun, "Conjuring the Modern in Africa"; and Tantala, "The Early History of Kitara in Western Uganda."

19. Feierman, *Peasant Intellectuals*.

20. Inside biomedical institutions, however, medical doctors are regularly referred to as *mganga* and the chief medical officer as the *mganga mkuu*. Part of the history of biomedicine, then, is in the translation of this term from a flexible, ethical evaluation of how knowledge is used to a title conferring authority and identity on the basis of mastery over facts.

21. Langwick, "Geographies of Medicine."

22. The work of "middles" (nurses, teachers, and others who translated practices for colonial institutions) and African field assistants was critical in shaping understandings of which practices constituted medicine and what forms of expertise legitimated a healer. See Hunt, *A Colonial Lexicon*, and Schumaker, *Africanizing Anthropology*.

23. Arnold, *Toxic Histories*. See especially chap. 2, pp. 41–77.

24. Flint, *Healing Traditions*, and Langwick, *Bodies, Politics, and African Healing*.

25. Arnold, *Toxic Histories*.

26. Brockway, *Science and Colonial Expansion*.

27. See Geissler and Prince, "Active Compounds and Atoms of Society." In addition, for a helpful articulation of thinking vegetal agency versus plants as responsible, see Hetherington, "Beans Before the Law."

28. See Mukharji, "Vishalyakarani as *Eupatorium Ayapana*," and Nappi, "Winter Worm, Summer Grass."

29. Feierman, "Struggles for Control."

30. See Feierman, *The Shamba Kingdom*; Feierman, *Peasant Intellectuals*; and Vaughan, *Curing Their Ills*. Additionally, two pieces that elaborate the behind-the-scenes, low-level debates generated because witchcraft ordinances did not reflect the complexity on the ground are as follows: Tilley, "An Anthropological Laboratory," and Langwick, "Witchcraft, Oracles, and Native Medicine."

31. See Langwick, *Bodies, Politics, and African Healing*.

32. For more on how colonial interventions into *uchawi* ruptured indigenous notions of justice, see Chanock, *Law, Custom, and Social Order*.

33. Vähäkangas, "Negotiating Religious Traditions: Babu Wa Loliondo's Theology of Healing," and Vähäkangas, "The Grandpa's Cup."

34. Mukharji, "Vishalyakarani as Eupatorium Ayapana."

35. The Nyerere government articulated "self-reliance" as a political goal and strategy for development. See Tanganyika African National Union, *The Arusha Declaration*. For the historical context in which to locate ways in which Nyerere's project drew on, and contributed to, international efforts to articulate African sovereignty, see Tilley, "Traditional Medicine Goes Global."

36. For more detail, see Mesaki, "Witchcraft and the Law in Tanzania."

37. A rich historical and anthropological literature has accounted for the ways that colonial and missionary clinics drew on healers in Africa for biomedical outreach and staffing. See Langwick, "Geographies of Medicine," and Hunt, *A Colonial Lexicon*. In the mid-1970s, the language of international public health policies, focused on primary health care, narrowed healers' ideal roles in biomedicine. See World Health Organization, "Declaration of Alma-Ata."

38. Hsu and Harris, eds., *Plants, Health and Healing*.

39. This collaboration was cut short very suddenly in the mid-1980s, unfortunately, when Deval was killed in a car accident.

40. Parliamentary Act No. 23 of 1979.

41. Nyerere, *Ujamaa-Essays on Socialism*.

42. Langwick, "Healers and Scientists." This essay also describes contests over the process of separating plants from healers and (re)embedding plants into laboratory research and state development projects.

43. Kasilo, Samson, and Ngenda, "An Overview of the Traditional Medicine Situation in the African Region."

44. In 2005, 27 percent of the African countries that responded to the WHO global survey had a national law or regulation that was focused on traditional medicine. See World Health Organization, *National Policy on Traditional Medicine and Regulation of Herbal Medicines*.

45. For more on the focus on herbals over pharmaceuticals at the Mabibo lab, see Langwick, "Partial Publics."

46. Latour, *We Have Never Been Modern*. The market does not necessarily share the state's commitment to engendering "the modern constitution."

47. It does, however, generate a politics of difference. Registration is one area that illustrates the work of the state to reinforce some forms of difference through traditional medicine and de-center others. The articulation of difference through the dualisms that generated traditional medicine in Africa has always been insufficient to account for the multiplicities of African therapeutic practice. Traditional medicine emerged as homogeneous only in relation to biomedicine. Magic, witchcraft, faith, and traditional practices have long all glossed the diverse engagements that have been refused by or slipped away from state-led projects in medicine and science. As postcolonial traditional medicine becomes a more elaborate apparatus for generating knowledge, practice, and value, it is implementing new ways of generating its subjects and objects.

48. Holbraad, Pedersen, and Viveiros de Castro, "The Politics of Ontology."

49. This became visible to John Ogondiek, a pharmacist working for the NIMR, who remembers being aware that references to *materia medica* grew consistently stronger and more frequent during these stakeholder meetings. Interview with John Ogondiek, May 2, 2013.

50. Eight focus group discussions, including a total of forty healers, were conducted in Meru, Arusha, in April 2013 and Rombo, Kilimanjaro, in February 2014. Open-ended questions explored healers' experiences with registration, drew out their thoughts on the benefits and drawbacks of registration, collected accounts of their own history of becoming a healer, and investigated how new knowledge is generated and gained by healers. Last, we elicited their opinions about scientific collaboration.

51. Feierman, "Struggles for Control."

52. Additionally, a few of the people who attended our focus group discussions, particularly in Meru, that included areas inhabited primarily by Maasai communities did not register because they refused the status of "healer." They argued that they "just knew a few specific medicines from the grandparents" and that they shared these with those in need, but they were not *waganga*.

53. *Kuota*, to dream, also means to germinate.

54. Langwick, "Partial Publics."

55. Povinelli, *The Cunning of Recognition*.

56. For a foundational argument about colonial misinterpretations of *uganga* and *uchawi*, specifically in relation to colonial state legal formations, see Chanock, *Law, Custom, and Social Order*.

57. Farquhar, *Knowing Practice*.

58. Agamben, *Nudities*, 113–14.

## 4. WORK OF TIME

1. Diagne, "The Time We Need."

2. Parsons, *The Social System*.

3. Berlant, "Slow Death (Sovereignty, Obesity, Lateral Agency)."

4. Berlant, "Slow Death," 754.

5. Escobar, *Encountering Development*.

6. For example, at the time, Ehrlich and Ehrlich's *The Population Bomb* was particularly influential.

7. Weisz and Olszynko-Gryn, "The Theory of Epidemiologic Transition."

8. Fox, "The Significance of the Milbank Memorial Fund for Policy"; Robine, "Life Course, Environmental Change, and Life Span"; and Susser, "Epidemiology in the United States after World War II."

9. Smallman-Raynor, "Late Stages of Epidemiological Transition."

10. Omran, "The Epidemiologic Transition Theory Revisited Thirty Years Later."

11. Gaylin and Kates, "Refocusing the Lens."

12. Institute for Health Metrics and Evaluation, "Tanzania."

13. Internationally, the Joint United Nations Program on HIV/AIDS (UNAIDS) has widely promoted the goal of ending HIV/AIDS by 2030. See UNAIDS, "2016 United Nations Political Declaration on Ending AIDS." In the United States, see Fauci et al., "Ending the HIV Epidemic."

14. World Health Organization, *Antiretroviral Drugs for Treating Pregnant Women and Preventing HIV Infection in Infants*.

15. Thompson and Abel, "The Work of Negotiating HIV as a Chronic Condition"; Mattes, "Caught in Transition."

16. Moyer, "The Anthropology of Life After AIDS."

17. Moyer and Hardon, "A Disease Unlike Any Other?"

18. Mattes, "Caught in Transition."

19. Morens, Folkers, and Fauci, "The Challenge of Emerging and Re-Emerging Infectious Diseases."

20. See, for example, Cavagna et al., "Blood Pressure-Lowering Medicines Implemented in 12 African Countries." See also Chiwandire et al., "Trends, Prevalence and Factors Associated with Hypertension and Diabetes Among South African Adults Living with HIV, 2005–2017."

21. Greene, *Prescribing by Numbers*.

22. Aronowitz, "The Converged Experience of Risk and Disease."

23. Smith-Morris, "The Chronicity of Life, the Acuteness of Diagnosis." Also, for the ways the ideology of cure is challenged by attending to issues of disability, see Clare, *Brilliant Imperfection*.

24. In the United States, when people have tried to talk to their doctor about getting off of statins, for example, they have at times faced incredulous clinicians who accuse them of wanting to commit suicide.

25. Dumit, *Drugs for Life*.

26. Biehl, "Pharmaceuticalization."

27. Cupit et al., "Overruling Uncertainty About Preventative Medications."

28. Paparini and Rhodes, "The Biopolitics of Engagement and the HIV Cascade of Care."

29. Setel, "Local Histories of Sexually Transmitted Diseases and AIDS in Western and Northern Tanzania."

30. National Tanzania Commission for AIDS (TACAIDS), Bureau of Statistics (NBS), and ORC Marco, *Tanzania HIV/AIDS Indicator Survey 2003–04*.

31. Davies, "Shattered Assumptions."

32. Marten, "Living with HIV as Donor Aid Declines in Tanzania."

33. De Klerk, "The Compassion of Concealment."

34. Ashforth, *Witchcraft, Violence, and Democracy in South Africa*. In Tanzania, see Mshana et al., "'She Was Bewitched and Caught an Illness Similar to AIDS.'"

35. Nguyen, "Antiretroviral Globalism, Biopolitics, and Therapeutic Citizenship."

36. Crane, *Scrambling for Africa*.

37. For more on the broad trend of pharmaceuticalization, see Petryna, "The Pharmaceutical Nexus"; Biehl, "Pharmaceuticalization: AIDS Treatment"; and Greene, "Making Medicines Essential."

38. I draw the concept of "project time" from Benton, Sangaramoorthy, and Kalofonos, "Temporality and Positive Living in the Age of HIV/AIDS." The impact of this temporal formation is a product of what Meinert and Whyte have called "projectification"; the counterresponse to the rapid spread of HIV, they argue, was the multiplication of projects addressing AIDS prevention, palliation, and treatment. See Meinert and Whyte, "Epidemic Projectification."

39. Kirigia et al., "Effects of Global Financial Crisis on Funding for Health Development in Nineteen Countries of the WHO African Region."

40. Prince, "Situating Health and the Public in Africa."

41. In a comment on Benton, Sangaramoorthy, and Kalofonos' article "Temporality and Positive Living," Whyte called these temporal logics that mark the emergence of the HIV/AIDS epidemics and international interventions "global health time" (471).

42. Geissler and Prince, *The Land Is Dying*.

43. The names of all TRMEGA members, as well as other clients, consumers, and patients, have been changed to maintain confidentiality.

44. Others have written on the cancer epidemic in Africa that is following in the wake of AIDS. See, for example, Livingston, *Improvising Medicine*, and Mika, *Africanizing Oncology*. In addition, as I write, emerging disease specialists are just beginning to frame their research question around HIV and COVID, suggesting that the SARS-CoV-2 virus has more opportunity to mutate in immune-compromised bodies and pointing to links between metabolic disorders and COVID.

45. Plants in this way might be seen as in conversation or in collaboration with techniques in African healing, described as "diagnosis by addition," in Feierman, "Explanation and Uncertainty in the Medical World of Ghaambo."

46. Anthropology's grappling with cultural relativism as a concept and method posed belief as an excuse and a conceptual program. Its impact on medical anthropology and studies of health and healing is well-trod territory. For a key example, see Good, "Medical Anthropology and the Problem of Belief." See also Pigg, "The Credible and the Credulous."

47. Guyer, "Prophecy and the Near Future."

48. Guyer, "Prophecy and the Near Future," 409.

49. Guyer, "Prophecy and the Near Future," 410.

50. Prince, "Situating Health and the Public in Africa."

51. Shifts in the structure of biomedical sciences shaped and were shaped by these political and economic changes. Public health programming bled increasingly often into clinical trials, and health emergencies offered a platform for governance. See Geissler, ed., *Para-States and Medical Science*.

52. Cohen, "Making Peasants Protestant and Other Projects."

53. This is not only an ethnographic question for "alternative treatments." McKay's *Medicine in the Meantime* has shown how the affective politics that motivate Africans to take up the goods of humanitarian initiatives draw diverse temporal sensibilities, folding the tempos of caring and dwelling of earlier times into the experience of the present.

54. *Cassia spectabilis* is originally from South America. It is used in Tanzania as an ornamental tree as well as for shade, firewood, charcoal, medicine, and tool handles.

55. For a study of this sort in which data collection focused on an area in Tanzania known for sisal production, see Paul, Kangalawe, and Mboera, "Land-Use Patterns and Their Implication on Malaria Transmission in Kilosa District, Tanzania."

56. "Asili" suggests the origin, a starting point, that from which others are derived.

57. Zhan, "Does It Take a Miracle?"; Chudakova, *Mixing Medicines*; Livingston, *Debility and the Moral Imagination*; Fullwiley, "The Biosocial Politics of Plants and People"; and Hannig, "Healing and Reforming."

58. Myhre, *Returning Life*.

59. Interestingly, the colleague who transcribed this interview translated Jane's comments as connoting that educated people "are stagnant"—they cannot hear or move with the idea that there may be a day that someone taking ART could stop taking the medication and remain strong.

60. Sanabria, "Endurance and Alterability."

61. Cousins, "A Mediating Capacity."

62. Whyte, "Food Sovereignty, Justice, and Indigenous Peoples," and Corntassel, "Re-envisioning Re-surgence."

63. Haraway, *Staying with the Trouble*. For a critique of the effects of Haraway's lack of explicit engagement with, and insufficient citation of, indigenous and Black feminist work, see Jegathesan, "Black Feminist Plots Before the Plantationocene and Anthropology's 'Regional Closets.'"

### 5. PROPERTIES OF HEALING

1. The forms of knowledge critical to articulating matters of appropriation (e.g., the biomedical properties of plants) also create zones of ignorance. Notably, here, appropriating arrangements fade into the background, or in Latour's influential terms, the politics

and practicalities through which these arrangements function are "blackboxed." In *Pandora's Hope*, Latour defines blackboxing as "the way scientific and technical work is made invisible by its own success. When a machine runs efficiently, when a matter of fact is settled, one need focus only on its inputs and outputs and not on its internal complexity. Thus, paradoxically, the more science and technology succeed, the more opaque and obscure they become" (304).

2. Marder, *Plant-Thinking*; Stark, "Deleuze and Critical Plant Studies"; Nealon, *Plant Theory*; Lawrence, "Listening to Plants"; and Irigaray and Marder, *Through Vegetal Being*. In addition, Brill and Rowman and Littlefield now both have an imprint for Critical Plant Studies.

3. See, for example, Jones, "Dying to Eat? Black Food Geographies of Slow Violence and Resilience," as well as the special issue edited by Hirsch and Jones, "Incontestable: Imagining Possibilities Through Intimate Black Geographies."

4. In "Ecology as Event," Marder describes our times as "the times of rampant economism, eclipsing all other modes of thinking, and of an unprecedented environmental crisis that provokes ecological thought as a reaction to the emergency of a planetary house on fire" (142).

5. Myhre, "Ngakuuriya Moo." For beer and rainmaking in Tanzania, see also Sanders, *Beyond Bodies*.

6. Murphy, "Alterlife and Decolonial Chemical Relations."

7. As Lock and Farquhar argue in *Beyond the Body Proper*, "The classic problematic of the relations between individual and society that still provides analytic tools for most of the social sciences seems to require a 'proper' body as a unit of individuality. This body proper, the unit that supports the individual from which societies are apparently assembled, has been treated as a skin-bound, rights bearing, communicating, experience-collecting, biomechanical entity" (40).

8. Clark, "Scale as a Force of Deconstruction."

9. A young male schoolteacher who was helping me with transcriptions translated *athirika* as "victimized" rather than "affected" to stress the point that he understood Rukia to be making.

10. Shortly after the passing of the 2002 Traditional and Alternative Medicines Act, government officials began convening a series of stakeholders' meetings around the country in an effort to conceive of a new system of registration. In 2011, when a new three-tiered registration system for healers, premises, and medicines was rolled out, a struggle emerged between the Baraza and TFDA over who should register medicines and how. By 2014, TFDA had hired their first graduate of the Master's Program on Traditional Medicine and Drug Development from the Institute of Traditional Medicine at Muhimbili to develop their engagements with herbal medicine regulation. By 2016, it was at least temporarily resolved that TFDA would register foreign herbal medicines entering the country as well as "processed" herbals from Tanzanian producers, while the Baraza would register traditional healers' "raw raw" therapies, in the words of the TFDA Complementary Medicines Program manager. Neither Dorcas nor her partner Rukia, however, have registered as a "healer." Dorkia Enterprises is a small enterprise, registered as a business and connected to the Small Industries Development Organization (SIDO). As a result, *kitarasa* does not move through the Baraza as a "raw" therapy, made by an individual

healer for her patients. Rather, commercialization means moving through TFDA as a drug or a food.

11. Barnes and Bloor, "Relativism, Rationalism and the Sociology of Knowledge."

12. Happi Emaga et al., "Effects of the Stage of Maturation and Varieties on the Chemical Composition of Banana and Plantain Peels"; Rodríguez-Ambriz et al., "Characterization of a Fibre-Rich Powder Prepared by Liquefaction of Unripe Banana Flour."

13. Someya, Yoshiki, and Okubo, "Antioxidant Compounds from Bananas (Musa Cavendish)."

14. Alkarkhi et al., "Comparing Physicochemical Properties of Banana Pulp and Peel Flours Prepared from Green and Ripe Fruits."

15. White, *Speaking with Vampires.*

16. Nguyen, "Antiretroviral Globalism, Biopolitics, and Therapeutic Citizenship."

17. Dilger, "Targeting the Empowered Individual."

18. Langwick, "Properties of (Dis)Possession."

19. Tanzania has a long history developing "traditional medicine" as a discrete category of knowledge and practice. See Feierman, "Struggles for Control," and Langwick, *Bodies, Politics, and African Healing.* This is true throughout Africa and other sites of European colonization, as empire explicitly remade relations between people and plants. See Brockway, *Science and Colonial Expansion*; Grove, *Green Imperialism*; Geissler and Prince, *The Land Is Dying*; and Hsu and Harris, eds., *Plants, Health and Healing.*

20. *Kitarasa* might be thought of as an "interscalar vehicle" in the sense that Hecht describes in "Interscalar Vehicles for an African Anthropocene."

21. As Chakrabarty has argued in "The Climate of History: Four Theses," the Anthropocene—or, more precisely, efforts to grapple with the ways that bodies and lands are emerging through one another given the toxicity of industrialization—"collapse[s] the age-old humanist distinction between natural history and human history." Starting with this premise, I wonder with activists, *dawa lishe* producers, and *kitarasa* what sort of account becomes possible in these ruins. Who and what can tell a history of the co-constitution of people and plants, bodies and land?

22. For instance, see Natural Justice's articulation on biocultural rights, especially their work with rooibos. National Khoisan Council and Cederberg Belt Indigenous Farmers Representatives, "The Khoikhoi People's Rooibos Biocultural Community Protocol." For more on rooibos, see Ives, *Steeped in Heritage.* For more on hoodia, see Foster, *Reinventing Hoodia.*

23. Bananas and humans have made each other what they are today. Similar arguments about the co-constitution are better elaborated between humans and animals. For instance, see Swanson, "Landscapes, by Comparison," and Tsai et al., "Golden Snail Opera."

24. Denham et al., "Origins of Agriculture at Kuk Swamp in the Highlands of New Guinea."

25. Lejju, Robertshaw, and Taylor, "Africa's Earliest Bananas?"

26. German colonial rule in the nineteenth century reified the identity of twenty to thirty loose confederations organizing social life and settlement there as "the Chagga." Perhaps only in the 1940s did Chagga emerge as a powerful self-declared political identity (Bender, "Being 'Chagga'").

27. These communities also developed markets as a mechanism to distribute goods across microenvironments even before political organization later reified them as "the Chagga." See Kimambo, "Environmental Control and Hunger."

28. Stump, "The Archaeology of Agricultural Intensification in Africa"; Silayo and Pikirayi, "Community-Based Approaches in the Construction and Management of Water Infrastructures Among the Chagga, Kilimanjaro, Tanzania."

29. Spear, *Mountain Farmers*.

30. Cadena, *Earth Beings*.

31. Marder, *Plant-Thinking*, 144.

32. See Marder, *Plant-Thinking*. The Matsutake Worlds Research Group also has taught us about the value of such material specificity to our theorizing. One of the generative frictions in Tsing's story of the *Mushroom at the End of the World* is that it has yet to be domesticated successfully. Matsutake resists the plantation, and this capacity for resistance shapes the ways that it articulates the edges of capitalism. The banana establishes a different kind of example.

33. Frison, Karamura, and Sharrock, "Banana Production Systems in Eastern and Southern Africa"; Tushemereirwe and Kubiriba, "Achieving Sustainable Cultivation of Bananas."

34. Tousignant, *Edges of Exposure*, 8.

35. Marder, "Ecology as Event."

## CONCLUSION: THERAPEUTIC SOVEREIGNTY

1. Silayo, "Land and Culture as Symbols of Remembrance, Ancestry, Rituals and Initiations."

2. For more on COVID in Tanzania, see Ciribassi, "The Temporal Politics of Ethnography, Heritability, and Contagion in Tanzania During Covid-19." See also Richey et al., "South-South Humanitarianism."

3. As discussed in greater detail in the Introduction and in chapter 4, the concept of collective continuance emerges from the scholarship of North American writers and environmental justice advocates. See, for example, Whyte, "Collective Continuance."

4. Haraway, "Tentacular Thinking: Anthropocene, Capitalocene, Chthulucene."

5. Foucault, *The Order of Things*.

6. Barad, *Meeting the Universe Halfway*.

7. For debates around this term, see Olsen and Sargent, eds., *African Medical Pluralism*.

8. Seth, "Pluriversal Problems, Revolutionary Remedies."

# Bibliography

Abrahamsson, Sebastian, and Filippo Bertoni. "Compost Politics: Experimenting with Togetherness in Vermicomposting." *Environmental Humanities* 4, no. 1 (2014): 125–48.

Afonso, Camila Bevilaqua, and Arthur Pereira Imbassahy. "Feeding the Earth: Composting and Compost in an Indigenous Garden in Rio de Janeiro." *Vibrant, Virtual Brazilian Anthropology* 20 (2023): 10. https://doi.org/10.1590/1809-43412023v20d905.

Agamben, Giorgio. *Nudities*. Translated by David Kishik and Stefan Pedatella. Palo Alto, CA: Stanford University Press, 2010.

Agard-Jones, Vanessa. "Chemical Kin/Esthesia." Paper presented at the Annual Meeting of the American Anthropological Association, November 18, 2016, Minneapolis, MN.

Ahmann, Chloe. *Futures After Progress: Hope and Doubt in Late Industrial Baltimore*. Chicago: University of Chicago Press, 2024.

Alaimo, Stacy. *Exposed: Environmental Politics and Pleasures in Posthuman Times*. Minneapolis: University of Minnesota Press, 2016.

Alam, Muzaffar, and Sanjay Subrahmanyam. "A View from Mecca: Notes on Gujarat, the Red Sea, and the Ottomans, 1517–39/923–946 H." *Modern Asian Studies* 51, no. 2 (2017): 268–318. https://doi.org/10.1017/S0026749X16000172.

Alkarkhi, Abbas F. M., Saifullah Bin Ramli, Yeoh Shin Yong, and Azhar Mat Easa. "Comparing Physicochemical Properties of Banana Pulp and Peel Flours Prepared from Green and Ripe Fruits." *Food Chemistry* 129, no. 2 (2011): 312–18. https://doi.org/10.1016/j.foodchem.2011.04.060.

Alpers, Edward A. "Gujarat and the Trade of East Africa, c. 1500–1800." *International Journal of African Historical Studies* 9, no. 1 (1976): 22. https://doi.org/10.2307/217389.

Altamirano-Jiménez, Isabel. "Indigenous Women Refusing the Violence of Resource Extraction in Oaxaca." *AlterNative: An International Journal of Indigenous Peoples* 17, no. 2 (2021): 215–23.

Anwar, Farooq, Sajid Latif, Muhammad Ashraf, and Anwarul Hassan Gilani. "*Moringa Oleifera*: A Food Plant with Multiple Medicinal Uses." *Phytotherapy Research* 21, no. 1 (2007): 17–25. https://doi.org/10.1002/ptr.2023.

Appleton, Judith. "'At My Age I Should Be Sitting Under That Tree': The Impact of AIDS on Tanzanian Lakeshore Communities." *Gender and Development* 8, no. 2 (2000): 19–27. https://doi.org/10.1080/741923627.

Archambault, Julie Soleil. "Taking Love Seriously in Human-Plant Relations in Mozambique: Toward an Anthropology of Affective Encounters." *Cultural Anthropology* 31, no. 2 (2016): 244–71. https://doi.org/10.14506/ca31.2.05.

Arnold, David. *Toxic Histories: Poison and Pollution in Modern India*. Science in History. New York: Cambridge University Press, 2016.

Aronowitz, Robert A. "The Converged Experience of Risk and Disease." *Milbank Quarterly* 87, no. 2 (2009): 417–42. https://doi.org/10.1111/j.1468-0009.2009.00563.x.

Ashforth, Adam. *Witchcraft, Violence, and Democracy in South Africa*. Chicago: University of Chicago Press, 2005.

Balick, Michael J., and Paul Alan Cox. *Plants, People, and Culture: The Science of Ethnobotany*. New York: Scientific American Library, 1996.

Barad, Karen Michelle. *Meeting the Universe Halfway: Quantum Physics and the Entanglement of Matter and Meaning*. Durham, NC: Duke University Press, 2007.

Bareetseng, Sechaba. "The Worldwide Herbal Market: Trends and Opportunities." *Journal of Biomedical Research and Environmental Sciences* 3, no. 5 (2022): 575–84. https://doi.org/10.37871/jbres1482.

Barnes, Barry, and David Bloor. "Relativism, Rationalism and the Sociology of Knowledge." In *Rationality and Relativism*, edited by Martin Hollis and Steven Lukes, 21–47. Oxford: Blackwell, 1993.

Baskin, Jeremy. "Paradigm Dressed as Epoch: The Ideology of the Anthropocene." *Environmental Values* 24, no. 1 (2015): 9–29. https://doi.org/10.3197/096327115X14183182353746.

Bayart, Jean-François. *The State in Africa: The Politics of the Belly*. 2nd ed. London: Longman, 1993.

Belem, Gisele. "Mining, Poverty Reduction, the Protection of the Environment, and the Role of the World Bank Group in Mali." In *Mining in Africa: Regulation and Development*, edited by Bonnie K. Campbell, 119–49. New York: Pluto Press, 2009.

Bender, Matthew V. "Being 'Chagga': Natural Resources, Political Activism, and Identity on Kilimanjaro." *Journal of African History* 54, no. 2 (2013): 199–220. https://doi.org/10.1017/S0021853713000273.

Benton, Adia, Thurka Sangaramoorthy, and Ippolytos Kalofonos. "Temporality and Positive Living in the Age of HIV/AIDS: A Multisited Ethnography." *Current Anthropology* 58, no. 4 (2017): 454–76. https://doi.org/10.1086/692825.

Berlant, Lauren. "Slow Death (Sovereignty, Obesity, Lateral Agency)." *Critical Inquiry* 33, no. 4 (2007): 754–80. https://doi.org/10.1086/521568.

Bhabha, Homi K. *The Location of Culture*. Routledge Classics. New York: Routledge, 2004.

Bicego, George, Shea Rutstein, and Kiersten Johnson. "Dimensions of the Emerging Orphan Crisis in Sub-Saharan Africa." *Social Science and Medicine* 56, no. 6 (2003): 1235–47. https://doi.org/10.1016/S0277-9536(02)00125-9.

Biehl, João Guilherme. "Pharmaceuticalization: AIDS Treatment and Global Health Politics." *Anthropological Quarterly* 80, no. 4 (2007): 1083–126. https://doi.org/10.1353/anq.2007.0056.

Bitar, Adrienne Rose. *Diet and the Disease of Civilization*. New Brunswick, NJ: Rutgers University Press, 2018.

Brockway, Lucile H. *Science and Colonial Expansion: The Role of the British Royal Botanic Gardens*. New Haven, CT: Yale University Press, 2002.

Cabnal, Lorena. "Acercamiento a la construcción del pensamiento epistémico de las mujeres indígenas feministas comunitarias de Abya Yala." In *Feminismos diversos: el feminismo comunitario*, by Lorena Cabnal and Asociación para la Cooperación con el Sur. Madrid: ACSUR–Las Segovias, 2010.

Cadena, Marisol de la. *Earth Beings: Ecologies of Practice across Andean Worlds*. Durham, NC: Duke University Press, 2015.

Campbell, Linda, D. G. Dixon, and R. E. Hecky. "A Review of Mercury in Lake Victoria, East Africa: Implications for Human and Ecosystem Health." *Journal of Toxicology and Environmental Health, Part B* 6, no. 4 (2003): 325–56. https://doi.org/10.1080/10937400306474.

Casey, Conerly. "Eco-Intimacy and Spirit Exorcism in the Nigerian Sahel." *The Senses and Society* 16, no. 2 (2021): 132–50. https://doi.org/10.1080/17458927.2020.1858651.

Castellano, Katey. "Provision Grounds Against the Plantation." *Small Axe: A Caribbean Journal of Criticism* 25, no. 1 (2021): 15–27. https://doi.org/10.1215/07990537-8912758.

Cavagna, Pauline, Jean Laurent Takombe, Jean Marie Damorou, et al. "Blood Pressure-Lowering Medicines Implemented in 12 African Countries: The Cross-Sectional Multination EIGHT Study." *BMJ Open* 11, no. 12 (2021): e049632. https://doi.org/10.1136/bmjopen-2021-049632.

Chakrabarty, Dipesh. "The Climate of History: Four Theses." *Critical Inquiry* 35, no. 2 (2009): 197–222. https://doi.org/10.1086/596640.

Chanock, Martin. *Law, Custom, and Social Order: The Colonial Experience in Malawi and Zambia*. Portsmouth, NH: Heinemann, 1998.

Chao, Sophie, and Dion Enari. "Decolonising Climate Change: A Call for Beyond-Human Imaginaries and Knowledge Generation." *eTropic* 20, no. 2 (2021): 32–54.

Chen, Mel Y. *Animacies: Biopolitics, Racial Mattering, and Queer Affect*. Durham, NC: Duke University Press, 2014.

Chiwandire, Nicola, Nompumelelo Zungu, Musawenkosi Mabaso, and Charles Chasela. "Trends, Prevalence and Factors Associated with Hypertension and Diabetes Among South African Adults Living with HIV, 2005–2017." *BMC Public Health* 21, no. 1 (2021): 462. https://doi.org/10.1186/s12889-021-10502-8.

Choy, Timothy K. *Ecologies of Comparison: An Ethnography of Endangerment in Hong Kong*. Experimental Futures: Technological Lives, Scientific Arts, Anthropological Voices. Durham, NC: Duke University Press, 2011.

Chudakova, Tatiana. *Mixing Medicines: Ecologies of Care in Buddhist Siberia*. Thinking from Elsewhere. New York: Fordham University Press, 2021.

Ciribassi, Rebekah. "The Temporal Politics of Ethnography, Heritability, and Contagion in Tanzania During Covid-19." *Platypus: The CASTAC Blog*, May 6, 2020. https://blog

.castac.org/2020/05/the-temporal-politics-of-ethnography-heritability-and-contagion
-in-tanzania-during-covid-19/.

Clapp, Jennifer. "The Toxic Waste Trade with Less-Industrialised Countries: Economic
Linkages and Political Alliances." *Third World Quarterly* 15, no. 3 (1994): 505–18.
https://doi.org/10.1080/01436599408420393.

Clare, Eli. *Brilliant Imperfection: Grappling with Cure*. Durham, NC: Duke University
Press, 2017.

Clark, Timothy. "Scale as a Force of Deconstruction." In *Eco-Deconstruction: Derrida
and Environmental Philosophy*, edited by Matthias Fritsch, 81–97. Groundworks: Ecological Issues in Philosophy and Theology. New York: Fordham University Press, 2018.

Cohen, Ed. "A Body Worth Having?: Or, a System of Natural Governance." *Theory, Culture and Society* 25, no. 3 (2008): 103–29. https://doi.org/10.1177/0263276408090660.

Cohen, Ed. "The Paradoxical Politics of Viral Containment; Or, How Scale Undoes Us
One and All." *Social Text* 29, no. 1 (2011): 15–35. https://doi.org/10.1215/01642472
-1210247.

Cohen, Lawrence. "Making Peasants Protestant and Other Projects: Medical Anthropol-
ogy and its Global Condition." In *Medical Anthropology at the Intersections: Histories,
Activisms, and Futures*, edited by Marcia Claire Inhorn and Emily A. Wentzell, 65–92.
Durham, NC: Duke University Press, 2012.

Comaroff, Jean, and John L. Comaroff. *Theory from the South, or, How Euro-America Is
Evolving Toward Africa*. New York: Routledge, 2012.

Cone, Marla. "Should DDT Be Used to Combat Malaria?" *Scientific American*, May 4,
2009. https://www.scientificamerican.com/article/ddt-use-to-combat-malaria.

Corntassel, Jeff. "Re-envisioning Re-surgence: Indigenous Pathways to Decolonization and Sustainable Self-determination." *Decolonization: Indigeneity, Education and
Society* 1, no. 1 (2012): 86–101.

Cousins, Thomas. "A Mediating Capacity." *Medicine Anthropology Theory* 2, no. 2 (2015).
https://doi.org/10.17157/mat.2.2.175.

Crane, Emma Shaw. "Lush Aftermath: Race, Labor, and Landscape in the Suburb." *Environment and Planning D: Society and Space* 41, no. 2 (2023): 210–30. https://doi.org
/10.1177/02637758231172202.

Crane, Johanna Tayloe. *Scrambling for Africa: AIDS, Expertise, and the Rise of American
Global Health Science*. Ithaca, NY: Cornell University Press, 2013.

Crutzen, Paul J. "Geology of Mankind." *Nature* 415, no. 6867 (2002): 23. https://doi.org
/10.1038/415023a.

Crutzen, Paul J., and Eugene F. Stoermer. "The 'Anthropocene.'" *Global Change Newslet-
ter* 41 (2000): 17–18.

Cupit, Caroline, Janet Rankin, Natalie Armstrong, and Graham P. Martin. "Overruling
Uncertainty About Preventative Medications: The Social Organisation of Healthcare
Professionals' Knowledge and Practices." *Sociology of Health and Illness* 42 (Au-
gust 2020): 114–29. https://doi.org/10.1111/1467-9566.12998.

*Daily Nation.* "Tanzania Sentences Albino Killers to Death." March 6, 2015.

Davies, Michele L. "Shattered Assumptions: Time and the Experience of Long-Term HIV
Positivity." *Social Science and Medicine* 44, no. 5 (1997): 561–71. https://doi.org/10
.1016/S0277-9536(96)00177-3.

De Klerk, Josien. "The Compassion of Concealment: Silence Between Older Caregivers and Dying Patients in the AIDS Era, Northwest Tanzania." *Culture, Health and Sexuality* 14 (November 2012): S27–38. https://doi.org/10.1080/13691058.2011.631220.

DeLoughrey, Elizabeth. "Yam, Roots, and Rot: Allegories of the Provision Grounds." *Small Axe: A Caribbean Journal of Criticism* 15, no. 1 (2011): 58–75. https://doi.org/10.1215/07990537-1189530.

Denham, T. P., S. G. Haberle, C. Lentfer, et al. "Origins of Agriculture at Kuk Swamp in the Highlands of New Guinea." *Science* 301, no. 5630 (2003): 189–93. https://doi.org/10.1126/science.1085255.

Diagne, Souleymane Bachir. *The Ink of the Scholars: Reflections on Philosophy in Africa.* Translated by Jonathan Adjemian. Codesria Book Series. Dakar: CODESRIA, Council for the Development of Social Science Research in Africa, 2016.

Diagne, Souleymane Bachir. "The Time We Need." In *The Ink of the Scholars: Reflections on Philosophy in Africa.* Translated by Jonathan Adjemian, 35–47. Dakar: CODESRIA, Council for the Development of Social Science Research in Africa, 2016.

Digby, Anne, Waltraud Ernst, and Projit B. Muhkarji, eds. *Crossing Colonial Historiographies: Histories of Colonial and Indigenous Medicines in Transnational Perspective.* Cambridge: Cambridge Scholars, 2010.

Dilger, Hansjörg. "Healing the Wounds of Modernity: Salvation, Community and Care in a Neo-Pentecostal Church in Dar Es Salaam, Tanzania." *Journal of Religion in Africa* 37, no. 1 (2007): 59–83. https://doi.org/10.1163/157006607X166591.

Dilger, Hansjörg. "Targeting the Empowered Individual: Transnational Policy Making, the Global Economy of Aid, and the Limitations of Biopower in Tanzania." In *Medicine, Mobility, and Power in Global Africa: Transnational Health and Healing,* edited by Hansjorg Dilger, Abdoulaye Kane, and Stacey Ann Langwick, 60–91. Bloomington: Indiana University Press, 2012.

Dumit, Joseph. *Drugs for Life: How Pharmaceutical Companies Define Our Health.* Experimental Futures. Durham, NC: Duke University Press, 2012.

Eglash, Ron. *African Fractals: Modern Computing and Indigenous Design.* New Brunswick, NJ: Rutgers University Press, 1999.

Ehrlich, Paul R., and Anne Howland Ehrlich. *The Population Bomb.* New York: Ballantine Books, 1968.

Escobar, Arturo. *Encountering Development: The Making and Unmaking of the Third World.* Princeton, NJ: Princeton University Press, 2012.

Evans, Ruth. "'We Are Managing Our Own Lives . . .': Life Transitions and Care in Sibling-Headed Households Affected by AIDS in Tanzania and Uganda." *Area* 43, no. 4 (2011): 384–96. https://www.jstor.org/stable/41406021.

Evich, Helena Bottemiller. "The Great Nutrient Collapse." *Politico,* September 13, 2017. http://politi.co/2zACS5k.

Farquhar, Judith. "Eating Chinese Medicine." *Cultural Anthropology* 9, no. 4 (1994): 471–97. https://doi.org/10.1525/can.1994.9.4.02a00020.

Farquhar, Judith. *Knowing Practice: The Clinical Encounter of Chinese Medicine.* New York: Routledge, 1994.

Fassin, Didier. *Humanitarian Reason: A Moral History of the Present Times.* Berkeley: University of California Press, 2012.

Fauci, Anthony S., Robert R. Redfield, George Sigounas, Michael D. Weahkee, and Brett P. Giroir. "Ending the HIV Epidemic: A Plan for the United States." *JAMA* 321, no. 9 (2019): 844. https://doi.org/10.1001/jama.2019.1343.

Feierman, Steven. "Explanation and Uncertainty in the Medical World of Ghaambo." *Bulletin of the History of Medicine* 74, no. 2 (2000): 317–44. https://doi.org/10.1353/bhm.2000.0070.

Feierman, Steven. *Peasant Intellectuals: Anthropology and History in Tanzania.* Madison: University of Wisconsin Press, 1990.

Feierman, Steven. "Struggles for Control: The Social Roots of Health and Healing in Modern Africa." *African Studies Review* 28, no. 2/3 (1985): 73–147. https://doi.org/10.2307/524604.

Feierman, Steven. *The Shambaa Kingdom: A History.* Madison: University of Wisconsin Press, 1974.

Fisher, Josh, Mary Mostafanezhad, Alex Nading, and Sarah Marie Wiebe. "Introduction: Pollution and Toxicity: Cultivating Ecological Practices for Troubled Times." *Environment and Society* 12, no. 1 (2021): 1–4. https://doi.org/10.3167/ares.2021.120101.

Fisher, Thomas J. "Chagga Elites and the Politics of Ethnicity in Kilimanjaro, Tanzania." PhD diss., University of Edinburgh, 2012.

Flint, Karen E. *Healing Traditions: African Medicine, Cultural Exchange, and Competition in South Africa, 1820–1948.* Edited by Jean Allman and Allen Isaacman. Athens: Ohio University Press, 2008.

Fortun, Kim. *Advocacy After Bhopal: Environmentalism, Disaster, New Global Orders.* Chicago: University of Chicago Press, 2001.

Fortun, Kim, and Mike Fortun. "Scientific Imaginaries and Ethical Plateaus in Contemporary U.S. Toxicology." *American Anthropologist* 107, no. 1 (2005): 43–54. https://doi.org/10.1525/aa.2005.107.1.043.

Foster, Laura A. *Reinventing Hoodia: Peoples, Plants, and Patents in South Africa.* Feminist Technosciences. Seattle: University of Washington Press, 2017.

Foucault, Michel. *The History of Sexuality.* Vol. 2, *The Use of Pleasure.* Translated by Robert Hurley. New York: Vintage Books, 1990.

Foucault, Michel. *The Order of Things: An Archaeology of the Human Sciences.* Vintage Books ed. New York: Vintage Books, 1994.

Fox, Daniel M. "The Significance of the Milbank Memorial Fund for Policy: An Assessment at Its Centennial." *Milbank Quarterly* 84, no. 1 (2006): 5–36. https://doi.org/10.1111/j.1468-0009.2006.00411.x.

Frison, E., D. A. Karamura, and S. Sharrock. "Banana Production Systems in Eastern and Southern Africa." In *Bananas and Food Security,* edited by Claudine Picq, Eric Fouré, and Émile A. Frison. Montpellier: INIBAP Montpellier, 1998.

Fuller, R. J. M. "Ethnobotany: Major Developments of a Discipline Abroad, Reflected in New Zealand." *New Zealand Journal of Botany* 51, no. 2 (2013): 116–38. https://doi.org/10.1080/0028825X.2013.778298.

Fullwiley, Duana. "The Biosocial Politics of Plants and People." In *The Encultured Gene: Sickle Cell Health Politics and Biological Difference in West Africa,* 77–118. Princeton, NJ: Princeton University Press, 2011.

Gaylin, Daniel S., and Jennifer Kates. "Refocusing the Lens: Epidemiologic Transition Theory, Mortality Differentials, and the AIDS Pandemic." *Social Science and Medicine* 44, no. 5 (1997): 609–21. https://doi.org/10.1016/S0277-9536(96)00212-2.

Geissler, P. Wenzel. "Public Secrets in Public Health: Knowing Not to Know While Making Scientific Knowledge." *American Ethnologist* 40, no. 1 (2013): 13–34. https://doi.org/10.1111/amet.12002.

Geissler, P. Wenzel, and Ruth J. Prince. "Active Compounds and Atoms of Society: Plants, Bodies, Minds and Cultures in the Work of Kenyan Ethnobotanical Knowledge." *Social Studies of Science* 39, no. 4 (2009): 599–634. https://www.jstor.org/stable/27793310.

Geissler, Wenzel, ed. *Para-States and Medical Science: Making African Global Health.* Critical Global Health. Durham, NC: Duke University Press, 2015.

Geissler, Wenzel, and Ruth Jane Prince. *The Land Is Dying: Contingency, Creativity and Conflict in Western Kenya.* New York: Berghahn Books, 2010.

Gerdes, Paulus. *Geometry from Africa: Mathematical and Educational Explorations.* Classroom Resource Materials. Washington, DC: Mathematical Association of America, 1999.

Ghosh, Amitav. *The Great Derangement: Climate Change and the Unthinkable.* The Randy L. and Melvin R. Berlin Family Lectures. Chicago: University of Chicago Press, 2016.

Giblin, James. "Land Tenure, Traditions of Thought about Land, and Their Environmental Implications in Tanzania." *Tanzania Zamani: A Journal of Historical Research and Writing* 4, nos. 1–2 (1998): 1–56.

Glover, Dominic, and Glenn Davis Stone. "Heirloom Rice in Ifugao: An 'Anti-Commodity' in the Process of Commodification." *Journal of Peasant Studies* 45, no. 4 (2018): 776–804. https://doi.org/10.1080/03066150.2017.1284062.

Gomez-Temesio, Veronica. "Outliving Death: Ebola, Zombies, and the Politics of Saving Lives: Outliving Death." *American Anthropologist* 120, no. 4 (2018): 738–51. https://doi.org/10.1111/aman.13126.

Good, Byron. "Medical Anthropology and the Problem of Belief." In *Medicine, Rationality and Experience: An Anthropological Perspective*, 1–24. New York: Cambridge University Press, 1994.

Good, Byron. *Medicine, Rationality, and Experience: An Anthropological Perspective.* The Lewis Henry Morgan Lectures 1990. Cambridge: Cambridge University Press, 1994.

Gopalakrishnan, Lakshmipriya, Kruthi Doriya, and Devarai Santhosh Kumar. "*Moringa Oleifera*: A Review on Nutritive Importance and Its Medicinal Application." *Food Science and Human Wellness* 5, no. 2 (2016): 49–56. https://doi.org/10.1016/j.fshw.2016.04.001.

Grand View Research. "Nutraceuticals Market Size, Share and Trends Analysis Report by Product (Dietary Supplements, Functional Foods, Functional Beverages), by Ingredient, by Application, by Region, and by Segment Forecast, 2024–2030." *Market Analysis Report*, 2023.

Greene, Jeremy A. "Making Medicines Essential: The Emergent Centrality of Pharmaceuticals in Global Health." *BioSocieties* 6, no. 1 (2011): 10–33. https://doi.org/10.1057/biosoc.2010.39.

Greene, Jeremy A. *Prescribing by Numbers: Drugs and the Definition of Disease.* Baltimore: Johns Hopkins University Press, 2007.

Grove, Richard. *Green Imperialism: Colonial Expansion, Tropical Island Edens and the Origins of Environmentalism, 1600–1860*. Studies in Environment and History. Cambridge, MA: Cambridge University Press, 1996.

Guyer, Jane I. "Prophecy and the Near Future: Thoughts on the Macroeconomic, Evangelical, and Punctuated Time." *American Ethnologist*, 34, no. 3 (2007): 409–21.

Hakansson, N. Thomas. "Politics, Cattle and Ivory: Regional Interaction and Changing Land-Use Prior to Colonialism." In *Culture, History, and Identity: Landscapes of Inhabitation in the Mount Kilimanjaro Area, Tanzania: Essays in Honour of Paramount Chief Thomas Lenana Mlanga Marealle II (1915–2007)*, edited by Timothy Clack and Thomas Lenana Mlanga Marealle, 141–54. BAR International Series 1966. Oxford: Archaeopress, 2009.

Hall, Kim Q. "Toward a Queer Crip Feminist Politics of Food." *philoSOPHIA* 4, no. 2 (2014): 177–96. https://muse.jhu.edu/pub/163/article/565882.

Hall, Laura. "My Mother's Garden: Aesthetics, Indigenous Renewal, and Creativity." In *Art in the Anthropocene: Encounters Among Aesthetics, Politics, Environments and Epistemologies*, edited by Heather Davis and Etienne Turpin. Critical Climate Change. London: Open Humanities Press, 2015.

Hamdy, Sherine. *Our Bodies Belong to God: Organ Transplants, Islam, and the Struggle for Human Dignity in Egypt*. Berkeley: University of California Press, 2012.

Hannig, Anita. "Healing and Reforming: The Making of the Modern Clinical Subject." In *Beyond Surgery: Injury, Healing, and Religion at an Ethiopian Hospital*, 149–75. Chicago: University of Chicago Press, 2017.

Happi Emaga, Thomas, Rado Herinavalona Andrianaivo, Bernard Wathelet, Jean Tchango Tchango, and Michel Paquot. "Effects of the Stage of Maturation and Varieties on the Chemical Composition of Banana and Plantain Peels." *Food Chemistry* 103, no. 2 (2007): 590–600. https://doi.org/10.1016/j.foodchem.2006.09.006.

Haraway, Donna Jeanne. *The Companion Species Manifesto: Dogs, People, and Significant Otherness*. Vol. 1. Chicago: Prickly Paradigm Press, 2003.

Haraway, Donna Jeanne. *Staying with the Trouble: Making Kin in the Chthulucene*. Experimental Futures: Technological Lives, Scientific Arts, Anthropological Voices. Durham, NC: Duke University Press, 2016.

Haraway, Donna Jeanne. "Tentacular Thinking: Anthropocene, Capitalocene, Chthulucene." *E-Flux*, no. 75 (September 2016). https://www.e-flux.com/journal/75/67125/tentacular-thinking-anthropocene-capitalocene-chthulucene/.

Hardin, Jessica A. *Faith and the Pursuit of Health: Cardiometabolic Disorders in Samoa*. Medical Anthropology. New Brunswick, NJ: Rutgers University Press, 2018.

Hatch, Anthony Ryan. *Blood Sugar: Racial Pharmacology and Food Justice in Black America*. Minneapolis: University of Minnesota Press, 2016.

Hecht, Gabrielle. "The African Anthropocene." *Aeon*, February 6, 2018. https://aeon.co/essays/if-we-talk-about-hurting-our-planet-who-exactly-is-the-we.

Hecht, Gabrielle. *Being Nuclear: Africans and the Global Uranium Trade*. Cambridge, MA: MIT Press, 2012.

Hecht, Gabrielle. "Interscalar Vehicles for an African Anthropocene: On Waste, Temporality, and Violence." *Cultural Anthropology* 33, no. 1 (2018): 109–41. https://doi.org/10.14506/ca33.1.05.

Hecht, Gabrielle, and Pamila Gupta, eds. "Toxicity, Waste, Detritus: An Introduction." In *Somatosphere, Toxicity, Waste, and Detritus in the Global South: Africa and Beyond*, October 10, 2017. http://somatosphere.net/2017/toxicity-waste-detritus-an-introduction.html/.

Hemp, Andreas. "The Banana Forests of Kilimanjaro: Biodiversity and Conservation of the Chagga Homegardens." *Biodiversity and Conservation* 15, no. 4 (2006): 1193–217. https://doi.org/10.1007/s10531-004-8230-8.

Hemp, Claudia. "The Chagga Home Gardens—Relict Areas for Endemic Saltatoria Species (Insecta: Orthoptera) on Mount Kilimanjaro." *Biological Conservation* 125, no. 2 (2005): 203–9. https://doi.org/10.1016/j.biocon.2005.03.018.

Hetherington, Kregg. "Beans Before the Law: Knowledge Practices, Responsibility, and the Paraguayan Soy Boom." *Cultural Anthropology* 28, no. 1 (2013): 65–85. https://doi.org/10.1111/j.1548-1360.2012.01173.x.

Hirsch, Lioba A., and Naya Jones. "Incontestable: Imagining Possibilities Through Intimate Black Geographies." *Transactions of the Institute of British Geographers* 46, no. 4 (2021): 796–800.

Hoffman, Danny. "Toxicity." In *Somatosphere, Toxicity, Waste, and Detritus in the Global South: Africa and Beyond*, edited by Pamila Gupta and Gabrielle Hecht, October 16, 2017. http://somatosphere.net/2017/toxicity.html/.

Holbraad, Martin, Morten Axel Pedersen, and Eduardo Viveiros de Castro. "The Politics of Ontology: Anthropological Positions." *Fieldsights*, January 13, 2014. https://culanth.org/fieldsights/the-politics-of-ontology-anthropological-positions.

Hong, Esther S., Hajo Zeeb, and Michael H. Repacholi. "Albinism in Africa as a Public Health Issue." *BMC Public Health* 6, no. 1 (2006): 212. https://doi.org/10.1186/1471-2458-6-212.

Horton, Richard, and Selina Lo. "Planetary Health: A New Science for Exceptional Action." *The Lancet* 386, no. 10007 (2015): 1921–22. https://doi.org/10.1016/S0140-6736(15)61038-8.

Hsu, Elisabeth, and Stephen Harris, eds. *Plants, Health and Healing: On the Interface of Ethnobotany and Medical Anthropology*. Epistemologies of Healing. New York: Berghahn Books, 2010.

Hunt, Nancy Rose. *A Colonial Lexicon of Birth Ritual, Medicalization, and Mobility in the Congo*. Durham, NC: Duke University Press, 1999.

Hustak, Carla, and Natasha Myers. "Involutionary Momentum: Affective Ecologies and the Sciences of Plant/Insect Encounters." *Differences* 23, no. 3 (2012): 74–118. https://doi.org/10.1215/10407391-1892907.

Institute for Health Metrics and Evaluation. "Tanzania." Health Data, September 9, 2015. https://www.healthdata.org/tanzania.

IPPmedia.com. "Two Cabinet Ministers Meet Loliondo Ex-Pastor." March 28, 2011.

Irigaray, Luce, and Michael Marder. *Through Vegetal Being: Two Philosophical Perspectives*. Critical Life Studies. New York: Columbia University Press, 2016.

Ives, Sarah Fleming. *Steeped in Heritage: The Racial Politics of South African Rooibos Tea*. New Ecologies for the Twenty-First Century. Durham, NC: Duke University Press, 2017.

Jain, Sarah S. Lochlann. *Malignant: How Cancer Becomes Us*. Berkeley: University of California Press, 2013.

Janzen, John M. *Lemba, 1650–1930: A Drum of Affliction in Africa and the New World*. New York: Garland, 1982.

Janzen, John M. *Ngoma: Discourses of Healing in Central and Southern Africa*. Comparative Studies of Health Systems and Medical Care. Berkeley: University of California Press, 1992.

Jegathesan, Mythri. "Black Feminist Plots Before the Plantationocene and Anthropology's 'Regional Closets.'" *Feminist Anthropology* 2, no. 1 (2021): 78–93. https://doi.org/10.1002/fea2.12037.

Johnson, Paul C., ed. *Spirited Things: The Work of "Possession" in Afro-Atlantic Religions*. Chicago: University of Chicago Press, 2014.

Jones, Bradley M. "(Com)Post-Capitalism: Cultivating a More-than-Human Economy in the Appalachian Anthropocene." *Environmental Humanities* 11, no. 1 (2019): 3–26.

Jones, Naya. "Dying to Eat? Black Food Geographies of Slow Violence and Resilience." *ACME: An International Journal for Critical Geographies* 18, no. 5 (2019): 1076–99.

Kasilo, Ossy M. J., Paul Samson, and Chris Mwikisa Ngenda. "An Overview of the Traditional Medicine Situation in the African Region." *African Health Monitor* 14 (2010): 7–15.

Kimambo, Isaria N. "Environmental Control and Hunger: In the Mountains and Plains of Northeastern Tanzania." In *Custodians of the Land: Ecology and Culture in the History of Tanzania*, edited by Gregory Maddox, James Leonard Giblin, and Isaria N. Kimambo, 71–95. Eastern African Studies. London: James Curry, 1996.

Kirigia, Joses M., Benjamin M. Nganda, Chris N. Mwikisa, and Bernardino Cardoso. "Effects of Global Financial Crisis on Funding for Health Development in Nineteen Countries of the WHO African Region." *BMC International Health and Human Rights* 11, no. 1 (2011): 4. https://doi.org/10.1186/1472-698X-11-4.

Kodesh, Neil. *Beyond the Royal Gaze: Clanship and Public Healing in Buganda*. Charlottesville: University of Virginia Press, 2010.

Kodesh, Neil. "History from the Healer's Shrine: Genre, Historical Imagination, and Early Ganda History." *Comparative Studies in Society and History* 49, no. 3 (2007): 527–52. https://doi.org/10.1017/S0010417507000618.

Krupar, Shiloh R. *Hot Spotter's Report: Military Fables of Toxic Waste*. Minneapolis: University of Minnesota Press, 2013.

Laist, Randy, ed. *Plants and Literature: Essays in Critical Plant Studies*. Amsterdam: Editions Rodopi, 2013.

Landecker, Hannah. "Being and Eating: Losing Grip on the Equation." *BioSocieties* 10, no. 2 (2015): 253–58. https://doi.org/10.1057/biosoc.2015.15.

Landecker, Hannah. "Food as Exposure: Nutritional Epigenetics and the New Metabolism." *BioSocieties* 6, no. 2 (2011): 167–94. https://doi.org/10.1057/biosoc.2011.1.

Landecker, Hannah. "Postindustrial Metabolism: Fat Knowledge." *Public Culture* 25, no. 3 (2013): 495–522. https://doi.org/10.1215/08992363-2144625.

Langston, Nancy. *Toxic Bodies: Hormone Disruptors and the Legacy of DES*. New Haven, CT: Yale University Press, 2010.

Langwick, Stacey Ann. *Bodies, Politics, and African Healing: The Matter of Maladies in Tanzania*. Bloomington: Indiana University Press, 2011.

Langwick, Stacey Ann. "Cultivating Vitality: A Photo Essay." *Anthropology News*, January 24, 2018.

Langwick, Stacey Ann. "Geographies of Medicine: Interrogating the Boundary Between 'Traditional' and 'Modern' Medicine in Colonial Tanganyika." In *Borders and Healers: Brokering Therapeutic Resources in Southeast Africa*, edited by Tracy J. Luedke and Harry G. West. Bloomington: Indiana University Press, 2006.

Langwick, Stacey Ann. "Healers and Scientists: The Epistemological Politics of Research about Medicinal Plants in Tanzania, or 'Moving Away from Traditional Medicine.'" In *Evidence, Ethos, and Experiment: The Anthropology and History of Medical Research in Africa*, edited by P. Wenzel Geissler and Catherine Molyneux. New York: Berghahn, 2011.

Langwick, Stacey Ann. "Healers and Their Intimate Becomings." In *Bodies, Politics, and African Healing: The Matter of Maladies in Tanzania*, 87–120. Bloomington: Indiana University Press, 2011.

Langwick, Stacey Ann. "Healing in the Anthropocene." In *The World Multiple: Politics of Knowing and Generating Entangled Worlds*, edited by Keiichi Omura, Atsuro Morita, Shiho Satsuka, and Grant Jun Otsuki, 155–72. New York: Routledge, 2018.

Langwick, Stacey Ann. "Partial Publics: The Political Promise of Traditional Medicine in Africa." *Current Anthropology* 56, no. 4 (2015): 493–514. https://doi.org/10.1086/682285.

Langwick, Stacey Ann. "A Politics of Habitability: Plants, Healing and Sovereignty in a Toxic World." *Cultural Anthropology* 33, no. 3 (2018): 415–43.

Langwick, Stacey Ann. "Properties of (Dis)Possession: Therapeutic Plants, Intellectual Property, and Questions of Justice in Tanzania." *Osiris* 36 (2021): 284–305. https://doi.org/10.1086/714263.

Langwick, Stacey Ann. "The Value of Secrets: Pragmatic Healers and Proprietary Knowledge." In *African Medical Pluralism*, edited by William Olsen and Carolyn Sargent, 31–49. Bloomington: Indiana University Press, 2017.

Langwick, Stacey Ann. "Witchcraft, Oracles, and Native Medicine." In *Bodies, Politics, and African Healing: The Matter of Maladies in Tanzania*. Bloomington: Indiana University Press, 2011.

Langwick, Stacey Ann, Victor Wiketye, and John Wilfred Ogondiek. "Working with Traditional Healers: Recognition and Knowledge Sharing." Paper presented at the Annual Meeting of the National Institute of Medical Research (NIMR), Dar es Salaam, April 22, 2014.

Last, Murray. "The Importance of Knowing about Not Knowing." *Social Science and Medicine. Part B: Medical Anthropology* 15, no. 3 (1981): 387–92. https://doi.org/10.1016/0160-7987(81)90064-8.

Latour, Bruno. *Pandora's Hope: Essays on the Reality of Science Studies*. Cambridge, MA: Harvard University Press, 1999.

Latour, Bruno. *We Have Never Been Modern*. Translated by Catherine Porter. Cambridge, MA: Harvard University Press, 1993.

Latour, Bruno, and Catherine Porter. *Facing Gaia: Eight Lectures on the New Climatic Regime*. Cambridge: Polity, 2017.

Lawrence, Anna M. "Listening to Plants: Conversations Between Critical Plant Studies and Vegetal Geography." *Progress in Human Geography* 46, no. 2 (2022): 629–51. https://doi.org/10.1177/03091325211062167.

Lejju, B. Julius, Peter Robertshaw, and David Taylor. "Africa's Earliest Bananas?" *Journal of Archaeological Science* 33, no. 1 (2006): 102–13. https://doi.org/10.1016/j.jas.2005.06 .015.

Liboiron, Max. *Pollution Is Colonialism*. Durham, NC: Duke University Press, 2021.

Littlewood, Roland, ed. *On Knowing and Not Knowing in the Anthropology of Medicine*. London: Routledge, 2007.

Livingston, Julie. *Debility and the Moral Imagination in Botswana*. African Systems of Thought. Bloomington: Indiana University Press, 2005.

Livingston, Julie. *Improvising Medicine: An African Oncology Ward in an Emerging Cancer Epidemic*. Durham, NC: Duke University Press, 2012.

Livingston, Julie. "In the Time of Beef." In *Self-Devouring Growth: A Planetary Parable as Told from Southern Africa*, 35–60. Critical Global Health: Evidence, Efficacy, Ethnography. Durham, NC: Duke University Press, 2019.

Livingston, Julie. *Self-Devouring Growth: A Planetary Parable as Told from Southern Africa*. Critical Global Health: Evidence, Efficacy, Ethnography. Durham, NC: Duke University Press, 2019.

Lock, Margaret M., and Judith Farquhar, eds. *Beyond the Body Proper: Reading the Anthropology of Material Life*. Durham, NC: Duke University Press, 2007.

Lyons, Kristina Marie. "Decomposition as Life Politics: Soils, Selva, and Small Farmers under the Gun of the U.S.–Colombia War on Drugs." *Cultural Anthropology* 31, no. 1 (2015): 56–81. https://doi.org/10.14506/ca31.1.04.

Maghimbi, Sam. "Recent Changes in Crop Patterns in the Kilimanjaro Region of Tanzania: The Decline of Coffee and the Rise of Maize and Rice." *African Study Monographs*, March 2007, 73–83.

Majumdar, Aparajita. "Recalcitrant Life Worlds: Decolonizing the History of Human-Plant Relations." PhD diss., Cornell University, 2024.

Malebo, Hamisi M., and Zakaria H. Mbwambo. *Technical Report on Miracle Cure Prescribed by Rev. Ambilikile Mwasupile in Samunge Village, Loliondo, Arusha*. Dar es Salaam: National Institute for Medical Research and Institute of Traditional Medicine, 2011.

Malhi, Yadvinder. "The Concept of the Anthropocene." *Annual Review of Environment and Resources* 42, no. 1 (2017): 77–104. https://doi.org/10.1146/annurev-environ -102016-060854.

Manji, Ambreena S. "The Case for Women's Rights to Land in Tanzania: Some Observations in the Context of AIDS." *UTAFITI* 3, no. 2 (1996): 11–38.

Marder, Michael. "Ecology as Event." In *Eco-Deconstruction: Derrida and Environmental Philosophy*, edited by Matthias Fritsch, Philippe Lynes, and David Wood, 141–64. New York: Fordham University Press, 2018.

Marder, Michael. *Plant-Thinking: A Philosophy of Vegetal Life*. New York: Columbia University Press, 2013.

Maro, Paul S. "Agricultural Land Management Under Population Pressure: The Kilimanjaro Experience, Tanzania." *Mountain Research and Development* 8, no. 4 (1988): 273–82. https://doi.org/10.2307/3673548.

Maron, Dina Fine. "Witchcraft Trade, Skin Cancer Pose Serious Threats to Albinos in Tanzania." *Scientific American*, October 11, 2013. https://www.scientificamerican.com /article/witchcraft-trade-skin/.

Marten, Meredith G. "Living with HIV as Donor Aid Declines in Tanzania." *Medical Anthropology* 39, no. 3 (2020): 197–210. https://doi.org/10.1080/01459740.2019.1644334.

Masquelier, Adeline Marie. *Prayer Has Spoiled Everything: Possession, Power, and Identity in an Islamic Town of Niger*. Body, Commodity, Text. Durham, NC: Duke University Press, 2001.

Mastnak, Tomaz, Julia Elyachar, and Tom Boellstorff. "Botanical Decolonization: Rethinking Native Plants." *Environment and Planning D: Society and Space* 32, no. 2 (2014): 363–80. https://doi.org/10.1068/d13006p.

Mattes, Dominik. "The Blood of Jesus and CD4 Counts. Dreaming, Developing and Navigating Therapeutic Options for Curing HIV/AIDS in Tanzania." In *Religion and AIDS-Treatment in Africa: Saving Souls, Prolonging Lives*, edited by Rijk van Dijk, Hansjörg Dilger, Marian Burchardt, and Thera Rasing. London: Ashgate, 2014.

Mattes, Dominik. "Caught in Transition: The Struggle to Live a 'Normal' Life with HIV in Tanzania." *Medical Anthropology* 33, no. 4 (2014): 270–87. https://doi.org/10.1080 /01459740.2013.877899.

McKay, Ramah. *Medicine in the Meantime: The Work of Care in Mozambique*. Critical Global Health: Evidence, Efficacy, Ethnography. Durham, NC: Duke University Press, 2018.

Meek, Laura. "Fugitive Science: Beer Brewing and Experiments with Pharmaceuticals in Tanzania." *Platypus: The CASTAC Blog*, February 16, 2021. https://blog.castac.org/2021 /02/fugitive-science-beer-brewing-experiments-with-pharmaceuticals-in-tanzania/.

Meinert, Lotte, and Susan Reynolds Whyte. "Epidemic Projectification: AIDS Responses in Uganda as Event and Process." *Cambridge Journal of Anthropology* 32, no. 1 (2014): 77–94. https://doi.org/10.3167/ca.2014.320107.

Mendenhall, Emily. *Rethinking Diabetes: Entanglements with Poverty, Trauma, and HIV*. Ithaca, NY: Cornell University Press, 2019.

Mesaki, Simon. "Witchcraft and the Law in Tanzania." *International Journal of Sociology and Anthropology* 1, no. 8 (2009): 132–38.

Mhando, David G. "Conflict as Motivation for Change: The Case of Coffee Farmers' Cooperatives in Moshi, Tanzania." *African Study Monographs*, October 2014, 137–54.

Mika, Marissa. *Africanizing Oncology: Creativity, Crisis, and Cancer in Uganda*. New African Histories. Columbus: Ohio University Press, 2021.

Million, Dian. "There Is a River in Me." In *Theorizing Native Studies*, edited by Audra Simpson and Andrea Smith. Durham, NC: Duke University Press, 2014.

Minde, Elizabeth Maro. "Law Reform and Land Rights for Women in Tanzania." *HIV/AIDS Policy and Law Review* 11, no. 2/3 (2006): 64–66.

Minter, Adam. "The Burning Truth Behind an E-Waste Dump in Africa." *Smithsonian Magazine*, January 13, 2016. https://www.smithsonianmag.com/science-nature /burning-truth-behind-e-waste-dump-africa-180957597.

Mitman, Gregg. "Reflections on the Plantationocene: A Conversation with Donna Haraway and Anna Tsing." Edge Effects, n.d. https://edgeeffects.net/haraway-tsing -plantationocene/.

Mitman, Gregg, Michelle Murphy, and Christopher Sellers, eds. "Landscapes of Exposure: Knowledge and Illness in Modern Environments." *Osiris* 19 (2004).

Mol, Annemarie. *The Body Multiple: Ontology in Medical Practice*. Science and Cultural Theory. Durham, NC: Duke University Press, 2002.

Mol, Annemarie. *Eating in Theory*. Durham, NC: Duke University Press, 2021.

Moran-Thomas, Amy. *Traveling with Sugar: Chronicles of a Global Epidemic*. Oakland: University of California Press, 2019.

Morens, David M., Gregory K. Folkers, and Anthony S. Fauci. "The Challenge of Emerging and Re-Emerging Infectious Diseases." *Nature* 430, no. 6996 (2004): 242–49. https://doi.org/10.1038/nature02759.

Moyer, Eileen. "The Anthropology of Life After AIDS: Epistemological Continuities in the Age of Antiretroviral Treatment." *Annual Review of Anthropology* 44, no. 1 (2015): 259–75. https://doi.org/10.1146/annurev-anthro-102214-014235.

Moyer, Eileen, and Anita Hardon. "A Disease Unlike Any Other? Why HIV Remains Exceptional in the Age of Treatment." *Medical Anthropology* 33, no. 4 (2014): 263–69. https://doi.org/10.1080/01459740.2014.890618.

Mshana, Gerry, Mary L. Plummer, Joyce Wamoyi, Zachayo S. Shigongo, David A. Ross, and Daniel Wight. "'She Was Bewitched and Caught an Illness Similar to AIDS': AIDS and Sexually Transmitted Infection Causation Beliefs in Rural Northern Tanzania." *Culture, Health and Sexuality* 8, no. 1 (2006): 45–58. https://doi.org/10.1080/13691050500469731.

Mukharji, Projit Bihari. "Vishalyakarani as *Eupatorium Ayapana*: Retro-Botanizing, Embedded Traditions, and Multiple Historicities of Plants in Colonial Bengal, 1890–1940." *Journal of Asian Studies* 73, no. 1 (2014): 65–87. https://doi.org/10.1017/S0021911813001733.

Munishi, Linus K., Anza A. Lema, and Patrick A. Ndakidemi. "Decline in Maize and Beans Production in the Face of Climate Change at Hai District in Kilimanjaro Region, Tanzania." *International Journal of Climate Change Strategies and Management* 7, no. 1 (2015): 17–26. https://doi.org/10.1108/IJCCSM-07-2013-0094.

Murphy, Michelle. "Alterlife and Decolonial Chemical Relations." *Cultural Anthropology* 32, no. 4 (2017): 494–503. https://doi.org/10.14506/ca32.4.02.

Murphy, Michelle. "The Experimental Otherwise." In *The Economization of Life*, 105–9. Durham, NC: Duke University Press, 2017.

Murphy, Michelle. *Sick Building Syndrome and the Problem of Uncertainty: Environmental Politics, Technoscience, and Women Workers*. Durham, NC: Duke University Press, 2006.

Muzale, Henry R. T., and Josephat M. Rugemalira. "Researching and Documenting the Languages of Tanzania." *Language Documentation and Conservation* 2, no. 1 (2008): 68–108.

Mwegoha, W. J. S., and C. Kihampa. "Heavy Metal Contamination in Agricultural Soils and Water in Dar Es Salaam City, Tanzania." *African Journal of Environmental Science and Technology* 4, no. 11 (2010): 763–69.

Mwita, Chagula A., Ken M. Hosea, and Masoud H. Muruke. "Assessment of Genetic Modification in Imported Maize (Zea Mays) Seeds and Processed Soybean (Glycine Max) Foods in Tanzania." *Journal of Chemical, Biological and Physical Sciences* 3, no. 4 (2013): 2809.

Myers, Natasha. "From the Anthropocene to the Planthroposcene: Designing Gardens for Plant/People Involution." *History and Anthropology* 28, no. 3 (2017): 297–301. https://doi.org/10.1080/02757206.2017.1289934.

Myhre, Knut Christian. "*Horu*: Channeling Bodies and Shifting Subjects in an Engaging World." In *Returning Life: Language, Life Force and History in Kilimanjaro*, 98–145. Methodology and History in Anthropology. New York: Berghahn, 2018.

Myhre, Knut Christian. "Ngakuuriya Moo: Returning Life, Affording Rain." In *Returning Life: Language, Life Force and History in Kilimanjaro*, 223–78. New York: Berghahn, 2018.

Myhre, Knut Christian. *Returning Life: Language, Life Force and History in Kilimanjaro*. Methodology and History in Anthropology. New York: Berghahn, 2018.

Nading, Alex M. "Living in a Toxic World." *Annual Review of Anthropology* 49, no. 1 (2020): 209–24. https://doi.org/10.1146/annurev-anthro-010220-074557.

Nappi, Carla. "Winter Worm, Summer Grass: Cordyceps, Colonial Chinese Medicine, and the Formation of Historical Objects." In *Crossing Colonial Historiographies: Histories of Colonial and Indigenous Medicines in Transnational Perspective*, edited by Anne Digby, Waltraud Ernst, and Projit B. Muhkarji. Cambridge: Cambridge Scholars Publishing, 2010.

National Khoisan Council and Cederberg Belt Indigenous Farmers Representatives. "The Khoikhoi People's Rooibos Biocultural Community Protocol." Natural Justice, 2019. https://naturaljustice.org/wp-content/uploads/2020/04/NJ-Rooibos-BCP -Web.pdf.

National Tanzania Commission for AIDS (TACAIDS), Bureau of Statistics (NBS), and ORC Marco. *Tanzania HIV/AIDS Indicator Survey 2003–04*. Calverton, MD: TACAIDS, NBS, and ORC Marco, 2005.

Ndlovu-Gatsheni, Sabelo J. *Epistemic Freedom in Africa: Deprovincialization and Decolonization*. Rethinking Development. New York: Routledge, 2018.

Nealon, Jeffrey T. *Plant Theory: Biopower and Vegetable Life*. Stanford, CA: Stanford University Press, 2016.

Ngowi, A. V. F., T. J. Mbise, A. S. M. Ijani, L. London, and O. C. Ajayi. "Smallholder Vegetable Farmers in Northern Tanzania: Pesticides Use Practices, Perceptions, Cost and Health Effects." *Crop Protection* 26, no. 11 (2007): 1617–24. https://doi.org/10 .1016/j.cropro.2007.01.008.

Nguyen, Vinh-Kim. "Antiretroviral Globalism, Biopolitics, and Therapeutic Citizenship." In *Global Assemblages: Technology, Politics, and Ethics as Anthropological Problems*, edited by Aihwa Ong and Stephen J. Collier, 124–44. Malden, MA: Blackwell, 2005.

Nguyen, Vinh-Kim. "Government-by-Exception: Enrolment and Experimentality in Mass HIV Treatment Programmes in Africa." *Social Theory and Health* 7, no. 3 (2009): 196–217. https://doi.org/10.1057/sth.2009.12.

Nguyen, Vinh-Kim. *The Republic of Therapy: Triage and Sovereignty in West Africa's Time of AIDS*. Body, Commodity, Text. Durham, NC: Duke University Press, 2010.

Nixon, Rob. *Slow Violence and the Environmentalism of the Poor*. Cambridge, MA: Harvard University Press, 2011.

Noe, Christine. "Reducing Land Degradation on the Highlands of Kilimanjaro Region: A Biogeographical Perspective." *Open Journal of Soil Science* 4, no. 13 (2014): 437–45. https://doi.org/10.4236/ojss.2014.413043.

Nott, John. "'How Little Progress'? A Political Economy of Postcolonial Nutrition." *Population and Development Review* 44, no. 4 (2018): 771–91. https://doi.org/10.1111/padr.12198.

Ntapanta, Samwel M. "'Lifescaping' Toxicants: Locating and Living with e-Waste in Tanzania." *Anthropology Today* 37, no. 4 (2021): 7–10.

Ntapanta, Samwel M. "Polarized Cityscapes: Gathering Electronic Waste and Its Malcontents in Dar es Salaam." *Norsk Antropologisk Tidsskrift* 3, nos. 3–4 (2023): 227–43.

Nyambedha, Erick Otieno, Simiyu Wandibba, and Jens Aagaard-Hansen. "Changing Patterns of Orphan Care Due to the HIV Epidemic in Western Kenya." *Social Science and Medicine* 57, no. 2 (2003): 301–11. https://doi.org/10.1016/S0277-9536(02)00359-3.

Nyerere, Julius K. *Ujamaa-Essays on Socialism.* Galaxy Book 359. London: Oxford University Press, 1977.

Oguamanam, Chidi. "Plant Breeders' Rights, Farmers' Rights and Food Security: Africa's Failure of Resolve and India's Wobbly Leadership." *Indian Journal of Law and Technology* 14, no. 2 (2018): 240–68.

O'kting'ati, A., J. A. Maghembe, E. C. M. Fernandes, and G. H. Weaver. "Plant Species in the Kilimanjaro Agroforestry System." *Agroforestry Systems* 2, no. 3 (1984): 177–86. https://doi.org/10.1007/BF00147032.

Olarinoye, Tunde, Jan Willem Foppen, William Veerbeek, Tlhoriso Morienyane, and Hans Komakech. "Exploring the Future Impacts of Urbanization and Climate Change on Groundwater in Arusha, Tanzania." *Water International* 45, no. 5 (2020): 497–511. https://doi.org/10.1080/02508060.2020.1768724.

Olsen, William, and Carolyn Sargent, eds. *African Medical Pluralism.* Bloomington: Indiana University Press, 2017.

Omran, Abdel R. "The Epidemiologic Transition: A Theory of the Epidemiology of Population Change." *Milbank Memorial Fund Quarterly* 49, no. 4 (1971): 509. https://doi.org/10.2307/3349375.

Omran, Abdel R. "The Epidemiologic Transition Theory Revisited Thirty Years Later." *World Health Statistics Quarterly* 53, nos. 2, 3, 4 (1998): 99–119.

Paparini, Sara, and Tim Rhodes. "The Biopolitics of Engagement and the HIV Cascade of Care: A Synthesis of the Literature on Patient Citizenship and Antiretroviral Therapy." *Critical Public Health* 26, no. 5 (2016): 501–17. https://doi.org/10.1080/09581596.2016.1140127.

Parsons, Talcott. *The Social System.* Glencoe, IL: Free Press, 1951.

Paul, Phillipo, Richard Y. M. Kangalawe, and Leonard E. G. Mboera. "Land-Use Patterns and Their Implication on Malaria Transmission in Kilosa District, Tanzania." *Tropical Diseases, Travel Medicine and Vaccines* 4, no. 1 (2018): 6. https://doi.org/10.1186/s40794-018-0066-4.

Petryna, Adriana. *Life Exposed: Biological Citizens After Chernobyl.* In-Formation Series. Princeton, NJ: Princeton University Press, 2002.

Petryna, Adriana. "The Pharmaceutical Nexus." In *Global Pharmaceuticals: Ethics, Markets, Practices,* edited by Adriana Petryna, Andrew Lakoff, and Arthur Kleinman, 1–32. Durham, NC: Duke University Press, 2006.

Pigg, Stacy Leigh. "The Credible and the Credulous: The Question of 'Villagers' Beliefs' in Nepal." *Cultural Anthropology* 11, no. 2 (1996): 160–201. https://doi.org/10.1525/can.1996.11.2.02a00020.

Pignarre, Philippe, and Isabelle Stengers. *Capitalist Sorcery: Breaking the Spell*. Translated by Andrew Goffey. New York: Palgrave Macmillan, 2011.

Povinelli, Elizabeth A. *The Cunning of Recognition: Indigenous Alterities and the Making of Australian Multiculturalism*, edited by George Steinmetz, Julia Adams, Nancy Rose Hunt, Webb Keane, and Fatma Müge Göcek. Durham, NC: Duke University Press, 2020.

Povinelli, Elizabeth A. *Geontologies: A Requiem to Late Liberalism*. Durham, NC: Duke University Press, 2016.

Povinelli, Elizabeth A. "Routes/Worlds." *E-Flux*, no. 27 (September 2011). https://www.e-flux.com/journal/27/67991/routes-worlds/.

Povinelli, Elizabeth A. "Toxic Late Liberalism." In *Between Gaia and Ground: Four Axioms of Existence and the Ancestral Catastrophe of Late Liberalism*, 36–59. Durham, NC: Duke University Press, 2021.

Prince, Ruth Jane. "Situating Health and the Public in Africa: Historical and Anthropological Perspectives." In *Making and Unmaking Public Health in Africa: Ethnographic and Historical Perspectives*, edited by Ruth Jane Prince and Rebecca Marsland, 1–51. Athens: Ohio University Press, 2014.

Puig De La Bellacasa, Maria. "Making Time for Soil: Technoscientific Futurity and the Pace of Care." *Social Studies of Science* 45, no. 5 (2015): 691–716. https://doi.org/10.1177/0306312715599851.

Ralph, Laurence. *Renegade Dreams: Living Through Injury in Gangland Chicago*. Durham, NC: Duke University Press, 2014.

Ralph, Laurence. "What Wounds Enable: The Politics of Disability and Violence in Chicago." *Disability Studies Quarterly* 32, no. 3 (2012). https://doi.org/10.18061/dsq.v32i3.3270.

Redfern, Paul. "EU, US Dumping Toxic Waste in Africa." *East African*, July 5, 2010. http://www.theeastafrican.co.ke/news/EU-US-dumping-toxic-waste-in-Africa/2558-951790-385m1f/index.html.

Redfield, Peter. "Bioexpectations: Life Technologies as Humanitarian Goods." *Public Culture* 24, no. 1 (2012): 157–84. https://doi.org/10.1215/08992363-1443592.

Reetsch, Anika, Didas Kimaro, Karl-Heinz Feger, and Kai Schwärzel. "Traditional and Adapted Composting Practices Applied in Smallholder Banana-Coffee-Based Farming Systems: Case Studies from Kagera and Morogoro Regions, Tanzania." In *Organic Waste Composting through Nexus Thinking: Practices, Policies, and Trends*, edited by Hiroshan Hettiarachchi, Serena Caucci, and Kai Schwärzel, 165–84. Cham, Switzerland: Springer Open, 2020.

Research and Markets. "Herbal Medicines—Global Market Trajectory and Analytics." GlobeNewswire, September 21, 2021. https://www.globenewswire.com/en/news-release/2022/09/21/2519931/28124/en/The-Worldwide-Herbal-Medicines-Industry-is-Projected-to-Reach-178-4-Billion-by-2026.html.

Richey, Lisa Ann, Line Engbo Gissel, Opportuna Kweka, et al. "South-South Humanitarianism: The Case of COVID-Organics in Tanzania." *World Development* 141 (2021): 105375.

Roberts, Elizabeth F. S. "Food Is Love: And So, What Then?" *BioSocieties* 10, no. 2 (2015): 247–52. https://doi.org/10.1057/biosoc.2015.18.

Roberts, Elizabeth F. S. "What Gets Inside: Violent Entanglements and Toxic Bound-aries in Mexico City." *Cultural Anthropology* 32, no. 4 (2017): 592–619. https://doi.org/10.14506/ca32.4.07.

Robine, Jean-Marie. "Life Course, Environmental Change, and Life Span." *Population and Development Review* 29 (2003): 229–38.

Rockwood, J. L., B. G. Anderson, and D. A. Casamatta. "Potential Uses of *Moringa Oleifera* and an Examination of Antibiotic Efficacy Conferred by M. Oleifera Seed and Leaf Extracts Using Crude Extraction Techniques Available to Underserved Indigenous Populations." *International Journal of Phytotherapy Research* 3, no. 2 (2013): 61–71.

Rodríguez-Ambriz, S. L., J. J. Islas-Hernández, E. Agama-Acevedo, J. Tovar, and L. A. Bello-Pérez. "Characterization of a Fibre-Rich Powder Prepared by Liquefaction of Unripe Banana Flour." *Food Chemistry* 107, no. 4 (2008): 1515–21. https://doi.org/10.1016/j.foodchem.2007.10.007.

Rogers, Susan G. "The Kilimanjaro Native Planters Association: Administrative Re-sponses to Chagga Initiatives in the 1920's." *Transafrican Journal of History* 4, nos. 1/2 (1974): 94–114.

Roitman, Janet L. *Anti-Crisis*. Durham, NC: Duke University Press, 2013.

Saas, Christine. "How a Garden Changed My Life." *Slow Food Foundation for Biodiversity* (blog), September 2, 2014. https://www.fondazioneslowfood.com/en/come-un-orto-mi-ha-cambiato-la-vita/.

Sanabria, Emilia. "Endurance and Alterability." Correspondences, *Fieldsights*, August 21, 2018. https://culanth.org/fieldsights/endurance-and-alterability.

Sanders, Todd. *Beyond Bodies: Rainmaking and Sense Making in Tanzania*. Anthropo-logical Horizons. Toronto: University of Toronto Press, 2008.

Scheid, Volker. "Promoting Free Flow in the Networks: Reimagining the Body in Early Modern Suzhou." *History of Science* 56, no. 2 (2018): 131–67. https://doi.org/10.1177/0073275317709406.

Schmidt, Charles W. "Unfair Trade E-Waste in Africa." *Environmental Health Perspec-tives* 114, no. 4 (2006): A232–35. https://doi.org/10.1289/ehp.114-a232.

Schoenbrun, David L. "Conjuring the Modern in Africa: Durability and Rupture in His-tories of Public Healing Between the Great Lakes of East Africa." *American Historical Review* 111, no. 5 (2006): 1403–39. https://doi.org/10.1086/ahr.111.5.1403.

Schoenbrun, David L., and Jennifer L. Johnson. "Introduction: Ethnic Formation with Other-Than-Human Beings." *History in Africa* 45 (June 2018): 307–45. https://doi.org/10.1017/hia.2018.11.

Schoenbrun, David Lee. *A Green Place, a Good Place: Agrarian Change, Gender, and Social Identity in the Great Lakes Region to the 15th Century*. Social History of Africa. Portsmouth, NH: Heinemann, 1998.

Schultes, Richard Evans, and Siri Von Reis, eds. *Ethnobotany: Evolution of a Discipline*. Portland, OR: Dioscorides Press, 1995.

Schumaker, Lyn. *Africanizing Anthropology: Fieldwork, Networks, and the Making of Cultural Knowledge in Central Africa*. Durham, NC: Duke University Press, 2001.

Scott, James C. *The Art of Not Being Governed: An Anarchist History of Upland Southeast Asia*. Yale Agrarian Studies Series. New Haven, CT: Yale University Press, 2009.

Scott-Smith, T. "Control and Biopower in Contemporary Humanitarian Aid: The Case of Supplementary Feeding." *Journal of Refugee Studies* 28, no. 1 (2015): 21–37. https://doi.org/10.1093/jrs/feu018.

Setel, Philip W. "Local Histories of Sexually Transmitted Diseases and AIDS in Western and Northern Tanzania." In *Histories of Sexually Transmitted Diseases and HIV/AIDS in Sub-Saharan Africa*, edited by Philip Setel, Milton James Lewis, and Maryinez Lyons. Westport, CT: Greenwood Press, 1999.

Seth, Suman. "Pluriversal Problems, Revolutionary Remedies." *Science, Technology and Society* 28, no. 1 (2022): 1–9. https://doi.org/10.1177/09717218221102602.

Shapiro, Nicholas. "Attuning to the Chemosphere: Domestic Formaldehyde, Bodily Reasoning, and the Chemical Sublime." *Cultural Anthropology* 30, no. 3 (2015): 368–93. https://doi.org/10.14506/ca30.3.02.

Shotwell, Alexis. *Against Purity: Living Ethically in Compromised Times*. Minneapolis: University of Minnesota Press, 2016.

Silayo, Valence M., and Innocent Pikirayi. "Community-Based Approaches in the Construction and Management of Water Infrastructures among the Chagga, Kilimanjaro, Tanzania." *Land* 12, no. 3 (2023): 570. https://doi.org/10.3390/land12030570.

Silayo, Valence Valerian Meriki Silayo. "Land and Culture as Symbols of Remembrance, Ancestry, Rituals and Initiations: The Case of Kihamba, Kyungu and Kifunyi Among the Chagga of Kilimanjaro, Tanzania." *African Journal on Land Policy and Geospatial Sciences* 5 (September 2022): 891–902. https://revues.imist.ma/index.php/AJLP-GS/index.

Simpson, Leanne Betasamosake. *As We Have Always Done: Indigenous Freedom Through Radical Resistance*. Minneapolis: University of Minnesota Press, 2021.

Slow Food Foundation for Biodiversity. "Arusha Stingless Bee Honey." https://www.fondazioneslowfood.com/en/slow-food-presidia/arusha-stingless-bee-honey/.

Smallman-Raynor, M. "Late Stages of Epidemiological Transition: Health Status in the Developed World." *Health and Place* 5, no. 3 (1999): 209–22. https://doi.org/10.1016/S1353-8292(99)00010-6.

Smith-Morris, Carolyn. "The Chronicity of Life, the Acuteness of Diagnosis." In *Chronic Conditions, Fluid States: Chronicity and the Anthropology of Illness*, edited by Lenore Manderson and Carolyn Smith-Morris, 21–37. Studies in Medical Anthropology. New Brunswick, NJ: Rutgers University Press, 2010.

Smith-Morris, Carolyn, Sylvia Rodriguez, Rose Soto, Morningstar Spencer, and Luigi Meneghini. "Decolonizing Care at Diagnosis: Culture, History, and Family at an Urban Inter-tribal Clinic." *Medical Anthropology Quarterly* 35, no. 3 (2021): 364–85. https://doi.org/10.1111/maq.12645.

Soini, Eija. "Changing Livelihoods on the Slopes of Mt. Kilimanjaro, Tanzania: Challenges and Opportunities in the Chagga Homegarden System." *Agroforestry Systems* 64, no. 2 (2005): 157–67. https://doi.org/10.1007/s10457-004-1023-y.

Soini, Eija. "Land Use Change Patterns and Livelihood Dynamics on the Slopes of Mt. Kilimanjaro, Tanzania." *Agricultural Systems* 85, no. 3 (2005): 306–23. https://doi.org/10.1016/j.agsy.2005.06.013.

Solomon, Harris. *Metabolic Living: Food, Fat and the Absorption of Illness in India*. Critical Global Health: Evidence, Efficacy, Ethnography. Durham, NC: Duke University Press, 2016.

Someya, Shinichi, Yumiko Yoshiki, and Kazuyoshi Okubo. "Antioxidant Compounds from Bananas (Musa Cavendish)." *Food Chemistry* 79, no. 3 (2002): 351–54. https://doi.org/10.1016/S0308-8146(02)00186-3.

Spear, Thomas T. *Mountain Farmers: Moral Economies of Land and Agricultural Development in Arusha and Meru*. Dar es Salaam, Tanzania: James Currey, 1997.

Stambach, Amy. *Lessons from Mount Kilimanjaro: Schooling, Community, and Gender in East Africa*. New York: Routledge, 2000.

Stark, Hannah. "Deleuze and Critical Plant Studies." In *Deleuze and the Non/Human*, edited by Hannah Stark and Jon Roffe, 180–96. New York: Palgrave Macmillan, 2015.

Stein, Serena, and Jessie Luna. "Toxic Sensorium: Agrochemicals in the African Anthropocene." *Environment and Society* 12, no. 1 (2021): 87–107. https://doi.org/10.3167/ares.2021.120106.

Stoller, Paul. *Embodying Colonial Memories: Spirit Possession, Power, and the Hauka in West Africa*. New York: Routledge, 1995.

Strathern, Marilyn. *Partial Connections*. New York: Rowman and Littlefield, 1991.

Street, Alice. "Food as Pharma: Marketing Nutraceuticals to India's Rural Poor." *Critical Public Health* 25, no. 3 (2025): 361–72.

Stump, Daryl. "The Archaeology of Agricultural Intensification in Africa." In *The Oxford Handbook of African Archaeology*, edited by Peter Mitchell and Paul Lane, 671–85. Oxford Handbooks. New York: Oxford University Press, 2013.

Subrahmanyam, Sanjay. "Between Eastern Africa and Western India, 1500–1650: Slavery, Commerce, and Elite Formation." *Comparative Studies in Society and History* 61, no. 4 (2019): 805–34. https://doi.org/10.1017/S0010417519000276.

Susser, Mervyn. "Epidemiology in the United States after World War II: The Evolution of Technique." *Epidemiologic Reviews* 7, no. 1 (1985): 147–77. https://doi.org/10.1093/oxfordjournals.epirev.a036280.

Swanson, Heather Anne. "Landscapes, by Comparison: Practices of Enacting Salmon in Hokkaido, Japan." In *The World Multiple: The Quotidian Politics of Knowing and Generating Entangled Worlds*, edited by Keiichi Ōmura, Grant Otsuki, Shiho Satsuka, and Atsurō Morita, 105–22. London: Routledge, 2020.

Tagseth, Mattias. "Oral History and the Development of Indigenous Irrigation. Methods and Examples from Kilimanjaro, Tanzania." *Norwegian Journal of Geography* 62, no. 1 (2008): 9–22. https://doi.org/10.1080/00291950701864898.

Tanganyika African National Union. *The Arusha Declaration and TANU's Policy on Socialism and Self Reliance*. Dar es Salaam, Tanzania: Publicity Section, TANU, 1967.

Tantala, Renee L. "The Early History of Kitara in Western Uganda: Process Models of Religious and Political Change." PhD diss., University of Wisconsin–Madison, 1989.

Tanzania National Bureau of Statistics. *National Sample Census of Agriculture 2019/20-Main Report*, Government of Tanzania, 2021. https://www.nbs.go.tz/index.php/en/census-surveys/agriculture-statistics/661-2019-20-national-sample-census-of-agriculture-main-report.

Taylor, Sunaura. "Age of Disability: On Living Well with Impaired Landscapes." *Orion Magazine*, November 2021. https://www.orionmagazine.org/article/age-of-disability/.

Thomas, Lynn M. "Beauty." In *Somatosphere, Toxicity, Waste, and Detritus in the Global South: Africa and Beyond*, edited by Pamila Gupta and Gabrielle Hecht, January 15, 2018. http://somatosphere.net/2018/beauty.html/.

Thompson, Lee, and Gillian Abel. "The Work of Negotiating HIV as a Chronic Condition: A Qualitative Analysis." *AIDS Care* 28, no. 12 (2016): 1571–76. https://doi.org/10.1080/09540121.2016.1191615.

Thornton, Thomas F., and Patricia M. Thornton. "The Mutable, the Mythical, and the Managerial: Raven Narratives and the Anthropocene." *Environment and Society* 6, no. 1 (2015): 66–86. https://doi.org/10.3167/ares.2015.060105.

Tilley, Helen. "An Anthropological Laboratory: Ethnographic Research, Imperial Administration, and Magical Knowledge." In *Africa as a Living Laboratory: Empire, Development, and the Problem of Scientific Knowledge, 1870–1950*. Chicago: University of Chicago Press, 2011.

Tilley, Helen. "Ecologies of Complexity: Tropical Environments, African Trypanosomiasis, and the Science of Disease Control in British Colonial Africa, 1900–1940." *Osiris* 19 (2004): 21–38.

Tilley, Helen. "Traditional Medicine Goes Global: Pan-African Precedents, Cultural Decolonization, and Cold War Rights/Properties." *Osiris* 36 (2021): 132–59.

Tilman, David, Joseph Fargione, Brian Wolff, et al. "Forecasting Agriculturally Driven Global Environmental Change." *Science* 292, no. 5515 (2001): 281–84. https://doi.org/10.1126/science.1057544.

Tironi, Manuel, and Israel Rodríguez-Giralt. "Healing, Knowing, Enduring: Care and Politics in Damaged Worlds." *Sociological Review* 65, no. 2 suppl. (2017): 89–109. https://doi.org/10.1177/0081176917712874.

Todd, Zoe. "Fish, Kin and Hope: Tending to Water Violations in *Amiskwaciwâskahikan* and Treaty Six Territory." *Afterall: A Journal of Art, Context and Enquiry* 43 (March 2017): 102–7. https://doi.org/10.1086/692559.

Tousignant, Noemi. *Edges of Exposure: Toxicology and the Problem of Capacity in Postcolonial Senegal*. Experimental Futures. Durham, NC: Duke University Press, 2018.

Tousignant, Noemi. "Toxic Residues of Senegal's Peanut Economy." *Anthropology Today* 36, no. 6 (2020): 5–8. https://doi.org/10.1111/1467-8322.12616.

Trapp, Micah M. "You-Will-Kill-Me-Beans: Taste and the Politics of Necessity in Humanitarian Aid." *Cultural Anthropology* 31, no. 3 (2016): 412–37. https://doi.org/10.14506/ca31.3.08.

Tsai, Yen-Ling, Isabelle Carbonell, Joelle Chevrier, and Anna Lowenhaupt Tsing. "Golden Snail Opera: The More-Than-Human Performance of Friendly Farming on Taiwan's Lanyang Plain." *Cultural Anthropology* 31, no. 4 (2016): 520–44. https://doi.org/10.14506/ca31.4.04.

Tsing, Anna Lowenhaupt. *The Mushroom at the End of the World: On the Possibility of Life in Capitalist Ruins*. Princeton, NJ: Princeton University Press, 2015.

Tsing, Anna Lowenhaupt, Andrew S. Mathews, and Nils Bubandt. "Patchy Anthropocene: Landscape Structure, Multispecies History, and the Retooling of Anthropology: An Introduction to Supplement 20." *Current Anthropology* 60 (August 2019): S186–97. https://doi.org/10.1086/703391.

Turner, Bethaney, and Wendy Somerville. "Composting with Cullunghutti: Experimenting with How to Meet a Mountain." *Journal of Australian Studies* 44, no. 2 (2020): 224–42. https://doi.org/10.1080/14443058.2020.1753223.

Tushemereirwe, W. K., and J. Kubiriba. "Achieving Sustainable Cultivation of Bananas Volume 1: Cultivation Techniques." In *Achieving Sustainable Cultivation of Bananas*, edited by Gert H. J. Kema and André Drenth. Cambridge: Burleigh Dodds Science, 2018.

UNAIDS. "2016 United Nations Political Declaration on Ending AIDS Sets World on the Fast-Track to End the Epidemic by 2030." UNAIDS, October 2016. http://www.unaids.org/sites/default/files/20160608_PS_HLM_Political_Declaration_final.pdf.

Under the Same Sun. "Reported Attacks RE: Persons with Albinism." Under that Same Sun, March 29, 2023.

Urassa, Mark, J. Ties Boerma, Japheth Z. L. Ng'weshemi, Raphael Isingo, Dick Schapink, and Yusufu Kumogola. "Orphanhood, Child Fostering and the AIDS Epidemic in Rural Tanzania." *Health Transition Review* 7 (1997): 141–53. https://www.jstor.org/stable/40652331.

Vähäkangas, Mika. "The Grandpa's Cup: A Tanzanian Healing Ritual as a Temporary Interreligious Platform." *Suomen Antropologi: Journal of the Finnish Anthropological Society* 41, no. 4 (2016): 14–28.

Vähäkangas, Mika. "Negotiating Religious Traditions: Babu Wa Loliondo's Theology of Healing." *Exchange* 45, no. 3 (2016): 269–97. https://doi.org/10.1163/1572543X-12341404.

Van Der Plas, Geert W., Stephen M. Rucina, Andreas Hemp, et al. "Climate-Human-Landscape Interaction in the Eastern Foothills of Mt. Kilimanjaro (Equatorial East Africa) during the Last Two Millennia." *The Holocene* 31, no. 4 (2021): 556–69. https://doi.org/10.1177/0959683620981694.

Vaughan, Megan. *Curing Their Ills: Colonial Power and African Illness*. Stanford, CA: Stanford University Press, 1991.

Venkat, Bharat Jayram. "Cures." *Public Culture* 28, no. 3 (2016): 475–97. https://doi.org/10.1215/08992363-3511502.

Verran, Helen. *Science and an African Logic*. Chicago: University of Chicago Press, 2001.

Vimalassery, Manu, Juliana Hu Pegues, and Alyosha Goldstein. "Introduction: On Colonial Unknowing." *Theory and Event* 19, no. 4 (2016): 891–93.

Weaver, Lesley Jo. *Sugar and Tension: Diabetes and Gender in Modern India*. Medical Anthropology: Health, Inequality, and Social Justice. New Brunswick, NJ: Rutgers University Press, 2018.

Weiss, Brad. *The Making and Unmaking of the Haya Lived World: Consumption, Commoditization, and Everyday Practice*. Durham, NC: Duke University Press, 1996.

Weiss, Robin A., and Anthony J. McMichael. "Social and Environmental Risk Factors in the Emergence of Infectious Diseases." *Nature Medicine* 10 (December 2004): S70–76. https://doi.org/10.1038/nm1150.

Weisz, George, and Jesse Olszynko-Gryn. "The Theory of Epidemiologic Transition: The Origins of a Citation Classic." *Journal of the History of Medicine and Allied Sciences* 65, no. 3 (2010): 287–326. https://doi.org/10.1093/jhmas/jrp058.

Wendland, Claire L. *A Heart for the Work: Journeys through an African Medical School.* Chicago: University of Chicago Press, 2010.

Wetsman, Nicole. "Air-Pollution Trackers Seek to Fill Africa's Data Gap." *Nature,* April 11, 2018. https://www.nature.com/articles/d41586-018-04330-x.

White, Luise. "Poisoned Food, Poisoned Uniforms, and Anthrax: Or, How Guerillas Die in War." *Osiris* 19 (2004): 220–33.

White, Luise. *Speaking with Vampires: Rumor and History in Colonial Africa.* Studies on the History of Society and Culture. Berkeley: University of California Press, 2000.

Whitehead, Alfred North. *Process and Reality: An Essay in Cosmology.* 1929. New York: Free Press, 1978.

Whyte, Kyle Powys. "Collective Continuance." In *50 Concepts for a Critical Phenomenology,* edited by Gail Weiss, Ann V. Murphy, and Gayle Salamon, 53–59. Evanston, IL: Northwestern University Press, 2020. https://doi.org/10.2307/j.ctvmx3j22.50.

Whyte, Kyle Powys. "Food Sovereignty, Justice, and Indigenous Peoples: An Essay on Settler Colonialism and Collective Continuance." In *Food Sovereignty and Indigenous Peoples,* edited by Anne Barnhill, Mark Budolfson, and Tyler Doggett, 345–66. Oxford: Oxford University Press, 2018.

World Health Organization. *Antiretroviral Drugs for Treating Pregnant Women and Preventing HIV Infection in Infants: Recommendations for a Public Health Approach.* Geneva: WHO, 2010. https://www.ncbi.nlm.nih.gov/books/NBK304945/#sectioniii.s5.

World Health Organization. "Declaration of Alma-Ata." Geneva: WHO, 1978. https://www.who.int/publications/almaata_declaration_en.pdf.

World Health Organization. *National Policy on Traditional Medicine and Regulation of Herbal Medicines: Report of a WHO Global Survey.* Geneva: WHO, 2005.

World Health Organization and Pierpaolo Mudu. *Ambient Air Pollution and Health in Accra, Ghana.* Geneva: WHO, 2021.

Wulfhorst, Ellen. "Tanzania Criticizes Film Documenting Attacks on Albinos for Witchcraft." Thomas Reuters Foundation News, October 16, 2015. http://www.trust.org/item/20151016100346-zywal/?source=dpagerel.

Yates-Doerr, Emily. *The Weight of Obesity: Hunger and Global Health in Postwar Guatemala.* California Studies in Food and Culture 57. Oakland: University of California Press, 2015.

Yusoff, Kathryn. *A Billion Black Anthropocenes or None.* Forerunners: Ideas First from the University of Minnesota Press 53. Minneapolis: University of Minnesota Press, 2018.

Zárate, Salvador. "Maintenance." Society for Cultural Anthropology, March 29, 2018. https://culanth.org/fieldsights/maintenance.

Zhan, Mei. "Does It Take a Miracle? Negotiating Knowledges, Identities, and Communities of Traditional Chinese Medicine." *Cultural Anthropology* 16, no. 4 (2001): 453–80. https://doi.org/10.1525/can.2001.16.4.453.

Zimmerman, Andrew. "'What Do You Really Want in German East Africa, Herr Professor?' Counterinsurgency and the Science Effect in Colonial Tanzania." *Comparative Studies in Society and History* 48, no. 2 (2006): 419–61. https://doi.org/10.1017/S0010417506000168.

# Index

baobab fruit (*ubuyu*), 94, 99–100

Barad, Karen, 230–31

Baraza la Tiba Asili na Tiba Mbadala (Council of Traditional Healing and Alternative Healing), 124, 137–39, 140, 212, 213, 260n10

Barnes, Barry, 206

Bayart, Jean-François, 90–91

bees and beekeeping, 33, 34, 53, 55, 68

being in a body, 19, 91, 92, 93, 115, 229; chronic pain and, 97, 166; individuality and, 197–98, 260n7; moving through the world and, 154–56, 177

Berlant, Lauren, 162

Bertoni, Filippo, 70–71

Bhabha, Homi, 250n8

biodiversity, 66, 90, 221; health and, 14, 245n38; in *vihamba* (home gardens), 38–39, 69, 220

biological indicators, 169

biomedicine, 7, 79, 153, 188, 231, 251n16; bodiliness and, 26, 91, 92, 197; chronicity and temporality of, 35, 167, 170, 187, 189–90; failure of, 182; *mganga* (medical doctors), 131, 255n20; symptomatology and, 98; traditional medicine and, 86–88, 136, 138, 140–41, 150, 151, 255n37, 256n47; Western, 87; work of TRMEGA and, 173–74. *See also* chronic diseases; pharmaceuticals

bitterness, 49, 81–82, 106, 113, 251nn9–10

blackboxing, 194, 259–60n1

Bloor, David, 206

body/bodies: cleansing or purifying, 5–6, 11; depleted or exhausted, 2, 16, 20, 42, 54, 90, 101, 109; dispersed, 45, 93, 115; ecologies and, 1, 24, 42, 106, 118, 152, 206; extraction of parts, 122–23, 254n9; political phenomenology of, 80, 88–90; soil and, 1, 23–24, 70, 115; temporalities and rhythms of, 160, 163, 190–91; toxins, 5, 58, 245n36; vulnerability of, 24, 34, 35, 50, 160, 190, 193. *See also* being in a body; chronic diseases

body-land relations, 13, 92, 118, 153, 163, 173, 198; bananas and, 195, 218–19; being in the world and, 154–56; *dawa lishe* and, 4, 8, 41–43, 58, 90, 128, 132, 192, 229–33; depletion of, 2, 16, 160; environmental change and, 14, 245n36, 261n21; healing, 50, 51–52,

54, 62, 111, 154, 164, 196, 252n39; *kihamba* (home garden) and, 227–29; plant remedies and, 73–74, 80; reconfiguring and coming together of, 234; separating or dividing, 119, 120, 138, 226; shifting relations of, 38, 178; soil and, 1, 23–24, 70, 115

Bomang'ombe, 80, 81, 96; location and town center, *29*, 83–84, 102

botanical studies, 58, 69, 148; ethnobotany, 119, 120, 133, 137

Business Registrations and Licensing Agency (BRELA), 139

Cabnal, Lorena, 252n39

care, modes of, 1, 4, 15, 44, 45, 69, 101; COVID-19 and, 228; eating and cooking and, 91, 100, 103–4, 201–2; HIV/AIDS treatment and, 170–73, 210; rhythms and temporalities and, 163, 259n53

*Carissa edulis* (*mugariga*), 127, 133, 148

*Cassia spectabilis*, 259n54

Catholic Church, 51, 63, 210

cattle, 37, 38, 84, 85

Chagga, 27, 81, 184, 261n26, 262n27; *vihamba* (home gardens), 35–36, 54, 69, 219–20, 227

Chakrabarty, Dipesh, 261n21

chickens, 12, 34, 62, 91, 202

childbirth, 227; attendants, 26; rates, 164. *See also* midwives

Chinese medicine, 78, 82

chronic diseases, 10, 35, 74; appetite and, 114, 115; concept in traditional medicine, 167–68; COVID-19 and, 50; "cup of life" remedy for, 126–27; epidemiological theory of, 164–65; exhaustion as, 41–42, 51–52, 90, 140; HIV/AIDS as, 160–61, 165–67, 172–73, 174, 178, 188–89; malaria as, 181–84, 202; rise in, 4, 7, 14, 15, 43, 79, 167, 178; at risk for, 168–70, 258n24

chronicity: geopolitics of, 160, 165–67; pharmaceutical adherence and, 161, 164, 170, 172–73, 187, 189; "removing," 182–83, 190; as the time of slow violence, 162, 232

citational practices, 232–33

Clare, Eli, 79–80

Clark, Timothy, 198

climate change, 9, 12, 15, 113–14, 161; human health and, 14, 246n39, 246n41 coffee, 69, 231; as a cash crop, 248n1; exports, 84, 247n67; home garden cultivation, 36–38

colonialism, 15, 20, 153, 261n19; attempts to control healers, 8, 21, 118, 134–35, 143; indentured laborers, 66; land tenure, 37–38, 40; lushness and, 59, 248n13; native medicine and witchcraft and, 21, 44, 118–19, 123, 129–32, 134, 212–13, 255n30; toxicity and remedy and, 8–9, 16–17. *See also* dispossession; German colonial rule; imperial pharmakon

compost/composting, 7, 32, 43, 44, 49, 250n37; at TRMEGA garden, 65, 70–72, 72, 75, 90

continuance, collective, 71, 191, 229, 262n3. *See also* ongoingness

corn, 12, 34, 72, 200, 219, 221, 245n31

Cornell University, 155, 246n40

Cousins, Thomas, 188

COVID-19, 35–36, 49–50, 73, 234, 258n44; *kihamba* (home garden) and, 228

Crane, Emma Shaw, 59, 248n13

critical plant studies, 69, 194–95, 250n41, 260n2

Croatia, 155

crossroads, 85, 86

Crutzen, Paul, 13

"cup of life" cure, 122, 126–28, 133, 148

cure, ideology of, 79, 169, 257n23

Dadi, Binti, 25–26, 27

dandelion greens, 42, 81–82, 93, 106, 113, 115

Dar es Salaam, 26–28, 32, 78, 108, 170; NIMR Mabibo lab, 138, 204

data collection, 25, 205, 209

*dawa lishe*, 32, 33, 84, 88, 144, 162, 195, 210, 211; appetite and, 79, 80, 93, 115; bananas and, 33, 196–98, 214, 220–22; body-land relations and, 90, 91, 229–33; categories of recognition and, 150–51; chronicity and chronic disease and, 188–91; as creative infidelity, 193–94; eating and nourishment and, 111, 186, 187, 188, 199, 205; healing as land relations and, 39–40, 229, 233; lushness and, 112, 114–15; meaning and usage, 2–9, 15, 20, 22–24, 42–43, 56, 225–27; politics of purity and, 70; precarity of, 223; sovereignty and, 23, 45, 51, 192; time

and temporalities and, 35, 116, 128, 160, 164; toxicity and, 22, 24, 57–58, 248n11; traditional medicine and, 44, 117–18, 120, 132–33, 134, 153

decomposers, 72, 75, 219

Deval, Margaret, 136, 255n39

diabetes, 7, 15, 79, 169, 189, 202–3, 204

Diagne, Souleymane Bachir, 19, 44, 159–60

diagnostics, 95, 98, 170, 252n40, 258n45

diet, 94, 111, 205, 206, 250n7. *See also* eating

difference, politics of, 256n47

diseases. *See* chronic diseases; HIV/AIDS

dispossession, 3, 11, 50, 64, 140, 162, 164, 233; attachment and, 114; bodies and land and, 41, 42, 120, 206; invisibility of, 22, 25, 84, 132, 150, 231; "lush aftermath" of, 59, 248n13; racial capitalism and, 13, 16, 190; slow violence of, 4, 17, 35, 163, 190, 211; toxicity and, 18, 230; traditional medicine and, 118–19; vulnerabilities and, 180, 215

divination and dreaming (*kupiga bao* and *ramli*), 145–46, 147–48, 150, 153

Dorkia Enterprises, 5, 163, 191, 193, 194, 227; founding and operations, 198–200; *kitarasa* flour production, 197, 198, 201–7, 214–15, 217; at Slow Food's Arc of Taste, 216, 221; testimonials for *kitarasa*, 208–9, 211

double-bind, 1, 6, 80, 99, 128, 177, 229

dwelling, modes of, 1, 8, 51, 114, 230, 234; bananas and, 196, 219, 220–21; Kagera, 55; *kihamba* as, 35, 37, 38, 69, 227; medicine and agriculture and, 226; plants and, 78, 181, 182, 194, 195

eating: body and politics, 88–91, 115–16; chronic pain when, 97–98; daily gestures of, 154–55, 156; detoxifying and, 200; flavor and, 80–82, 113, 251nn9–10; habits, 94–95, 205, 206; HIV/AIDS and, 103–4, 108, 110–11, 114, 187; household dynamics and cooking and, 96, 98–100; *kitarasa*, 201–3, 208–9, 215; therapeutic practices and rituals of, 91–93; vegetarianism, 250n7. *See also* nourishment

ecology-economy relationship, 51, 58, 184, 231, 260n4; *dawa lishe* and, 194–96, 208; *kitarasa* flour production and, 206, 207, 212, 220–21, 222–23

institutionalization: *dawa lishe* and, 150, 153, 196; regulation of food and drugs, 203–7; of traditional medicine, 81, 118, 119, 120, 121, 133–40, 149, 150. *See also* registration system for healers, medicines, and premises

intellectual property, 6, 32, 137, 139, 151, 212

International Monetary Fund (IMF), 125

*isale* (*Dracaena fragrans*), 227

Jameson, Fredric, 60

Kagera, 40, 53, 55, 110, 117, 170, 248n2

Kenya–Uganda railway, 66

Kibona, Dorcas, 30, 32, 223; *dawa lishe* phrase and, 225; founding of Dorkia Enterprises, 198–200; *kitarasa* flour production and registration struggles, 201–7, 215–16, *217*, 222; outreach and gathering testimonials, 208–9, 211, 215, 220; trip to India, 155–57

kidney disease, 167, 169, 189, 199

*kihamba/vihamba* (home garden/home gardens), 66, 194, 247n79; banana and coffee cultivation, 36–39; COVID-19 and, 228–29; healing as land relations and, 35–36, 39–40, 229; history and description of, 54–55, 219–20, 227–28; lushness of, 68–69; remembering and, 206, 215

Kikwete, Jakaya, 123

Kilimanjaro, Mount, 27, 54–55, 83–84, 99; banana plants, 12, 200, 201, 214, 215, 216, 218–20; elders, 86; map showing location of, *29. See also* Chagga; *kihamba/vihamba*

Kilimanjaro Christian Medical Center (KCMC), 28, *29*, 208; Uzima Project, 234

Kilimanjaro Women's Information Exchange and Consultancy Organization (KWIECO), 38

*kitarasa*, 12, 33, 44–45, 261nn20–21; childhood narratives of, 201–2, 207, 215; circulation of, 214–15, 223; *dawa lishe* and, 197–98, 214, 220–22; difficulties registering, 203–7, 212, 215–16, 260n10; healing properties of, 202–3, *203*, 222; testimonials for, 208–9, 211

knowledge: biomedical, 92, 93, 98, 255n20, 259n1; bodily relations of, 154–57; citational practices and, 232–33; colonial, 16, 130; *dawa lishe*, 193, 196, 197, 234;

economies, 14, 15, 25, 48, 51, 88, 90; forgetting and, 81; gardening, 62–63; healers/healing, 51–52, 85, 91, 129–31, 140, 145–46, 252n32; local or indigenous, 68, 213, 231; mobilizing multiple forms of, 29–30, 87–88, 110, 193; networks and, 29–30, 31; nonknowledge, 153, 178; nutritional, 94, 96; politics and, 144, 160; registers of, 152–53, 194; scientific, 28, 135, 138–39, 148, 206, 208, 251n16; subject and object of, 210, 223; therapeutic plant, 29–30, 84–85, 91, 114, 120, 131, 180–84; traditional medicine as a modern category of, 44, 117–20, 127, 132, 149–50; uneven production of, 10, 13. *See also* unknowing

labels and therapeutic claims, 205–6, 207, 212

labor and land, 7, 118, 119, 229, 248–49n13; alienation of, 41, 59, 79, 132, 197, 206, 229

Lake Zone, 124, 254n13

Landecker, Hannah, 88–89, 92

land relations. *See* body-land relations

land rights, 38, 40–41, 53, 216, 248n2

languages, 168, 243n1; dialects and translation, 19–22, 53

Latour, Bruno, 11, 121, 250n2, 251n16, 259n1

lemongrass (*mchaichai*), 2, 33, 56, 99, 199; medicinal properties, *5*, 5–6, 12, 176–77, 184, 200

lemon trees, 36, 228

lineage and land, 41, 43, 99, 132, 163, 173, 196, 215; *kihamba* (home garden) and, 36–37, 63, 220

livelihoods and liveliness, 64, 104, 193, 215, 222, 226–27; bananas and, 196, 197, 221

Livingston, Julie, 15

lushness, 32, 43, 45, 74, 75, 84, 177, 233; alternative forms of, 80, 209; colonialism and, 59, 118, 248n13; of composting, 72; *dawa lishe* and, 33, 59–60, 70, 112, 134; flavor and, 113, 215; futures for, 51, 234; healing and, 62, 129, 140, 162, 227; of TRMEGA gardens, 47–49, 52, 54, 56, 66; of *vihamba* (home gardens), 37, 68, 220, 228

Maasai, 54–55, 84–85, 125, 133, 256n52

Machange, Rose, 31–32, 54–55, 65, 68, 73, 155, 234

noncultivated plants, 69
nongovernmental organization (NGO), 7,
    28, 112, 121, 141, 152, 210, 229; advocacy
    for albinism, 122, 254n11; for elderly
    home care, 208, 220; entrepreneurship
    and, 4, 30–31, 153. *See also* Training,
    Research, Monitoring and Evaluation on
    Gender and AIDS (TRMEGA); Women
    Development for Science and Technology
    (WODSTA)
nourishment, 50, 53, 71, 79, 88; antiretrovirals
    (ARVs) and, 185–88; *dawa lishe* and, 33, 51,
    80, 132, 186, 188–90, 199; interrupting, 98,
    104; of life force, 43, 70, 111, 115
nutraceuticals, 6, 195, 196, 208
nutrient value, 2, 66, 67, 246n39; of bananas,
    208; caloric logics, 89; of EdenMark
    remedies, 94, 95; of TRMEGA remedies,
    56–57
nutritional science and research, 56–57, 88,
    90, 92
Nyerere, Julius, 54, 134, 135, 136, 255n35
*nyumbani kabisa* (natal home), 64, 200

Ogondiek, John, *89*, 107, 138, 185, 256n49;
    about, 27–30, 32; exchange with TFDA
    staff member, 204–5; interest in TRMEGA
    plant remedies, 174, 177; new registra-
    tion system for healers and, 141–42, 144,
    147–48; trip to India, 155–56
Omran, Abdel, 164–65
One Health, 14, 246n40
ongoingness, 50, 59, 75, 109, 132, 191, 244n11;
    of composting, 71; *dawa lishe* and, 115, 221,
    222, 230, 231; of the *kihamba* (home gar-
    den), 37, 227–28; lushness that supports,
    227; of the physical body, 186–87
ontology/ontics, 23, 79, 87, 191, 222, 227,
    248n11, 252n38; of the body, 4, 8, 19, 43,
    181; colonialism and imperial pharmakon
    and, 20, 23, 44, 70, 130, 131, 132, 233; com-
    posting and, 72; *dawa lishe* and, 60, 117,
    121, 186, 231; of food and drug regulations,
    204, 206, 207; registration of healers and,
    134, 147, 148, 149
organic crop management. *See* sustainable
    agriculture
organic matter. *See* compost/composting

pandemics, 164–65, 258n44. *See also*
    COVID-19
periurban, 173, 231; areas, 41, 42, 44, 62–65, 233
permaculture, 4, 35, 65, 153, 228, 234
pesticides and herbicides, 12, 202, 231,
    245n30
pharmaceuticalization, 45, 169, 172, 209
pharmaceuticals, 74, 137, 169, 189; biological
    body and, 105–6; companies, 6; drug trials,
    207; HIV/AIDS treatment and, 31, 160–61,
    166–67, 171–72, 186, 187, 190; patents, 120,
    134; plants medicines and, 163, 181, 183;
    toxicities of, 44, 199–200, 202
pharmacology, 10, 12, 132, 151, 152, 178, 185
philanthropy, 171, 172
phytochemistry, 22, 113, 132, 150, 207; NIMR
    Ngongongare lab and, 28, 138, 141
planetary health, 14, 246n41, 260n4
plantation crops, 59, 66, 69, 230, 262n32;
    bananas, 221; failure of moringa (*mlonge*),
    67–68; *Ficus elastica*, 249n13; sisal, 180–82
poisons, 10, 16, 17, 23, 130, 221, 243n7; medi-
    cines and, 109, 177, 178, 202; removing,
    184, 200
pollution, invisibility of, 10–11, 244n18
polygamy, 36, 37
population growth, 99, 164–65
Povinelli, Elizabeth, 18, 151, 197–98
power, forms of, 90–91, 92, 96, 152, 194;
    God's will as, 125–26, 127, 133
Prince, Ruth J., 173, 253n3
property, 25, 45, 51, 124, 140, 153, 227; enclo-
    sure of land, 37, 59, 113, 115; plants as, 28,
    48, 194–95, 196–97, 212, 213–14, 216
provincializing/deprovincializing, 17, 19, 82,
    244n9
public health, 28, 88, 106, 146, 246n40,
    255n37, 259n51; appetites and, 44, 79, 93,
    115; chronic diseases and, 165, 167, 169;
    HIV/AIDS and, 166, 171, 190, 210; neoliberal
    reforms and, 180, 259n51; scholars and
    studies, 10, 15, 27, 64, 167–68; structural
    obstacles, 161–62; traditional medicine
    and, 17, 184. *See also* global health; national
    health care system
purity/purification, 67, 190, 244n19, 245n21,
    250n8; Latour on, 11, 121; metaphysics of,
    70, 91, 249n27; politics of, 11, 12, 70

Taasisi ya Dawa za Asili (Institute of Traditional Medicine, or ITM), 135–37, 139, 151–52, 260n10

Tanganyika, 16, 118, 129, 130

Tanzania Bureau of Standards (TBS), 139, 203, 215–16

Tanzania Food and Drugs Authority (TFDA), 138, 139; *kitarasa* flour and, 203–7, 209, 211, 215; raw plant-based remedies and, 212, 213; registration for herbal medicines, 151–52, 260n10

Tanzanian Commission on AIDS (TACAIDS), 171

tea: black, 6; lemongrass, 56, 73, 176, 184, 199; lemon leaf, 36, 228; medicinal, 63, 126–27, 133, 176–77, 189

testing and testimony, 112, 207–12, 217, 222

theory/theorization, 3, 19, 194, 246n62; compost and, 70, 250n37; *dawa lishe* as, 2, 41, 44, 117, 234; epidemiological, 160, 164–67, 168; permaculture, 35

time and temporality, 74, 159, 164, 191, 232, 259n53; of the body, 88, 91, 115–16; of *dawa lishe*, 35, 189; HIV/AIDS treatment and, 160–61, 170–73, 178–80, 258n41; *kitarasa* and, 202, 215; of plants, 163, 183, 190, 192; power and, 168; remaking, 51; spaces of healing and, 85, 112; toxicities and, 44, 222. *See also* chronicity

Tousignant, Noémi, 10, 222

toxicity and remedy relationship, 1, 10, 12–13, 19, 177, 198, 216; climate crisis and, 14–15; colonialism and, 8–9, 16–17, 130, 132; compost and, 70, 72; *dawa lishe* and, 22, 24, 45, 57–58, 226–27, 230, 248n11; plant-based remedies and, 57, 151, 180–81, 185; side effects of medicines and, 199–200; time and temporality and, 160; TRMEGA's work and, 51, 73–75

toxicology, 10, 11, 17, 132, 151, 185, 222

toxic waste, 10–11, 12, 222, 244n18; human body and, 14, 245n36

Traditional and Alternative Medicines Act (2002), 137–38, 212, 213; new registration system, 121, 134, 140, 151, 260n10

traditional foods, 110, 133, 185–86, 201–2

traditional healing (*uganga asili*), meaning, 146

traditional medicine, 28, 194; biomedicine and, 17, 86–88; chronic diseases and, 167–68, 182–83; *dawa lishe* and, 44, 117–18, 132–33, 134; integration of, 17–18; Maasai knowledge of, 84–85; as a modern category of knowledge and practice, 4, 26, 30, 117–20, 132, 153, 261n19; "native medicine" precursor, 119, 129–32; registration and institutionalization of, 81, 121–22, 123–28, 133–40, 149–50, 151–52, 254n13, 256n47; standardization of, 181

Training, Research, Monitoring and Evaluation on Gender and AIDS (TRMEGA), 44, 91, 191, 234, 248n2; apprehending appetites, 49, 79–80, 93, 112–16; approach to HIV stigma, 60–62; composting practice, 70–72, *72*; *dawa lishe* and, 227, 229, 230, 233; EdenMark strategy, 30, 31, 48–49, *49*, 78–79; *imarisha* remedy, 30, 48–49, *49*, 56–58, 73, 75; Maji ya Chai garden, 2, *3*, 47–51, *48*, 53–56, 60, 65–66, 70–72, 173–74, 180; moringa (*mlonge*) cultivation, 56–57, *57*, 61–62, 66–67, *67*; origin point for, 51–52; periurban transplanting projects, 62–65, *179*, 233; plant-based remedies for chronic conditions and, 163, 173–78, 181–84, 189–90; registration of healers and, 117–18, 121; social connections and international reach, 31, 55–56, 112

Translating Vitalities group, 155

translation, 19–21

transplanting, 32, 228; bananas, 218–19; TRMEGA's projects, 2, *3*, 53, 62–63, 176, *179*

Tsing, Anna, 65, 250n42, 262n32

*uganga* and *uchawi* (healing and harming), 59, 91, 110–11, 186, 196, 221; colonial science and law and, 129–30, 133, 145; *kihamba* (home garden), 35; meaning and translation, 20–22, 50, 129, 226

*uji* (cornmeal porridge), 103–4

ulcers, 101, 187; peptic, 97–99, 167; skin, 175–76, 180

Umoja Waganga wa Tanzania (UWATA), 135

unknowing, 18, 195, 206, 227, 253n3; colonial, 17, 25, 81, 118, 131–32; traditional medicine and, 120, 128, 145, 150

urbanization, 64, 99
urinary tract infections (UTIs), 183–84
Uroki, Alex, 29–31, 78, *89*, 152, 155, 252n40;
    business model, 106, 253n48; *dawa lishe*
    phrase and, 225; Dorkia Enterprises and,
    198; on flavor and appetite, 80–82, 88,
    94, 106, 111, 113; gifting medicines, 102,
    105; healing and plant remedies, 84–88,
    117–18, 210
Uzima Project, 234

vascular plant species, 69
*vihamba. See kihamba/vihamba*
village markets, 39, 68, 98, 108–9; negotiating
    at, 104–5, 112, 162
violence, 145, 161, 171, 182; cure and, 80, 88;
    of dispossession, 4, 17, 25, 35, 163, 190, 211,
    248n13; ecological and social, 1; against
    people with albinism, 122–24, 128, 143,
    254n11, 254n13; state, 143; structural, 51,
    107, 183, 189. *See also* slow violence
vitamins and supplements, 6, 7, 56, 94, 95,
    184, 212; Omega Wash, 188
vulnerability, 65, 104–5, 107, 112, 122, 180, 215;
    bodily, 24, 34, 35, 50, 160, 190, 193

*wachawi. See* witchcraft
*waganga. See* healers

*wagonjwa* (people who are ill), 126, 211
Watoto Foundation, 63
wealth, 222, 251n22, 253n48, 254n9;
    accumulation, 41, 70, 90, 193
weeds/weeding, 68–69, 99, 232
Wendland, Claire, 86–87
whole-plant products, 212–14
Whyte, Susan Reynolds, 172, 258n38,
    258n41
widows, 41, 108–9
Wiketye, Victor, *89*, 107, 138, 174, 185; about,
    27–30, 32; new registration system for
    healers and, 141–42, 144–45, 147–48
wild cucumber, 69
witchcraft, 20–21, 44, 90, 123, 133, 151, 226
    256n47; cemeteries and, 254n9; *kupiga bao*
    and, 145; and "native medicine" distinction,
    119, 130–32; ordinances, 21, 124, 130–31,
    134, 212, 255n30
Women Development for Science and
    Technology (WODSTA), 31, 54, 55, 65, 191;
    *dawa lishe* and, 193, 194, 225, 227, 229, 230
women's rights, 38, 41, 248n2
World Health Organization (WHO), 10, 166,
    171, 244n15, 246n40, 253n6; Program on
    Traditional Medicine, 137, 256n44
world-making, 15, 18, 60, 221
Wynter, Sylvia, 247n79